# BUILDING CONSTRUCTION
# HANDBOOK

# BUILDING CONSTRUCTION HANDBOOK

## R. Chudley
MCIOB

NEWNES

Newnes
An imprint of Butterworth-Heinemann Ltd
Linacre House, Jordan Hill, Oxford OX2 8DP

 PART OF REED INTERNATIONAL BOOKS

OXFORD    LONDON    BOSTON
MUNICH    NEW DELHI    SINGAPORE    SYDNEY
TOKYO    TORONTO    WELLINGTON

First published 1988
Reprinted 1988, 1989 (twice), 1990 (three times), 1991

**British Library Cataloguing in Publication Data**
Chudley, R.
   Building construction handbook
   1. Building
   I. Title
   690      TH145

ISBN 0 7506 0115 9

Printed and bound in Great Britain by
Billing and Sons Ltd, Worcester

# CONTENTS

**Preface**

# PREFACE

This book presents the basic concepts and techniques of building construction, mainly by means of drawings illustrating typical construction details, processes and concepts. I have chosen this method because it reflects the primary means of communication on site between building designer and building contractor – the construction drawing or detail. It must be stressed that the drawings used here represent typical details, chosen to illustrate particular points of building construction or technology; they do not constitute the alpha and omega of any building design, detail or process. The principles they illustrate must therefore, in reality, be applied to the data of the particular problem or situation encountered.

Readers who want to pursue to greater depth any of the topics treated here will find many useful sources of information in specialist textbooks, research reports, manufacturer's literature, codes of practice and similar publications. One such subject is building services, which are dealt with here only in so far as they are applicable to domestic dwellings. A comparable but much wider treatment of services is given in Hall's *Essential Building Services and Equipment*, also published by Heinemann, and there also exist many excellent advanced reference works on particular aspects of building services.

In conclusion, I hope that this book will not only itself prove useful and helpful to the reader, but will act as a stimulus to the observation of actual buildings and the study of works in progress. In this way, the understanding gained here will be continually broadened and deepened by experience.

R.C.

# 1 GENERAL

BUILT ENVIRONMENT
THE STRUCTURE
PRIMARY AND SECONDARY ELEMENTS
CONSTRUCTION ACTIVITIES
CONSTRUCTION DOCUMENTS
CONSTRUCTION DRAWINGS
MODULAR COORDINATION
CONSTRUCTION REGULATIONS
BUILDING REGULATIONS
BRITISH STANDARDS
CI/SfB SYSTEM OF CODING

Environment = surroundings which can be natural, man-made or a combination of these.

Built Environment = created by man with or without the aid of the natural environment.

grasses and wild flowers

deciduous and coniferous trees →

shrubs and bushes

rock outcrops

waterways and lakes

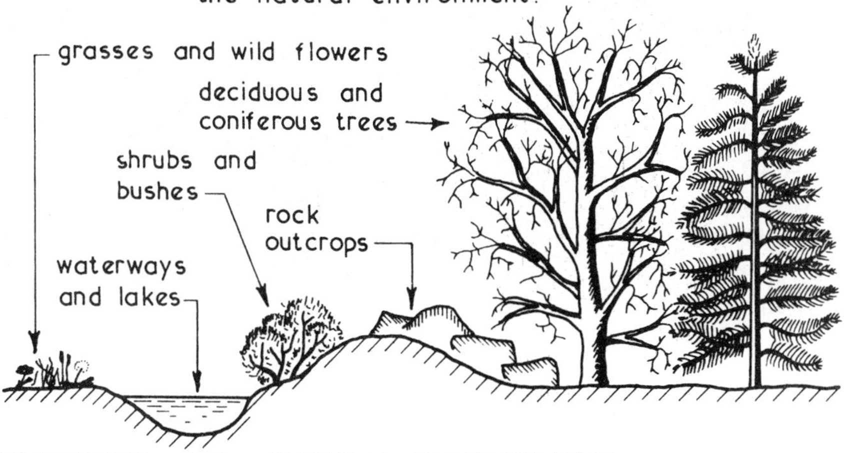

ELEMENTS of the NATURAL ENVIRONMENT

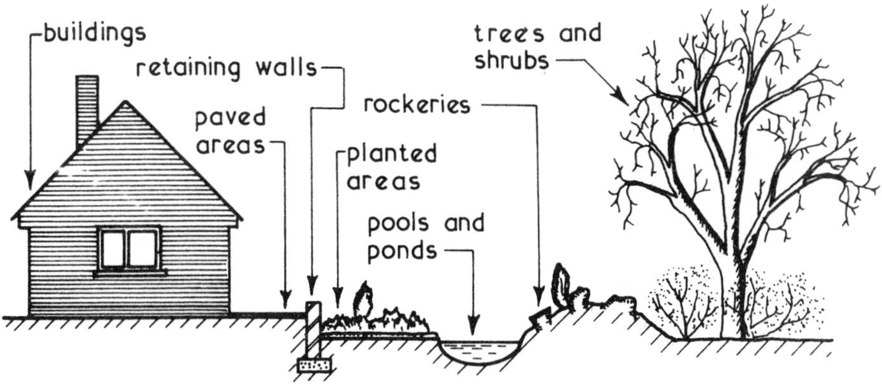

buildings

retaining walls

trees and shrubs

paved areas

rockeries

planted areas

pools and ponds

ELEMENTS of the BUILT ENVIRONMENT (EXTERNAL)

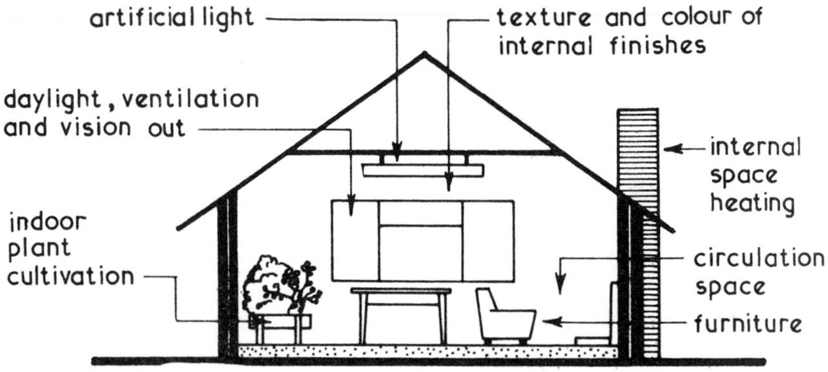

artificial light

texture and colour of internal finishes

daylight, ventilation and vision out

internal space heating

indoor plant cultivation

circulation space

furniture

ELEMENTS of the BUILT ENVIRONMENT (INTERNAL)

Environmental Considerations
1. Planning requirements.
2. Building Regulations.
3. Land restrictions by vendor or lessor.
4. Availability of services.
5. Local amenities including transport.
6. Subsoil conditions.
7. Levels and topography of land.
8. Adjoining buildings or land.
9. Use of building.
10. Daylight and view aspects.

gales

cold winds

N

gales

mild winds

longest day    shortest    day

ORIENTATION ASPECTS

Examples:~

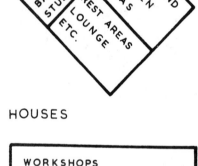

ENTRANCE BATHROOM STUDIOS ETC.

DINING AND KITCHEN AREAS

REST AREAS LOUNGE ETC.

HOUSES

STUDIOS LABORATORIES ART ROOMS

HANDICRAFT ROOMS

WORKSHOPS

LIBRARY

CLASSROOMS
STAFF ROOMS    OFFICES

SCHOOLS

WORKSHOPS
MACHINE SHOPS
STORAGE AREAS

LIGHT ASSEMBLY WORK
AND SIMILAR ACTIVITIES

OFFICES

FACTORIES

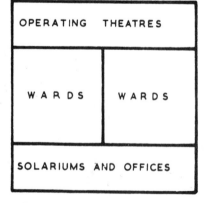

OPERATING    THEATRES

WARDS

WARDS

SOLARIUMS AND OFFICES

HOSPITALS

Physical Considerations

1. Natural contours of land.
2. Natural vegetation and trees.
3. Size of land and/or proposed building.
4. Shape of land and/or proposed building.
5. Approach and access roads and footpaths.
6. Services available.
7. Natural waterways, lakes and ponds.
8. Restrictions such as rights of way tree preservation and ancient buildings.
9. Climatic conditions created by surrounding properties, land or activities.
10. Proposed future developments.

Examples:~

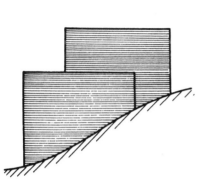

Split level construction to form economic shape.

Shape determined by existing trees.

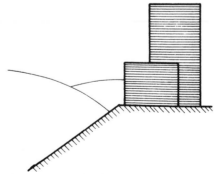

Plateau or high ground solution giving dry site conditions on sloping sites.

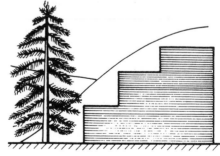

Stepped elevation or similar treatment to blend with the natural environment.

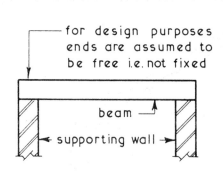

SIMPLY SUPPORTED BEAM

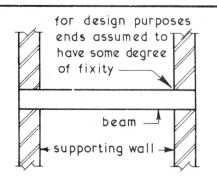

BUILT- IN BEAM

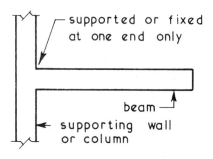

CANTILEVER BEAM

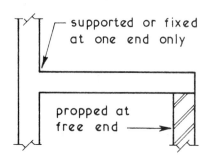

PROPPED CANTILEVER

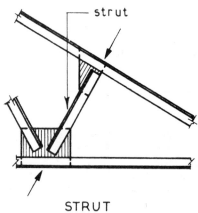

STRUT

structural member which is subjected mainly to compression forces

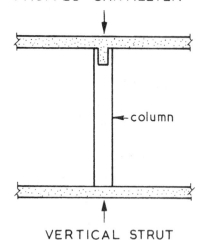

VERTICAL STRUT

usually called a column stanchion or pier

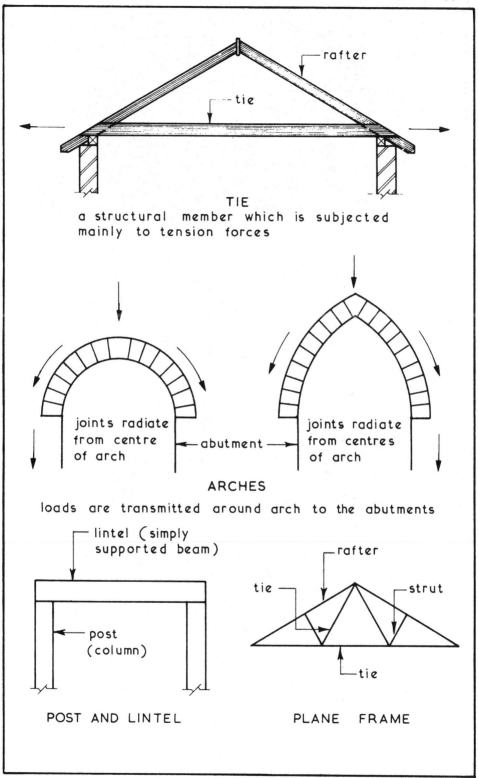

TIE

a structural member which is subjected
mainly to tension forces

joints radiate
from centre
of arch

←— abutment —→

joints radiate
from centres
of arch

ARCHES

loads are transmitted around arch to the abutments

lintel (simply
supported beam)

post
(column)

POST AND LINTEL

rafter

tie

strut

tie

PLANE FRAME

7

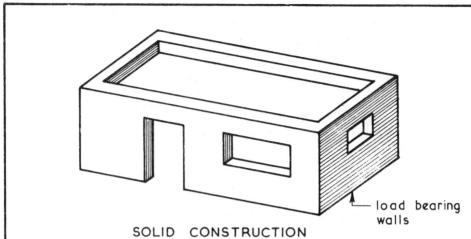

load bearing
walls

## SOLID CONSTRUCTION

structurally limited confined usually to buildings of
low height and short spans

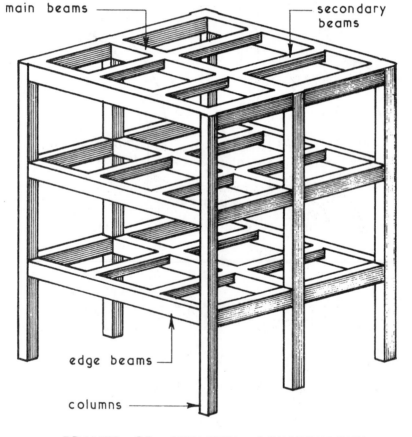

main beams ─── ┐      ┌─── secondary
beams

edge beams ───

columns ───→

## FRAMED OR SKELETAL CONSTRUCTION

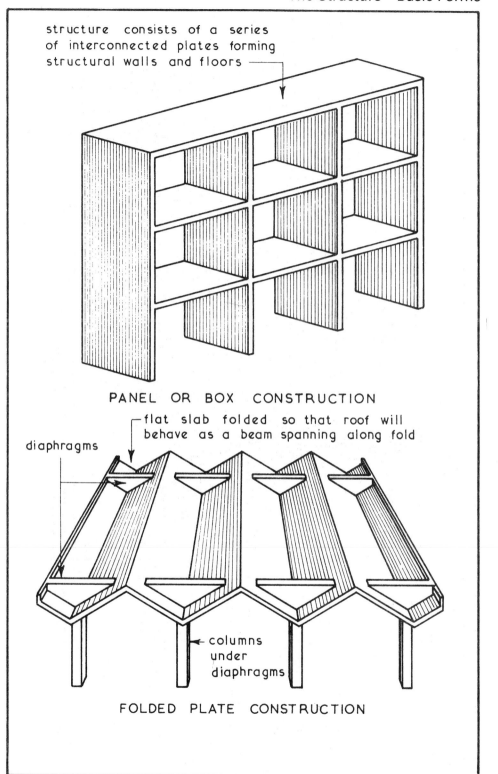

structure consists of a series
of interconnected plates forming
structural walls and floors

PANEL OR BOX CONSTRUCTION

flat slab folded so that roof will
behave as a beam spanning along fold

diaphragms

columns
under
diaphragms

FOLDED PLATE CONSTRUCTION

9

Shell Roofs~ these are formed by a structural curved skin covering a given plan shape and area.

Examples ~

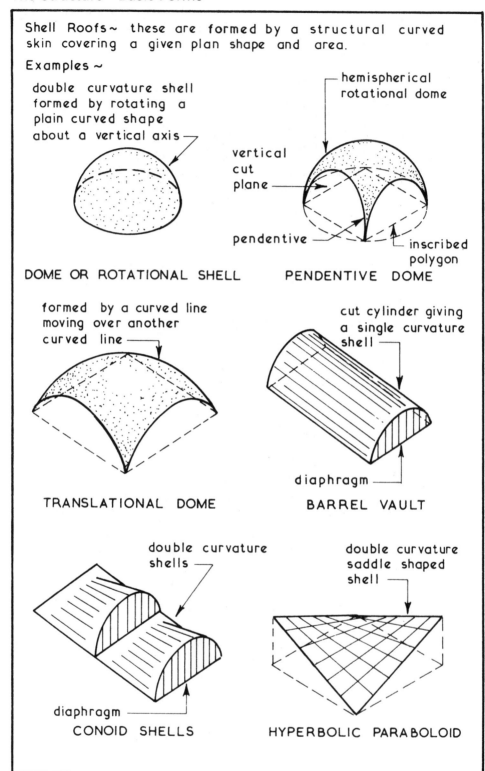

double curvature shell formed by rotating a plain curved shape about a vertical axis

hemispherical rotational dome

vertical cut plane

pendentive

inscribed polygon

DOME OR ROTATIONAL SHELL

PENDENTIVE DOME

formed by a curved line moving over another curved line

cut cylinder giving a single curvature shell

diaphragm

TRANSLATIONAL DOME

BARREL VAULT

double curvature shells

double curvature saddle shaped shell

diaphragm

CONOID SHELLS

HYPERBOLIC PARABOLOID

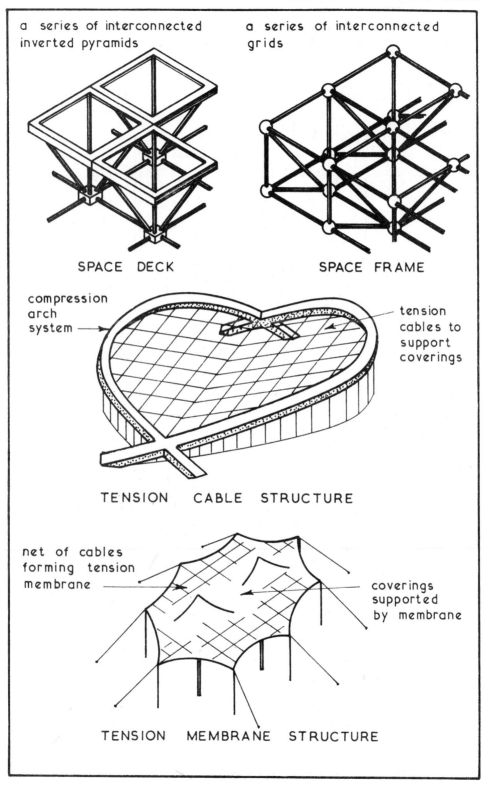

a series of interconnected
inverted pyramids

a series of interconnected
grids

SPACE DECK

SPACE FRAME

compression
arch
system

tension
cables to
support
coverings

TENSION CABLE STRUCTURE

net of cables
forming tension
membrane

coverings
supported
by membrane

TENSION MEMBRANE STRUCTURE

# Substructure

Substructure~ can be defined as all structure below the superstructure which in general terms is considered to include all structure below ground level but including the ground floor bed.

Typical Examples~

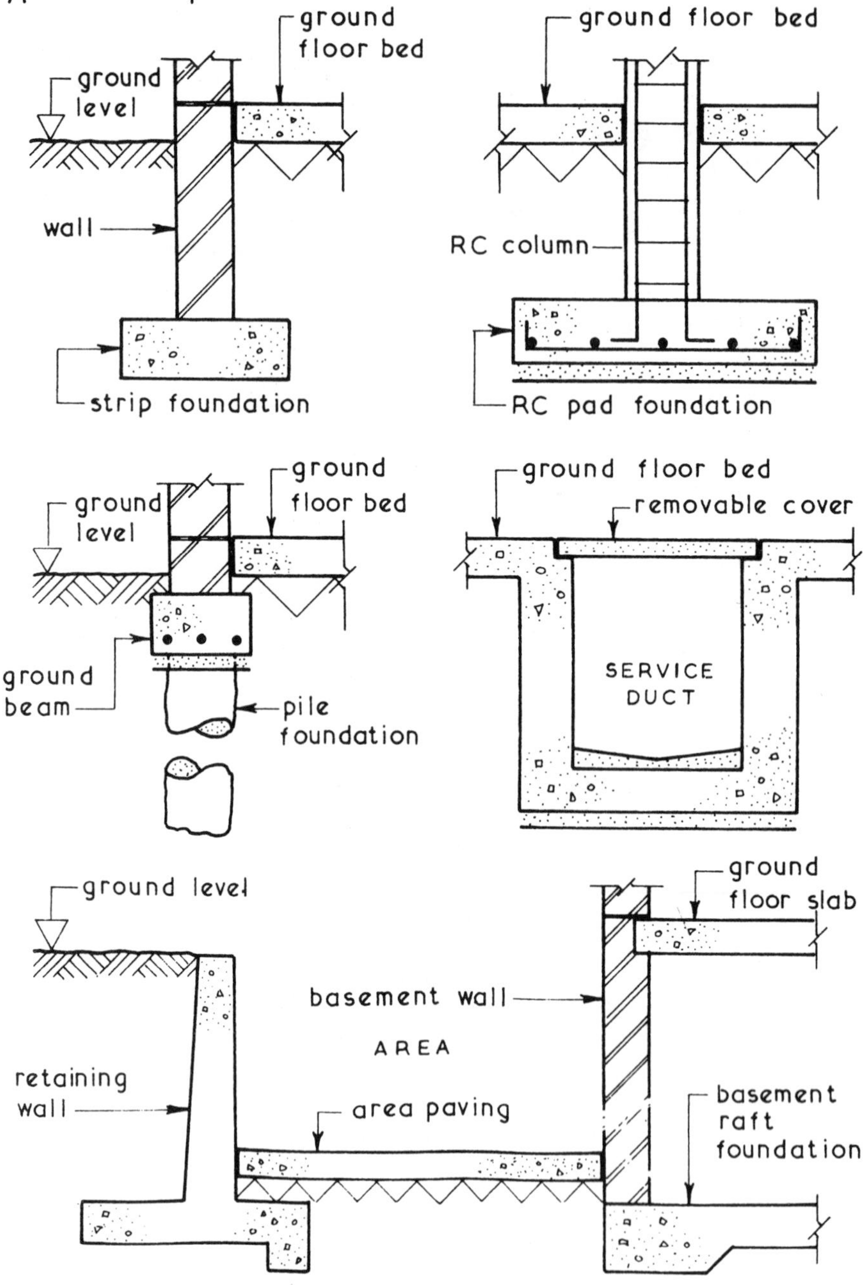

Fig 14   Substructure

Superstructure~ can be defined as all structure above substructure both internally and externally.

Primary Elements~ basically components of the building carcass above the substructure excluding secondary elements, finishes, services and fittings.

Typical Examples~

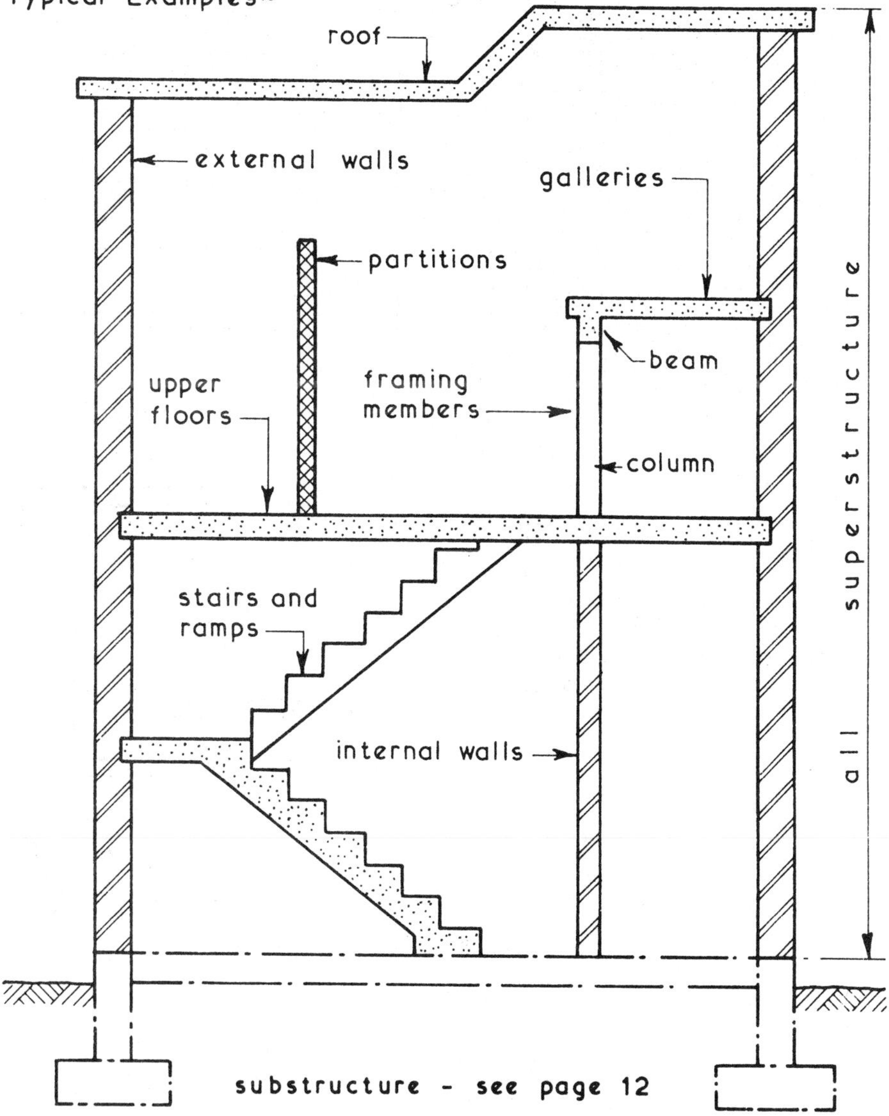

roof

external walls

galleries

partitions

upper floors

framing members

beam

column

superstructure

stairs and ramps

internal walls

all

substructure - see page 12

Secondary Elements~ completion of the structure including completion around and within openings in primary elements.

Typical Examples~

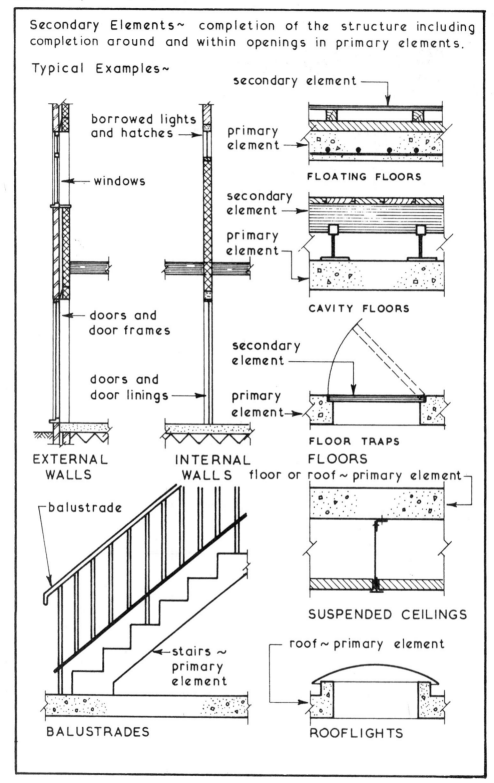

borrowed lights and hatches ──→

windows ──

doors and door frames

doors and door linings ──→

EXTERNAL WALLS

INTERNAL WALLS

primary element ──

primary element

primary element──→

secondary element ──

FLOATING FLOORS

secondary element ──→

primary element ──

CAVITY FLOORS

secondary element ──

primary element──→

FLOOR TRAPS

FLOORS

floor or roof ~ primary element ─

SUSPENDED CEILINGS

─balustrade

stairs ~ primary element

BALUSTRADES

roof ~ primary element

ROOFLIGHTS

Finish~ the final surface which can be self finished as with a trowelled concrete surface or an applied finish such as floor tiles.

Typical Examples~

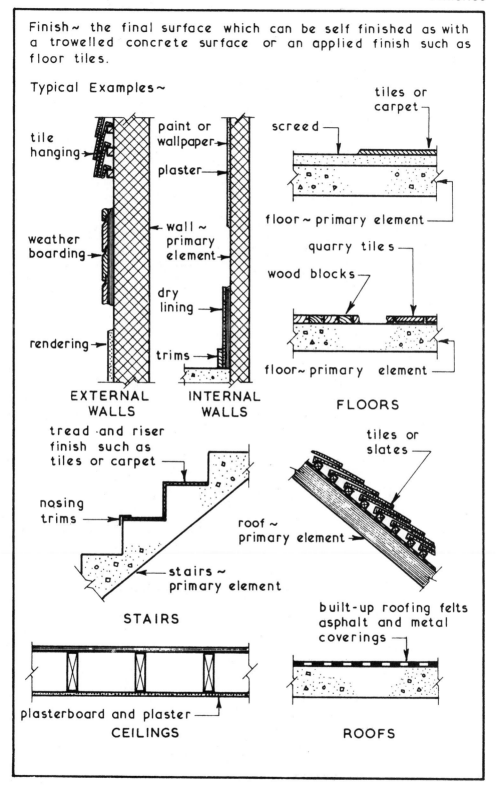

tile hanging

paint or wallpaper

plaster

weather boarding

wall ~ primary element

dry lining

rendering

trims

**EXTERNAL WALLS**

**INTERNAL WALLS**

tiles or carpet

screed

floor~ primary element

quarry tiles

wood blocks

floor~ primary element

**FLOORS**

tread and riser finish such as tiles or carpet

nosing trims

tiles or slates

roof ~ primary element

stairs ~ primary element

**STAIRS**

built-up roofing felts asphalt and metal coverings

plasterboard and plaster
**CEILINGS**

**ROOFS**

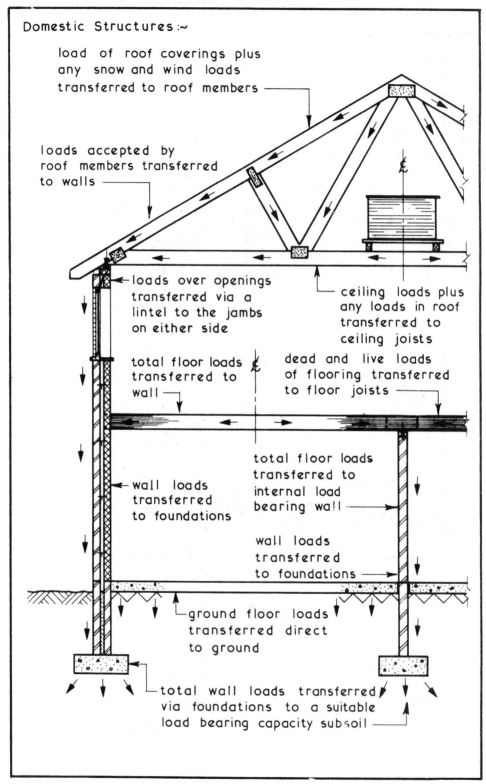

Domestic Structures :~

load of roof coverings plus any snow and wind loads transferred to roof members

loads accepted by roof members transferred to walls

loads over openings transferred via a lintel to the jambs on either side

ceiling loads plus any loads in roof transferred to ceiling joists

total floor loads transferred to wall

dead and live loads of flooring transferred to floor joists

wall loads transferred to foundations

total floor loads transferred to internal load bearing wall

wall loads transferred to foundations

ground floor loads transferred direct to ground

total wall loads transferred via foundations to a suitable load bearing capacity subsoil

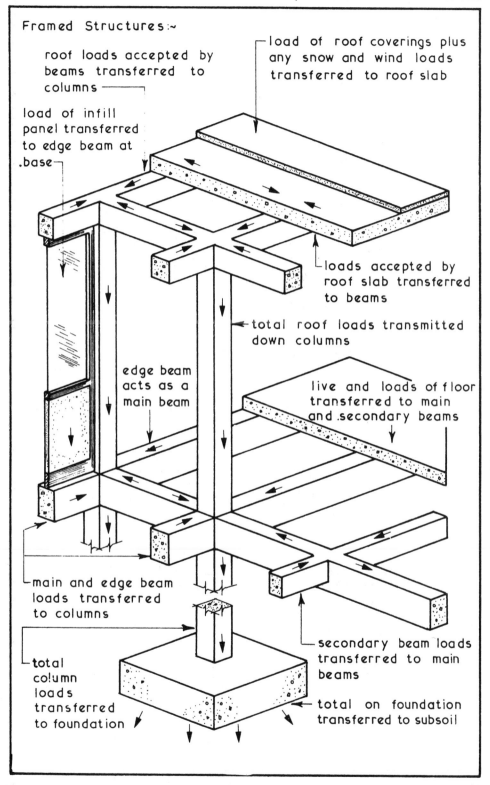

Framed Structures:~

roof loads accepted by beams transferred to columns

load of roof coverings plus any snow and wind loads transferred to roof slab

load of infill panel transferred to edge beam at .base

loads accepted by roof slab transferred to beams

total roof loads transmitted down columns

edge beam acts as a main beam

live and loads of floor transferred to main and .secondary beams

main and edge beam loads transferred to columns

secondary beam loads transferred to main beams

total column loads transferred to foundation

total on foundation transferred to subsoil

External Envelope ~ consists of the materials and components which form the external shell or enclosure of a building. These may be load bearing or non-load bearing according to the structural form of the building.

Primary Functions :~

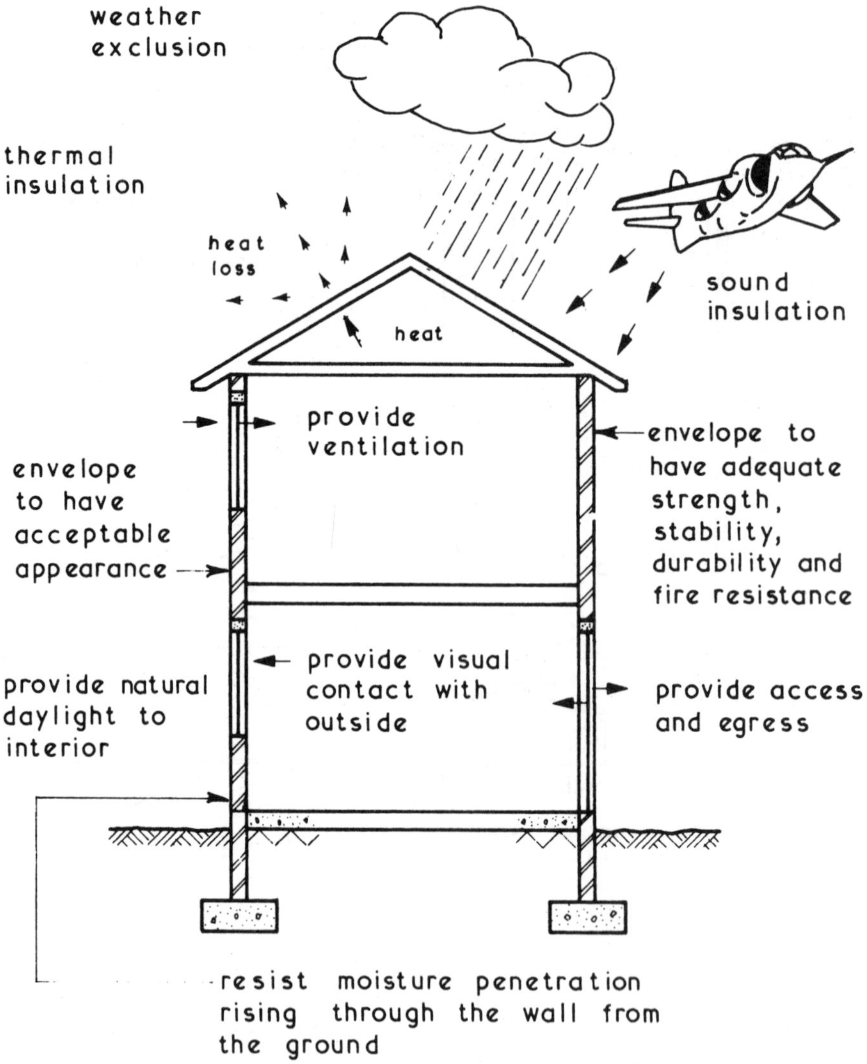

weather exclusion

thermal insulation

heat loss

heat

sound insulation

provide ventilation

envelope to have acceptable appearance --➤

envelope to have adequate strength, stability, durability and fire resistance

provide natural daylight to interior

provide visual contact with outside

provide access and egress

resist moisture penetration rising through the wall from the ground

A Building or Construction Site can be considered as a temporary factory employing the necessary resources to successfully fulfil a contract.

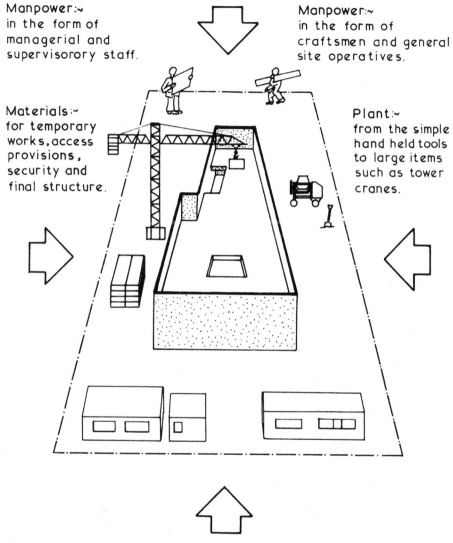

Manpower:~
in the form of managerial and supervisorory staff.

Manpower:~
in the form of craftsmen and general site operatives.

Materials:~
for temporary works, access provisions, security and final structure.

Plant:~
from the simple hand held tools to large items such as tower cranes.

Money :~
in the form of capital investment from the building owner to pay for the land, design team fees and a building contractor who uses his money to buy materials, buy or hire plant and hire labour to enable the project to be realised.

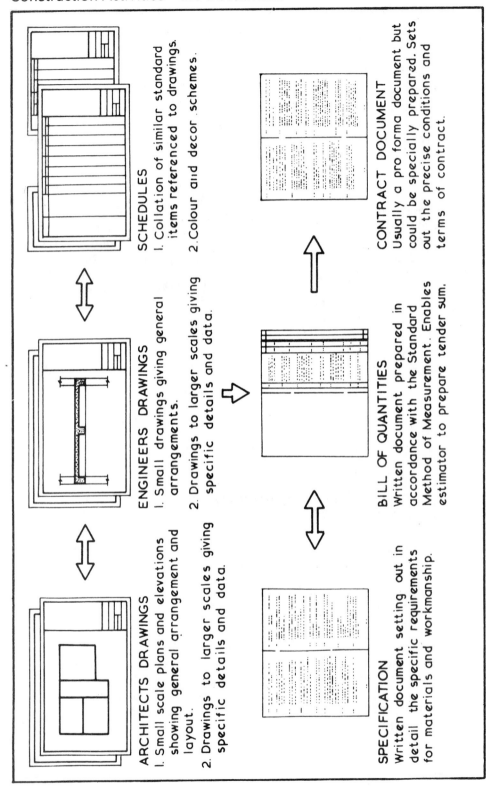

ARCHITECTS DRAWINGS
1. Small scale plans and elevations showing general arrangement and layout.
2. Drawings to larger scales giving specific details and data.

ENGINEERS DRAWINGS
1. Small drawings giving general arrangements.
2. Drawings to larger scales giving specific details and data.

SCHEDULES
1. Collation of similar standard items referenced to drawings.
2. Colour and decor schemes.

SPECIFICATION
Written document setting out in detail the specific requirements for materials and workmanship.

BILL OF QUANTITIES
Written document prepared in accordance with the Standard Method of Measurement. Enables estimator to prepare tender sum.

CONTRACT DOCUMENT
Usually a pro forma document but could be specially prepared. Sets out the precise conditions and terms of contract.

PRELIMINARY SKETCH

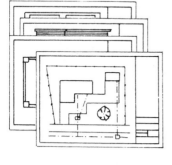

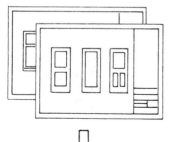

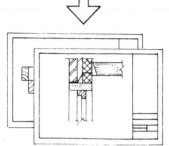

Location Drawings~

Site Plans - used to locate site, buildings, define site levels, indicate services to buildings, identify parts of site such as roads, footpaths and boundaries and to give setting out dimensions for the site and buildings as a whole. Suitable scale: not less than 1:2500

Floor Plans - used to identify and set out parts of the building such as rooms, corridors, doors, windows, etc., Suitable scale not less than 1:100

Elevations - used to show external appearance of all faces and to identify doors and windows. Suitable scale not less than 1:100

Sections - used to provide vertical views through the building to show method of construction. Suitable scale not less than 1:50

Component Drawings~

used to identify and supply data for components to be supplied by a manufacturer or for components not completely covered by assembly drawings. Suitable scale range 1:100 to 1:1

Assembly Drawings~

used to show how items fit together or are assembled to form elements. Suitable scale range 1:20 to 1:5

All drawings should be fully annotated, fully dimensioned and crossed referenced.

Sketch ~ this can be defined as a draft or rough outline of an idea, it can also be a means of depicting a three-dimensional form in a two-dimensional guise. Sketches can be produced free-hand or using rules and set squares to give basic guide lines.

All sketches should be clear, show all the necessary detail and above all be in the correct proportions.

Sketches can be drawn by observing a solid object or they can be produced from conventional orthographic views but in all cases can usually be successfully drawn by starting with an outline 'box' format giving length, width and height proportions and then building up the sketch within the outline box.

Example~ Square Based Chimney Pot.

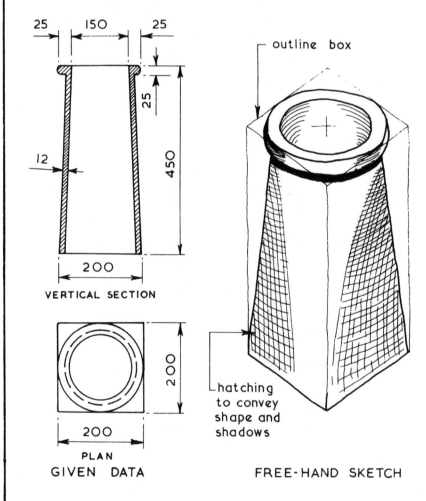

outline box

25  150  25

25

12

450

200

**VERTICAL SECTION**

200

200

200

**PLAN**

hatching
to convey
shape and
shadows

**GIVEN DATA**

**FREE-HAND SKETCH**

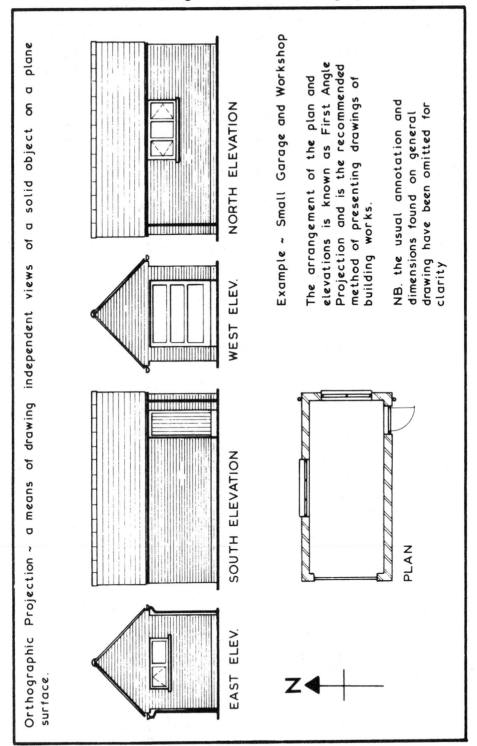

Orthographic Projection ~ a means of drawing independent views of a solid object on a plane surface.

NORTH ELEVATION

WEST ELEV.

SOUTH ELEVATION

EAST ELEV.

PLAN

N

Example ~ Small Garage and Workshop

The arrangement of the plan and elevations is known as First Angle Projection and is the recommended method of presenting drawings of building works.

NB. the usual annotation and dimensions found on general drawing have been omitted for clarity

Isometric Projection ~ a pictorial projection of a solid object on a plane surface drawn so that all vertical lines remain vertical and of true scale length, all horizontal lines are drawn at an angle of 30° and are of true scale length therefore scale measurements can be taken on the vertical and 30° lines but cannot be taken on any other inclined line.

A similar drawing can be produced using an angle of 45° for all horizontal lines and is called an Axonometric Projection

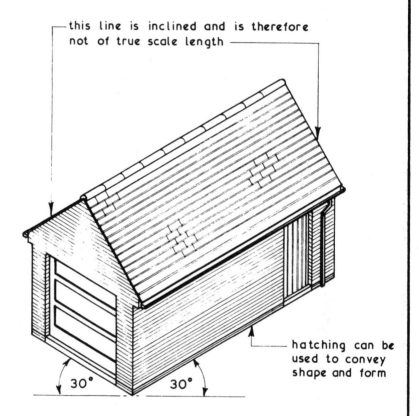

this line is inclined and is therefore not of true scale length

hatching can be used to convey shape and form

30°      30°

ISOMETRIC PROJECTION SHOWING SOUTH AND WEST ELEVATIONS OF SMALL GARAGE AND WORKSHOP ILLUSTRATED ON PAGE 23

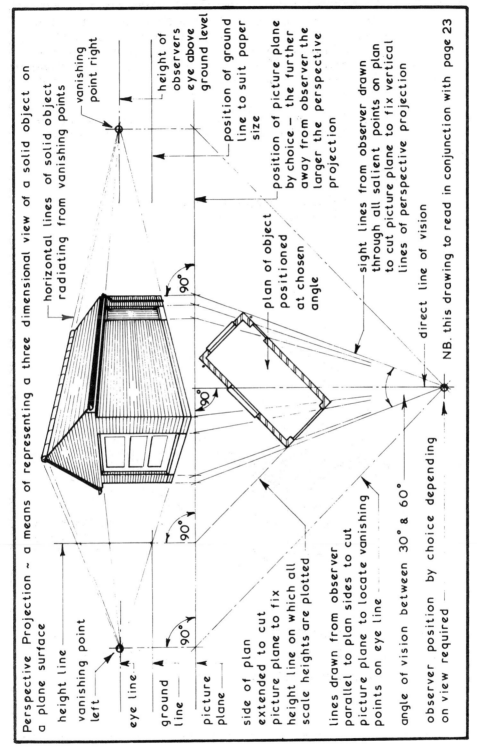

Perspective Projection ~ a means of representing a three dimensional view of a solid object on a plane surface

height line

vanishing point left

eye line

ground line

picture plane

side of plan extended to cut picture plane to fix height line on which all scale heights are plotted

lines drawn from observer parallel to plan sides to cut picture plane to locate vanishing points on eye line

angle of vision between 30° & 60°

observer position by choice depending on view required

horizontal lines of solid object radiating from vanishing points

vanishing point right

height of observers eye above ground level

position of ground line to suit paper size

position of picture plane by choice — the further away from observer the larger the perspective projection

plan of object positioned at chosen angle

sight lines from observer drawn through all salient points on plan to cut picture plane to fix vertical lines of perspective projection

direct line of vision

NB. this drawing to read in conjunction with page 23

90°

90°

90°

90°

25

Drawings~ these are the major means of communication between the designer and the contractor as to what, where and how the proposed project is to be constructed.

Drawings should therefore be clear, accurate, contain all the necessary information and be capable of being easily read.

To achieve these objectives most designers use the symbols and notations recommended in BS 1192  BS 308  and PD 6479 to which students should refer for full information.

Typical Examples~

| | |
|---|---|
| ———————— | outlines |
| ———————— | dimension and hatching lines |
| — — — — — — | hidden detail |
| —·———·— | drain and pipe lines |
| —··———··— | centre lines |
| ⊢————⊣ | modular and coordinating dimension lines |
| ⊢←———→⊣ | work size dimension lines |
| ⑫—·——— | controlling and grid lines |

**LINES**

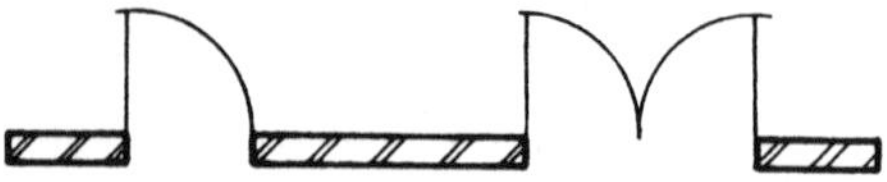

single door
single swing

double door
single swing

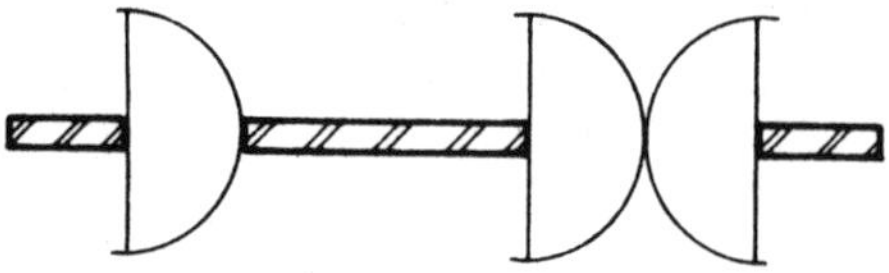

single door
double swing

double door
double swing

folding doors
side hung

folding doors
centre hung

**DOORS**

Hatchings~ the main objective is to differentiate between the materials being used thus enabling rapid recognition and location. Whichever hatchings are chosen they must be used consistently throughout the whole set of drawings.
In large areas it is not always necessary to hatch the whole area.

Symbols~ these are graphical representations and should wherever possible be drawn to scale but above all they must be consistent for the whole set of drawings and clearly drawn.

Typical Examples~

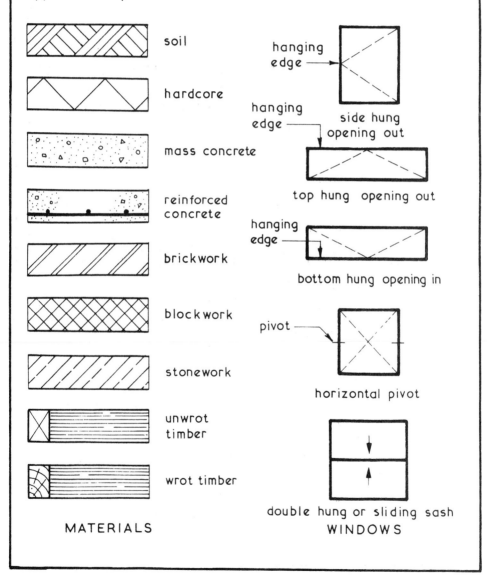

MATERIALS

WINDOWS

soil

hardcore

mass concrete

reinforced concrete

brickwork

blockwork

stonework

unwrot timber

wrot timber

side hung opening out

top hung opening out

bottom hung opening in

horizontal pivot

double hung or sliding sash

## Drawings—Hatchings, Symbols and Notations

| Name | Symbol | Name | Symbol |
|---|---|---|---|
| Rainwater pipe | ◯ R W P | Distribution board | ▢ |
| Gully | ▢ G | Electricity meter | (symbol) |
| Inspection chambers | —▢— soil or foul IC / —◯— surface water IC | Switched socket outlet | (symbol) |
| Boiler | ▢ B | Switch | (symbol) |
| Sink | S | Two way switch | (symbol) |
| Bath | (symbol) | Pendant switch | (symbol) |
| Wash basin | W B | Filament lamp | ◯ |
| Shower unit | S | Fluorescent lamp | (symbol) |
| Urinal | stall    bowl | Bed | (symbol) |
| Water closet | (symbol) | Table and chairs | (symbol) |
| TYPICAL COMPONENT, FITMENT AND ELECTRICAL SYMBOLS | | | |

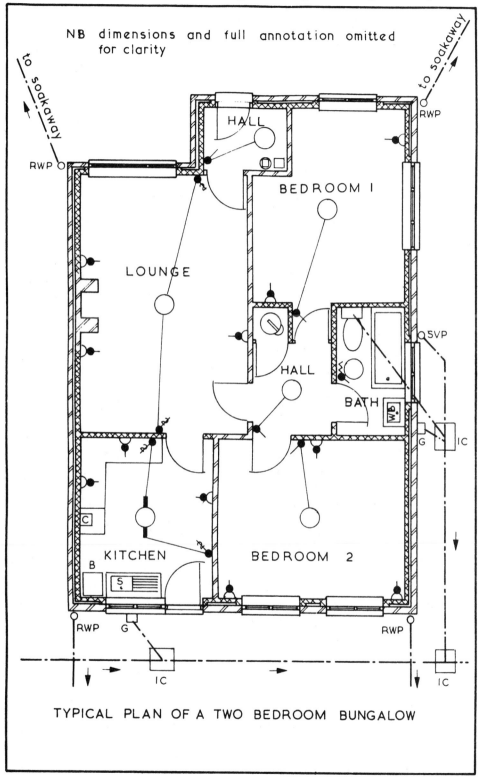

NB dimensions and full annotation omitted for clarity

to soakaway

to soakaway

RWP

RWP

HALL

BEDROOM 1

LOUNGE

SVP

HALL

BATH

WB

G

IC

RWP

G

IC

KITCHEN

C

B

S

BEDROOM 2

RWP

RWP

IC

IC

TYPICAL PLAN OF A TWO BEDROOM BUNGALOW

# Modular Coordination

Modular Coordination ~ a module can be defined as a basic dimension which could for example form the basis of a planning grid in terms of multiples and submultiples of the standard module.

Typical Modular Coordinated Planning Grid ~

Let M = the standard module

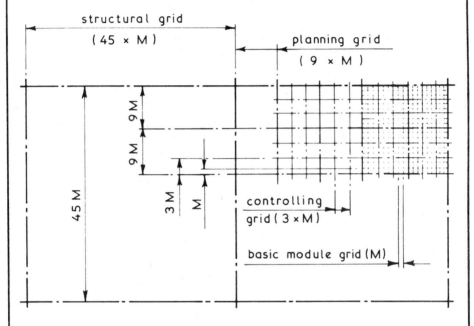

Structural Grid – used to locate structural components such as beams and columns.

Planning Grid – based on any convenient modular multiple for regulating space requirements such as rooms.

Controlling Grid – based on any convenient modular multiple for location of internal walls, partitions etc.

Basic Module Grid – used for detail location of components and fittings.

All the above grids, being based on a basic module, are contained one within the other and are therefore interrelated. These grids can be used in both the horizontal and vertical planes thus forming a three dimensional grid system. If a first preference numerical value is given to M dimensional coordination **is established - see page 31**

Dimensional Coordination ~ the practical aims of this concept are to :-

1. Size components so as to avoid the wasteful process of cutting and fitting on site.

2. Obtain maximum economy in the production of components.

3. Reduce the need for the manufacture of special sizes.

4. Increase the effective choice of components by the promotion of interchangeability.

BS 6570 specifies the increments of size for coordinating dimensions of building components thus :-

| Preference | 1st | 2nd | 3rd | 4th |
|---|---|---|---|---|
| Size (mm) | 300 | 100 | 50 | 25 |

the 3rd. and 4th. preferences having a maximum of 300mm

Dimensional Grids – the modular grid network as shown on page 30 defines the space into which dimensionally coordinated components must fit. An important factor is that the component must always be undersized to allow for the joint which is sized by the obtainable degree of tolerance and site assembly :-

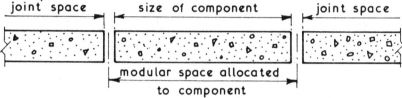

Controlling Lines, Zones and Controlling Dimensions – these terms can best be defined by example :-

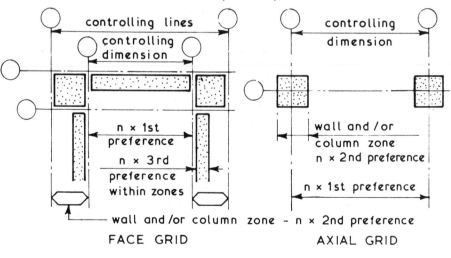

FACE GRID　　　　　　AXIAL GRID

Construction Regulations ~ these are Statutory Instruments made under the Factories Acts of 1937 and 1961 and come under the umbrella of the Health and Safety at Work etc., Act 1974. They set out the minimum legal requirements for construction works and relate primarily to the health, safety and welfare of the work force. The requirements contained within these documents must therefore be taken into account when planning construction operations and during the actual construction period. Reference should be made to the relevant document for specific requirements but the broad areas covered can be shown thus :-

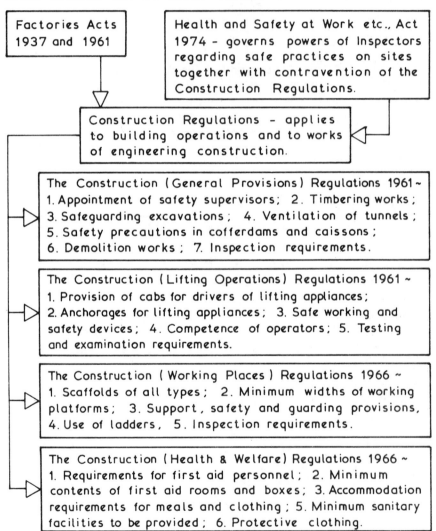

Factories Acts 1937 and 1961

Health and Safety at Work etc., Act 1974 - governs powers of Inspectors regarding safe practices on sites together with contravention of the Construction Regulations.

Construction Regulations - applies to building operations and to works of engineering construction.

The Construction (General Provisions) Regulations 1961 ~
1. Appointment of safety supervisors; 2. Timbering works;
3. Safeguarding excavations; 4. Ventilation of tunnels;
5. Safety precautions in cofferdams and caissons;
6. Demolition works; 7. Inspection requirements.

The Construction (Lifting Operations) Regulations 1961 ~
1. Provision of cabs for drivers of lifting appliances;
2. Anchorages for lifting appliances; 3. Safe working and safety devices; 4. Competence of operators; 5. Testing and examination requirements.

The Construction (Working Places) Regulations 1966 ~
1. Scaffolds of all types; 2. Minimum widths of working platforms; 3. Support, safety and guarding provisions,
4. Use of ladders, 5. Inspection requirements.

The Construction (Health & Welfare) Regulations 1966 ~
1. Requirements for first aid personnel; 2. Minimum contents of first aid rooms and boxes; 3. Accommodation requirements for meals and clothing; 5. Minimum sanitary facilities to be provided; 6. Protective clothing.

The Building Regulations – this is a Statutory Instrument which sets out the minimum performance standards for the design and construction of buildings and where applicable to the extension of buildings. The regulations are supported by other documents which generally give guidance on how to achieve the required performance standards. The relationship of these and other documents is set out below :-

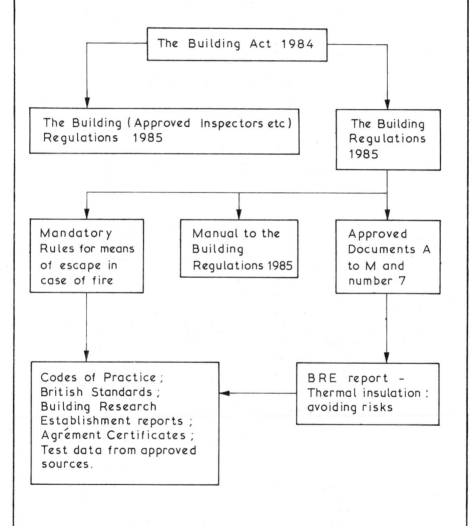

NB. the Building Regulations 1985 apply to England and Wales but not to Scotland and Northern Ireland which have separate systems of control.

## Building Regulations

Approved Documents ~ these are non-statutory publications supporting the Building Regulations prepared by the Department of the Environment, approved by the Secretary of State and issued by Her Majesty's Stationery Office. The Approved Documents (ADs) have been compiled to give practical guidance to comply with the performance standards set out in the various regulations. They are not mandatory but in the event of a dispute they will be seen in court as tending to show compliance with the requirements of the Building Regulations. If other solutions are used to satisfy the requirements the Regulations the burden of proving compliance rests with the applicant or designer. The various Approved Documents and their contents are set out below :-

**Part II**

Approved Document to support Regulation 7
Reg. 7 - Materials and workmanship.

**Schedule 1**

Approved Document A - STRUCTURE - Loading and ground movement; Disproportionate collapse.

Approved Document B - FIRE - Internal and external fire spread.

Approved Document C - SITE PREPARATION AND RESISTANCE TO MOISTURE - Site preparation and contaminants; Resistance to weather and ground moisture.

Approved Document D - TOXIC SUBSTANCES - Cavity insulation.

Approved Document E - SOUND - Airborne and impact sound.

Approved Document F - VENTILATION - Means of ventilation; Condensation.

Approved Document G - HYGIENE - Bathrooms; Hot water storage; Sanitary and washing accommodation.

Approved Document H - DRAINAGE AND WASTE DISPOSAL - Sanitary pipework and drainage; Cesspools and tanks; Rainwater drainage; Solid waste storage.

Approved Document J - HEAT PRODUCING APPLIANCES.

Approved Document K - STAIRS, RAMPS AND GUARDS.

Approved Document L - CONSERVATION OF FUEL AND POWER.

Approved Document M - ACCESS FOR DISABLED PEOPLE.

Example in the Use of Approved Documents

Problem :- the sizing of suspended upper floor joists to be spaced at 400 mm centres with a clear span of 3·600 m for use in a two storey domestic dwelling.

Building Regulation A1 :- states that the building shall be so constructed that the combined dead, imposed and wind loads are sustained and transmitted to the ground —

(a) safely, and

(b) without causing such deflection or deformation of any part of the building, or such movement of the ground, as will impair the stability of any part of another building.

Approved Document A :- this gives a series of tables the sizing of certain timber members for single family houses of not more than three storeys high. The table which can be applied to the above problem is Table B3 assuming a strength class 3 softwood timber is to be used.

Solution :-

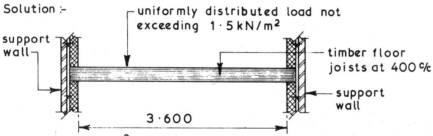

Dead load ($kN/m^2$) supported by joist excluding mass of joist :-

| | |
|---|---|
| Floor finish - carpet | — 0·03 ⎫ weights of |
| Flooring - 20 mm thick particle board | — 0·15 ⎬ materials |
| Ceiling - 9·5 mm thick plasterboard | — 0·08 ⎪ from BS 648 |
| Ceiling finish - 3 mm thick plaster | — 0·04 ⎭ |

total dead load $= \underline{\underline{0·30}}$ $kN/m^2$

Dead loading is therefore in the 0·25 to 0·50 $kN/m^2$ band

From Table B3 of AD 'A' suitable joist sizes are :-

38 × 200;  47 × 175;  50 × 175;  63 × 175  and  75 × 150

Final choice of section to be used will depend upon cost; availability; practical considerations and / or personal preference.

Building Control ~ unless the applicant has opted for control by a private approved inspector under The Building ( Approved Inspectors etc.) Regulations 1985 the control of building works in the contex of the Building Regulations is vested in the Local Authority. There are two systems of control namely the Building Notice and the Deposit of Plans. The sequence of systems is shown below :-

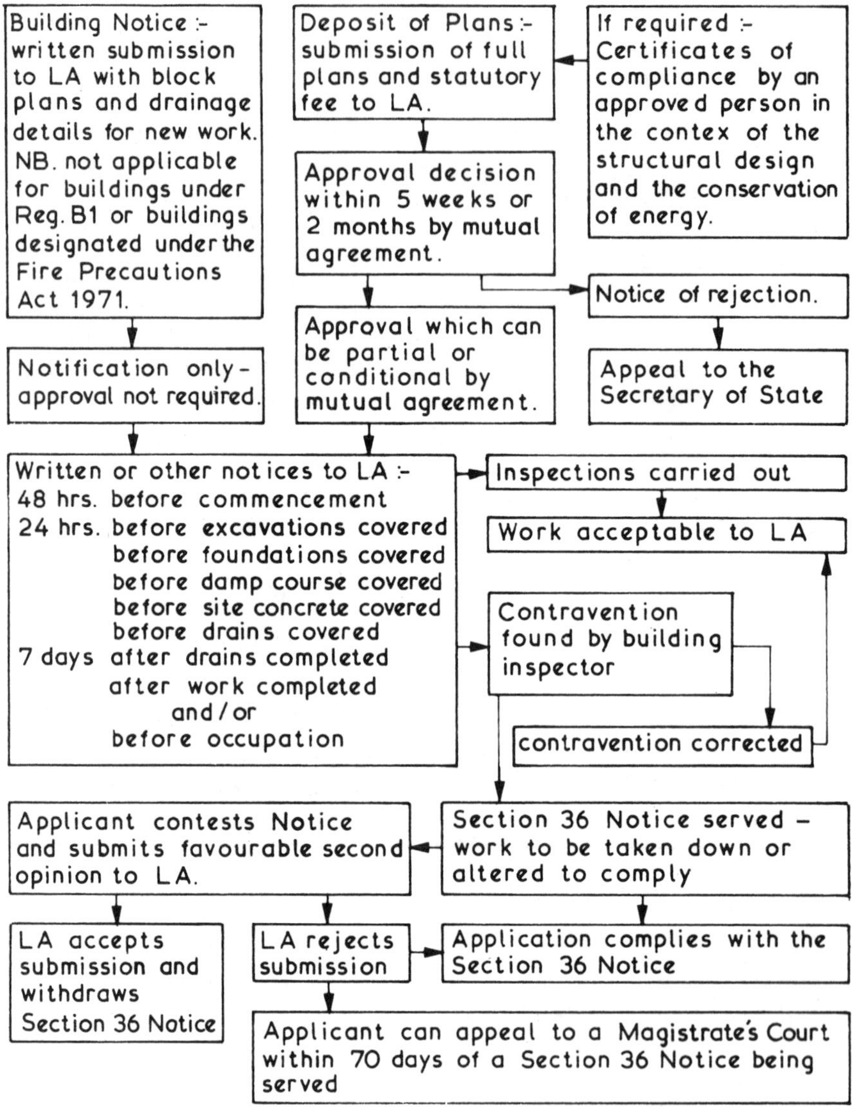

Building Notice :- written submission to LA with block plans and drainage details for new work. NB. not applicable for buildings under Reg. B1 or buildings designated under the Fire Precautions Act 1971.

Deposit of Plans :- submission of full plans and statutory fee to LA.

If required :- Certificates of compliance by an approved person in the contex of the structural design and the conservation of energy.

Approval decision within 5 weeks or 2 months by mutual agreement.

Notification only - approval not required.

Approval which can be partial or conditional by mutual agreement.

Notice of rejection.

Appeal to the Secretary of State

Written or other notices to LA :-
48 hrs. before commencement
24 hrs. before excavations covered
before foundations covered
before damp course covered
before site concrete covered
before drains covered
7 days after drains completed
after work completed
and / or
before occupation

Inspections carried out

Work acceptable to LA

Contravention found by building inspector

contravention corrected

Applicant contests Notice and submits favourable second opinion to LA.

Section 36 Notice served — work to be taken down or altered to comply

LA accepts submission and withdraws Section 36 Notice

LA rejects submission

Application complies with the Section 36 Notice

Applicant can appeal to a Magistrate's Court within 70 days of a Section 36 Notice being served

NB. In some stages of the above sequence statutory fees are payable as set out in The Building (Prescribed Fees) Regulations 1985

British Standards ~ these are publications issued by the British Standards Institution which give recommended minimum standards for materials, components, design and construction practises. These recommendations are not legally enforceable but some of the Building Regulations refer directly to specific British Standards and accept them as deemed-to-satisfy provisions. All materials and components complying with a particular British Standard are marked with the British Standards kitemark thus :- ⊕ together with the appropriate BS number.

This symbol assures the user that the product so marked has been produced and tested in accordance with the recommendations set out in that specific standard. British Standards can be purchased from the Sales Department at 101 Pentonville Road, London N1 9ND, as individual documents or in a combined form known as BS Handbook No. 3 which gives summaries of the basic data contained in over 1 200 of the British Standards applicable to building. British Standards are constantly under review and are amended, revised and rewritten as necessary, therefore a check should always be made to ensure that any standard being used is the current issue. There are over 1 500 British Standards which are directly related to the construction industry and these are prepared in four formats :-

1. British Standards - these give recommendations for the minimum standard of quality and testing for materials and components. Each standard number is prefixed BS.

2. Codes of Practice - these give recommendations for good practice relative to design, manufacture, construction, installation and maintenance with the main objectives of safety, quality, economy and fitness for the intended purpose. Each code of practice number is prefixed CP or BS.

3. Draft for Development - these are issued instead of a British Standard or Code of Practice when there is insufficient data or information to make firm or positive recommendations. Each draft number is prefixed DD.

4. Published Document - these are publications which cannot be placed into any one of the above categories. Each published document is numbered and prefixed PD.

CI/SfB System ~ this is a coded filing system for the classification and storing of building imformation and data. It was created in Sweden under the title of 'Samarbetskommittën för Byggnadsfrågor' and was introduced into this country in 1961 by the RIBA. In 1968 the CI (Construction Index) was added to the system which is used Nationally and recognised throughout the construction industry. The system consists of 5 sections called tables which are subdivided by a series of letters or numbers and these are listed in the CI/SfB index book to which reference should always be made in the first instance to enable an item to be correctly filed or retrieved.

### Table O - Physical Environment

This table contains ten sections O to 9 and deals mainly with the end product (i.e. the type of building.) Each section can be further subdivided (e.g. 21, 22 et seq.) as required.

### Table 1 - Elements

This table contains ten sections numbered (--) to (9-) and covers all parts of the structure such as walls, floors and services. Each section can be further subdivided (e.g. 31, 32 et seq,) as required.

### Table 2 - Construction Form

This table contains twenty five sections lettered A to Z (O being omitted) and covers construction forms such as excavation work, blockwork, cast insitu work etc., and is not subdivided but used in conjunction with Table 3

### Table 3 - Materials

This table contains twenty five sections lettered a to z (l being omitted) and covers the actual materials used in the construction form such as metal, timber, glass etc., and can be subdivided (e.g. n1, n2 et seq.) as required.

### Table 4 - Activities and Requirements

This table contains twenty five sections lettered (A) to (Z), (O being omitted) and covers anything which results from the building process such as shape, heat, sound, etc., Each section can be further subdivided [ (M1), (M2) et seq,] as required.

# 2 SITE WORKS

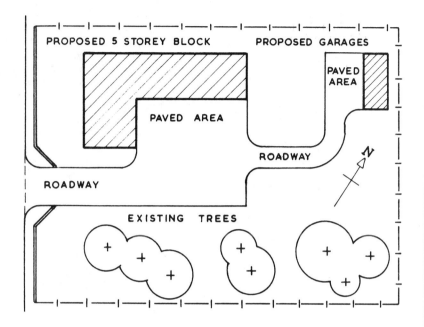

SITE INVESTIGATIONS

SOIL INVESTIGATION

SOIL ASSESSMENT AND TESTING

SITE LAYOUT CONSIDERATIONS

SITE SECURITY

SITE LIGHTING AND ELECTRICAL SUPPLY

SITE OFFICE ACCOMMODATION

MATERIALS STORAGE

MATERIALS TESTING

SETTING OUT

ROAD CONSTRUCTION

TUBULAR SCAFFOLDING AND SCAFFOLDING SYSTEMS

SHORING SYSTEMS

Site Investigation For New Works ~ the basic objective of this form of site investigation is to collect systematically and record all the necessary data which will be needed or will help in the design and construction processes of the proposed work. The collected data should be presented in the form of fully annotated and dimensioned plans and sections. Anything on adjacent sites which may affect the proposed works or conversely anything appertaining to the proposed works which may affect an adjacent site should also be recorded.

Typical Data Required ~

boundary hedges and /or fencing

property boundary lines and location of site

orientation

N

existing trees — type, girth, spread and height

existing buildings

trees and buildings on adjacent site

details of above ground obstructions such as transmisson lines

62·85
+
spot levels

contour lines     63·00

full data as to type, size, depth and location of all services such as gas, water, drains, electricity, telephone and relay services —

planning or similar restrictions on proposed building or structure

subsoil investigation data of soil types and properties together with ground water conditions

62·57

Bench Marks and OS levels

existing sewers

to Oldtown     A 3214     to Newtown

# Trial Pits and Hand Auger Holes

Purpose ~ primarily to obtain subsoil samples for identification, classification and ascertaining the subsoil's characteristics and properties. Trial pits and augered holes may also be used to establish the presence of any geological faults and the upper or lower limits of the water table.

Typical Details ~

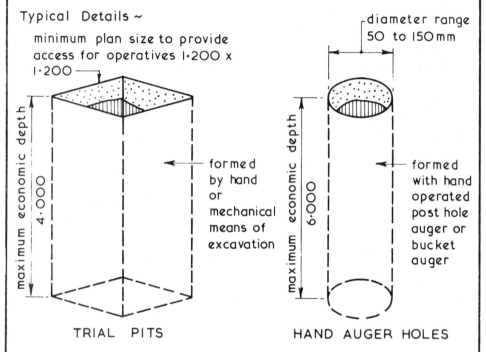

minimum plan size to provide access for operatives 1·200 x 1·200

diameter range 50 to 150 mm

maximum economic depth 4·000 — formed by hand or mechanical means of excavation

maximum economic depth 6·000 — formed with hand operated post hole auger or bucket auger

TRIAL PITS

HAND AUGER HOLES

General use ~
dry ground which requires little or no temporary support to sides of excavation.

Subsidiary use ~
to expose and /or locate underground services.

Advantages ~
subsoil can be visually examined insitu — both disturbed and undisturbed samples can be obtained.

General use ~
dry ground but liner tubes could be used if required to extract subsoil samples at a depth beyond the economic limit of trial holes.

Advantages ~
generally a cheaper and simpler method of obtaining subsoil samples than the trial pit method.

Trial pits and holes should be sited so that the subsoil samples will be representative but not interfering with works

Site Investigation ~ this is an all embracing term covering every aspect of the site under investigation.

Soil Investigation ~ specifically related to the subsoil beneath the site under investigation and could be part of or separate from the site investigation.

Purpose of Soil Investigation ~

1. Determine the suitability of the site for the proposed project.
2. Determine an adequate and economic foundation design.
3. Determine the difficulties which may arise during the construction process and period.
4. Determine the occurrence and /or cause of all changes in subsoil conditions.

The above purposes can usually be assessed by establishing the physical, chemical and general characteristics of the subsoil by obtaining subsoil samples which should be taken from positions on the site which are truly representative of the area but are not taken from the actual position of the proposed foundations. A series of samples extracted at the intersection points of a 20·000 square grid pattern should be adequate for most cases.

Soil Samples ~these can be obtained as disturbed or as undisturbed samples.

Disturbed Soil Samples ~ these are soil samples obtained from boreholes and trial pits. The method of extraction disturbs the natural structure of the subsoil but such samples are suitable for visual grading, establishing the moisture content and some laboratory tests. Disturbed soil samples should be stored in labelled air tight jars.

Undisturbed Soil Samples ~ these are soil samples obtained using coring tools which preserve the natural structure and properties of the subsoil. The extracted undisturbed soil samples are labelled and laid in wooden boxes for dispatch to a laboratory for testing. This method of obtaining soil samples is suitable for rock and clay subsoils but difficulties can be experienced in trying to obtain undisturbed soil samples in other types of subsoil.

The test results of soil samples are usually shown on a drawing which gives the location of each sample and the test results in the form of a hatched legend or section.

## Soil Investigation

Depth of Soil Investigation ~ before determining the actual method of obtaining the required subsoil samples the depth to which the soil investigation should be carried out must be established. This is usually based on the following factors-

1. Proposed foundation type.
2. Pressure bulb of proposed foundation.
3. Relationship of proposed foundation to other foundations.

Typical Examples ~

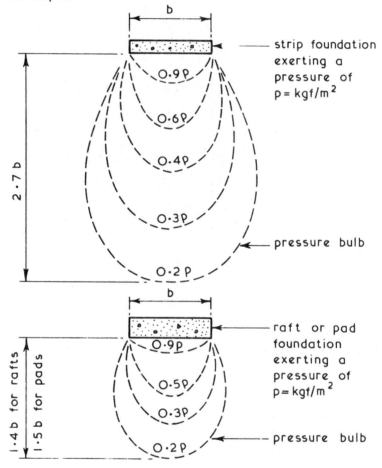

strip foundation exerting a pressure of $p = kgf/m^2$

pressure bulb

raft or pad foundation exerting a pressure of $p = kgf/m^2$

pressure bulb

Pressure bulbs of less than 20% of original loading at foundation level can be ignored — this applies to all foundation types.

**For further examples see page 45**

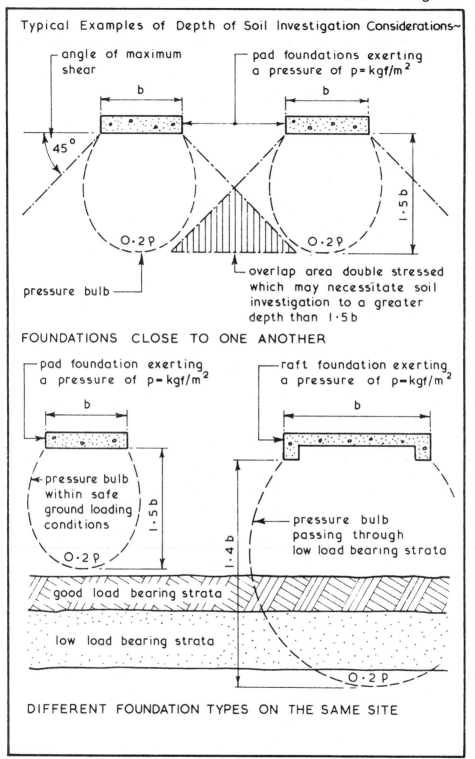

Typical Examples of Depth of Soil Investigation Considerations~

angle of maximum shear

pad foundations exerting a pressure of $p = kgf/m^2$

b

b

45°

1.5 b

O.2P

O.2P

pressure bulb

overlap area double stressed which may necessitate soil investigation to a greater depth than 1.5b

FOUNDATIONS CLOSE TO ONE ANOTHER

pad foundation exerting a pressure of $p = kgf/m^2$

raft foundation exerting a pressure of $p = kgf/m^2$

b

b

pressure bulb within safe ground loading conditions

1.5 b

O.2P

pressure bulb passing through low load bearing strata

1.4 b

good load bearing strata

low load bearing strata

O.2P

DIFFERENT FOUNDATION TYPES ON THE SAME SITE

# Soil Investigation

Soil Investigation Methods~ method chosen will depend on several factors –

1. Size of contract.
2. Type of proposed foundation.
3. Type of sample required.
4. Type of subsoils which may be encountered.

As a general guide the most suitable methods in terms of investigation depth are –

1. Foundations up to 3·000 deep – trial pits.
2. Foundations up to 30·000 deep – borings.
3. Foundations over 30·000 deep – deep borings and insitu examinations from tunnels and / or deep pits.

Typical Trail Pit Details ~

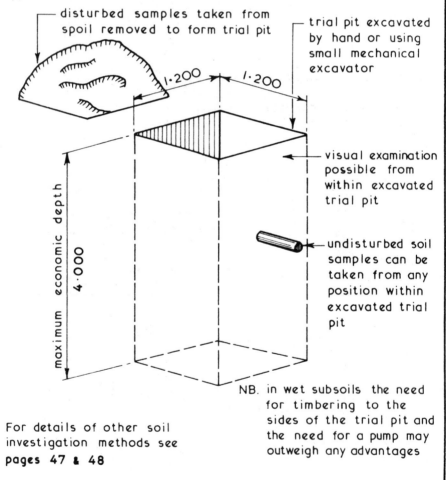

disturbed samples taken from spoil removed to form trial pit

trial pit excavated by hand or using small mechanical excavator

1·200  1·200

visual examination possible from within excavated trial pit

undististurbed soil samples can be taken from any position within excavated trial pit

maximum economic depth
4·000

NB. in wet subsoils the need for timbering to the sides of the trial pit and the need for a pump may outweigh any advantages

For details of other soil investigation methods see pages 47 & 48

Boring Methods to Obtain Disturbed Soil Samples ~

1. Hand or Mechanical Auger — suitable for depths up to 3·000 using a 150 or 200mm diameter flight auger.
2. Mechanical Auger — suitable for depths over 3·000 using a flight or Cheshire auger — a liner or casing is required for most granular soils and may be required for other types of subsoil.
3. Sampling Shells — suitable for shallow to medium depth borings in all subsoils except rock.

Typical Details ~

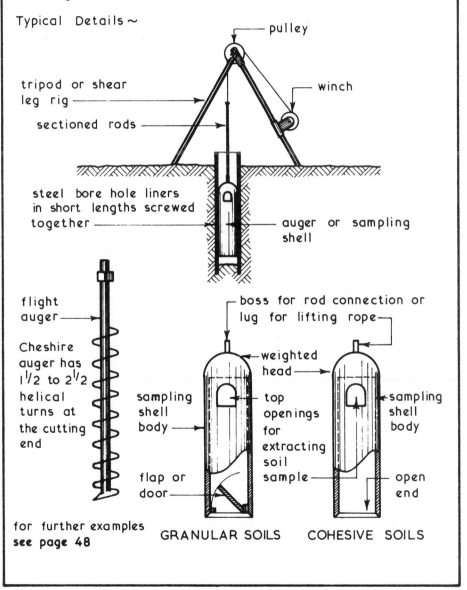

flight auger

Cheshire auger has 1$\frac{1}{2}$ to 2$\frac{1}{2}$ helical turns at the cutting end

for further examples **see page 48**

GRANULAR SOILS     COHESIVE SOILS

Wash Boring ~ this is a method of removing loosened soil from a bore hole using a strong jet of water or bentonite which is a controlled mixture of fullers earth and water. The jetting tube is worked up and down inside the bore hole, the jetting liquid disintegrates the subsoil which is carried in suspension up the annular space to a settling tank. The settled subsoil particles can be dried for testing and classification. This method has the advantage of producing subsoil samples which have not been disturbed by the impact of sampling shells however it is not suitable for large gravel subsoils or subsoils which contain boulders.

Typical Wash Boring Arrangement ~

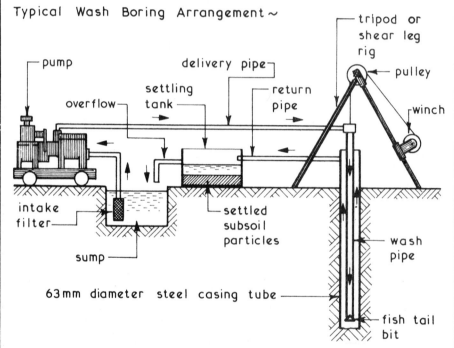

Mud-rotary Drilling ~ this is a method which can be used for rock investigations where bentonite is pumped in a continuous flow down hollow drilling rods to a rotating bit. The cutting bit is kept in contact with the bore face and the debris is carried up the annular space by the circulating fluid. Core samples can be obtained using coring tools.

Core Drilling ~ water or compressed air is jetted down the bore hole through a hollow tube and returns via the annular space. Coring tools extract continuous cores of rock samples which are sent in wooden boxes for laboratory testing.

Bore Hole Data ~ the information obtained from trial pits or bore holes can be recorded on a pro forma sheet or on a drawing showing the position and data from each trial pit or bore hole thus :-

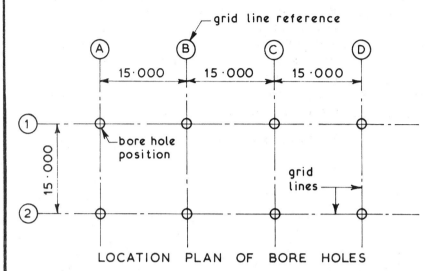

grid line reference

15·000    15·000    15·000

bore hole position

grid lines

15·000

LOCATION PLAN OF BORE HOLES

Bore holes can be taken on a 15·000 to 20·000 grid covering the whole site or in isolated positions relevant to the proposed foundation(s)

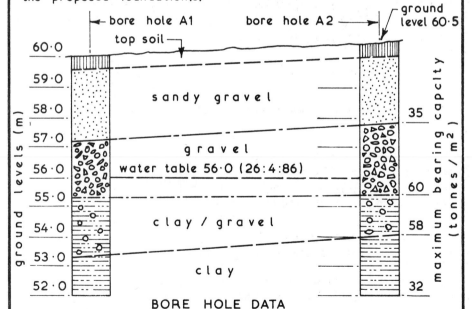

bore hole A1    bore hole A2    ground level 60·5

top soil

ground levels (m)

sandy gravel

gravel
water table 56·0 (26:4:86)

clay / gravel

clay

maximum bearing capcity (tonnes / m²)

35

60

58

32

BORE HOLE DATA

As a general guide the cost of site and soil investigations should not exceed 1% of estimated project costs

49

Soil Assessment ~ prior to designing the foundations for a building or structure the properties of the subsoil(s) must be assessed. These processes can also be carried out to confirm the suitability of the proposed foundations. Soil assessment can include classification, grading, tests to establish shear strength and consolidation. The full range of methods for testing soils is given in BS 1377.

Classification ~ soils may be classified in many ways such as geological origin, physical properties, chemical composition and particle size. It has been found that the partical size and physical properties of a soil are closely linked and are therefore of particular importance and interest to a designer.

Particle Size Distribution ~ this is the percentages of the various particle sizes present in a soil sample as determined by sieving or sedimentation. BS 1377 divides particle sizes into groups as follows :-

Gravel particles - over 2mm
Sand particles – between 2mm and 0·06mm
Silt particles – between 0·06mm and 0·002 mm
Clay particles – less than 0·002mm

The sand and silt classifications can be further divided thus:-

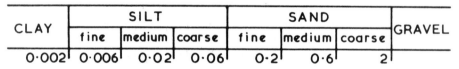

| CLAY | SILT | | | SAND | | | GRAVEL |
|------|------|--------|--------|------|--------|--------|--------|
|      | fine | medium | coarse | fine | medium | coarse |        |
| 0·002 | 0·006 | 0·02 | 0·06 | 0·2 | 0·6 | 2 | |

The results of a sieve analysis can be plotted as a grading curve thus :-

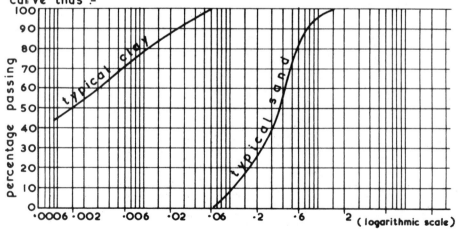

50

Site Soil Tests ~ these tests are designed to evaluate the density or shear strength of soils and are very valuable since they do not disturb the soil under test. Three such tests are the standard penetration test, the vane test and the unconfined compression test all of which are fully described in BS 1377.

Standard Penetration Test ~ this test measures the resistance of a soil to the penetration of a split spoon or split barrel sampler driven into the bottom of a bore hole. The sampler is driven into the soil to a depth of 150mm by a falling standard weight of 65 kg falling through a distance of 760 mm. The sampler is then driven into the soil a further 300mm and the number of blows counted up to a maximum of 50 blows. This test establishes the relative density of the soil.

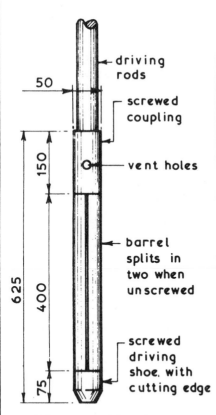

**TYPICAL SPLIT BARREL SAMPLER**

For other tests see pages 52 & 53

## TYPICAL RESULTS

Non cohesive soils:-

| No. of Blows | Relative Density |
|---|---|
| O to 4 | very loose |
| 4 to 10 | loose |
| 10 to 30 | medium |
| 30 to 50 | dense |
| 50 + | very dense |

Cohesive soils :-

| No. of Blows | Relative Density |
|---|---|
| O to 2 | very soft |
| 2 to 4 | soft |
| 4 to 8 | medium |
| 8 to 15 | stiff |
| 15 to 30 | very stiff |
| 30 + | hard |

The results of this test in terms of number of blows and amounts of penetration will need expert interpretation.

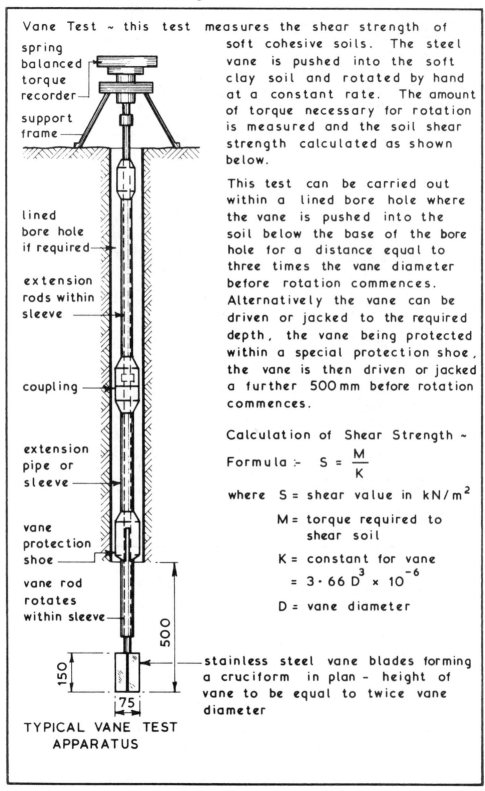

Vane Test ~ this test measures the shear strength of soft cohesive soils. The steel vane is pushed into the soft clay soil and rotated by hand at a constant rate. The amount of torque necessary for rotation is measured and the soil shear strength calculated as shown below.

This test can be carried out within a lined bore hole where the vane is pushed into the soil below the base of the bore hole for a distance equal to three times the vane diameter before rotation commences. Alternatively the vane can be driven or jacked to the required depth, the vane being protected within a special protection shoe, the vane is then driven or jacked a further 500 mm before rotation commences.

Calculation of Shear Strength ~

Formula :-  $S = \dfrac{M}{K}$

where  S = shear value in kN/m$^2$

   M = torque required to shear soil

   K = constant for vane
   $= 3 \cdot 66\, D^3 \times 10^{-6}$

   D = vane diameter

**Labels (left of figure):**

spring balanced torque recorder

support frame

lined bore hole if required

extension rods within sleeve

coupling

extension pipe or sleeve

vane protection shoe

vane rod rotates within sleeve

500

150

75

TYPICAL VANE TEST APPARATUS

stainless steel vane blades forming a cruciform in plan - height of vane to be equal to twice vane diameter

Unconfined Compression Test ~ this test can be used to establish the shear strength of a non-fissured cohesive soil sample using portable apparatus either on site or in a laboratory. The 75 mm long × 38 mm diameter soil sample is placed in the apparatus and loaded in compression until failure occurs by shearing or lateral bulging. For accurate reading of the trace on the recording chart a transparent viewfoil is placed over the trace on the chart.

Typical Apparatus Details ~

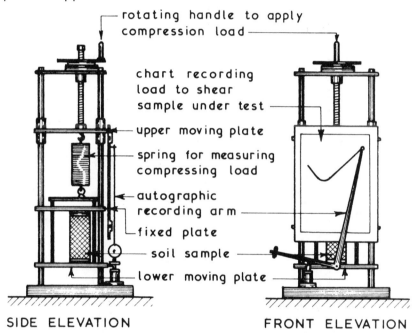

rotating handle to apply compression load

chart recording load to shear sample under test

upper moving plate

spring for measuring compressing load

autographic recording arm

fixed plate

soil sample

lower moving plate

SIDE ELEVATION                    FRONT ELEVATION

Typical Results ~ showing compression strengths of clays :-

Very soft clay   —   less than 25 kN/m$^2$

Soft clay        —   25 to 50 kN/m$^2$

Medium clay      —   50 to 100 kN/m$^2$

Stiff clay       —   100 to 200 kN/m$^2$

Very stiff clay  —   200 to 400 kN/m$^2$

Hard clay        —   more than 400 kN/m$^2$

NB. the shear strength of clay soils is only half of the compression strength values given above.

# Soil Assessment and Testing

Laboratory Testing ~ tests for identifying and classifying soils with regard to moisture content, liquid limit, plastic limit, particle size distribution and bulk density are given in BS 1377.

Bulk Density ~ this is the mass per unit volume which includes mass of air or water in the voids and is essential information required for the design of retaining structures where the weight of the retained earth is an important factor.

Shear Strength ~ this soil property can be used to establish its bearing capacity and also the pressure being exerted on the supports in an excavation. The most popular method to establish the shear strength of cohesive soils is the Triaxial Compression Test. In principle this test consists of subjecting a cyclindrical sample of undisturbed soil (75 mm long × 38 mm diameter) to a lateral hydraulic pressure in addition to a vertical load. Three tests are carried out on three samples (all cut from the same large sample) each being subjected to a higher hydraulic pressure before axial loading is applied. The results are plotted in the form of Mohr's circles.

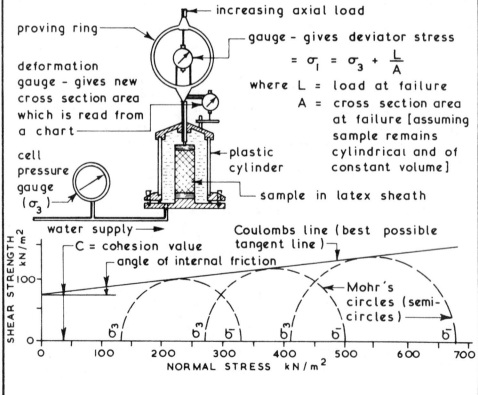

Shear Strength ~ this can be defined as the resistance offered by a soil to the sliding of one particle over another. A simple method of establishing this property is the Shear Box Test in which the apparatus consists of two bottomless boxes which are filled with the soil sample to be tested. A horizontal shearing force (S) is applied against a vertical load (W) causing the soil sample to shear along a line between the two boxes.

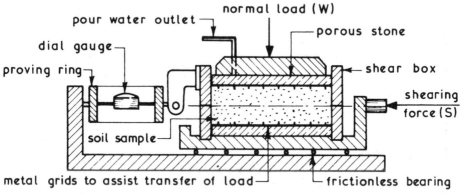

Typical Results ~

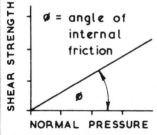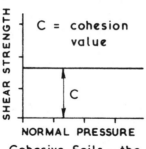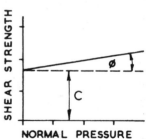

| Granular Soils – as load increases friction between particles increases therefore shear strength is increased. | Cohesive Soils – the very small particles develop no friction therefore shear strength remains constant. | Mixture of Soils – small angle of internal friction is developed as load increases. |
|---|---|---|

Consolidation of Soil ~ this property is very important in calculating the movement of a soil under a foundation. The laboratory testing apparatus is called an Oedometer.

75 mm dia. × 18 mm thick soil sample placed in a metal ring and capped with porous discs then placed in water filled tray and subjected to load

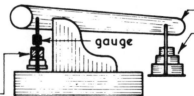

lever arm

weights – load increased every 24 hrs and time / settlement curve drawn

## Site Layout Considerations

General Considerations ~ before any specific considerations and decisions can be made regarding site layout a general appreciation should be obtained by conducting a thorough site investigation at the pre-tender stage and examining in detail the drawings, specification and Bill of Quantities to formulate proposals of how the contract will be carried out if the tender is successful. This will involve a preliminary assessment of plant, materials and manpower requirements plotted against the proposed time scale in the form of a bar chart.

Access Considerations ~ this must be considered for both on and off site access. Routes to and from the site must be checked as to the suitability for transporting all the requirements for the proposed works. Access on site for deliveries and general circulation must also be carefully considered.

Typical Site Access Considerations ~

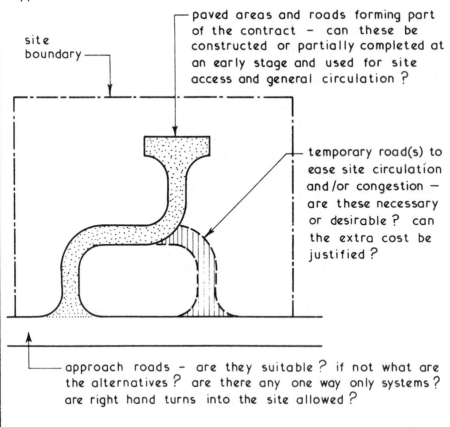

site boundary ⎯

paved areas and roads forming part of the contract – can these be constructed or partially completed at an early stage and used for site access and general circulation ?

temporary road(s) to ease site circulation and /or congestion — are these necessary or desirable ? can the extra cost be justified ?

approach roads – are they suitable ? if not what are the alternatives ? are there any one way only systems ? are right hand turns into the site allowed ?

Storage Considerations ~ amount and types of material to be stored , security and weather protection requirements, allocation of adequate areas for storing materials and allocating adequate working space around storage areas as required, siting of storage areas to reduce double handling to a minimum without impeding the general site circulation and/or works in progress.

Accommodation Considerations ~ number and type of site staff anticipated , calculate size and select units of accommodation and check to ensure compliance with the minimum requirements of the Construction (Health and Welfare) Regulations 1966, select siting for offices to give easy and quick access for visitors but at the same time giving a reasonable view of the site, select siting for messroom and toilets to reduce walking time to a minimum without impeding the general site circulation and/or works in progress.

Temporary Services Considerations ~ what, when and where are they required ? possibility of having permanent services installed at an early stage and making temporary connections for site use during the construction period, coordination with the various service undertakings is essential.

Plant Considerations ~ what plant, when and where is it required ? static or mobile plant ? if static select the most appropriate position and provide any necessary hard standing, if mobile check on circulation routes for optimum efficiency and suitability, provision of space and hard standing for on site plant maintenance if required.

Fencing and Hoarding Considerations ~ what is mandatory and what is desirable ? local vandalism record, type or types of fence and/or hoarding required, possibility of using fencing which is part of the contract by erecting this at an early stage in the contract.

Safety and Health Considerations ~ check to ensure that all the above conclusions from the considerations comply with the minimum requirements set out in the various Construction Regulations and in the Health and Safety at Work etc., Act 1974.

For a typical site layout example see page 58

## Site Layout Considerations

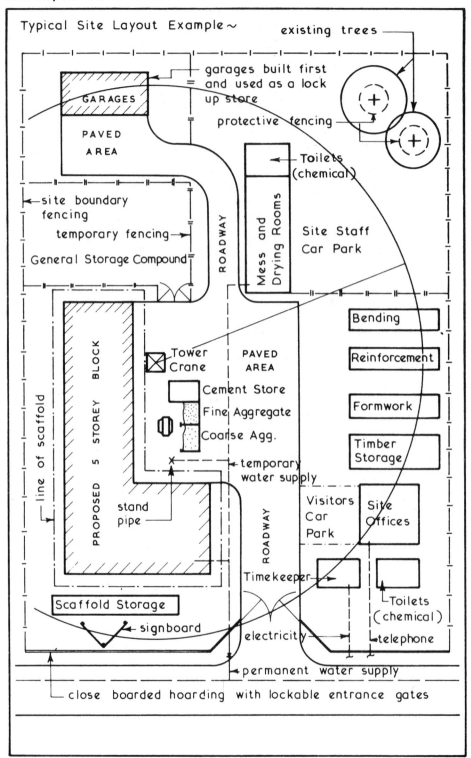

Typical Site Layout Example~

existing trees

garages built first and used as a lock up store

GARAGES

PAVED AREA

protective fencing

Toilets (chemical)

site boundary fencing

temporary fencing

General Storage Compound

ROADWAY

Mess and Drying Rooms

Site Staff Car Park

Bending

Reinforcement

Tower Crane

PAVED AREA

Formwork

Cement Store

Fine Aggregate

Coarse Agg.

Timber Storage

PROPOSED 5 STOREY BLOCK

line of scaffold

temporary water supply

stand pipe

Visitors Car Park

Site Offices

ROADWAY

Scaffold Storage

signboard

Timekeeper

Toilets (chemical)

electricity

telephone

permanent water supply

close boarded hoarding with lockable entrance gates

58

Site Security ~ the primary objectives of site security are-
1. Security against theft.
2. Security from vandals.
3. Protection from innocent trespassers.

The need for and type of security required will vary from site to site according to the neighbourhood, local vandalism record and the value of goods stored on site. Perimeter fencing, internal site protection and night security may all be necessary.

Typical Site Security Provisions ~

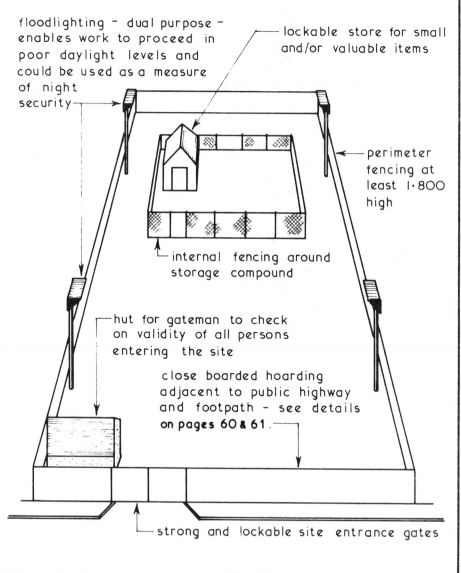

floodlighting – dual purpose – enables work to proceed in poor daylight levels and could be used as a measure of night security

lockable store for small and/or valuable items

perimeter fencing at least 1·800 high

internal fencing around storage compound

hut for gateman to check on validity of all persons entering the site

close boarded hoarding adjacent to public highway and footpath – see details on pages 60 & 61.

strong and lockable site entrance gates

# Hoardings

Hoardings ~ under the Highways Act 1959 a close boarded fence hoarding must be erected prior to the commencement of building operations if such operations are adjacent to a public footpath or highway. The hoarding needs to be adequately constructed to provide protection for the public, resist impact damage, resist anticipated wind pressures and adequately lit at night. Before a hoarding can be erected a licence or permit must be obtained from the local authority who will usually require 10 to 20 days notice. The licence will set out the minimum local authority requirements for hoardings and define the time limit period of the licence.

Typical Hoarding Details ~

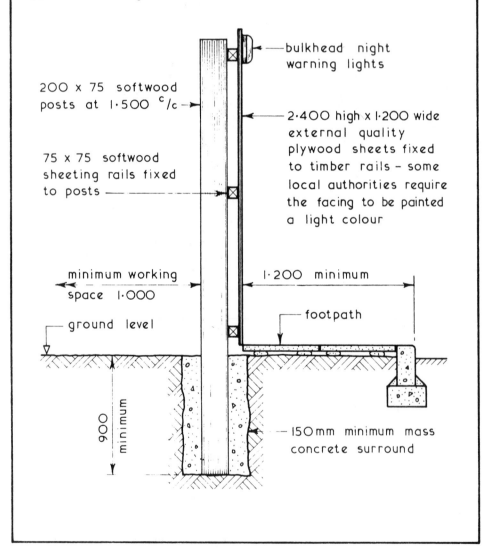

200 x 75 softwood posts at 1·500 $^c/_c$ →

75 x 75 softwood sheeting rails fixed to posts ———————

bulkhead night warning lights

2·400 high x 1·200 wide external quality plywood sheets fixed to timber rails – some local authorities require the facing to be painted a light colour

minimum working space 1·000

ground level

900 minimum

1·200 minimum

footpath

150mm minimum mass concrete surround

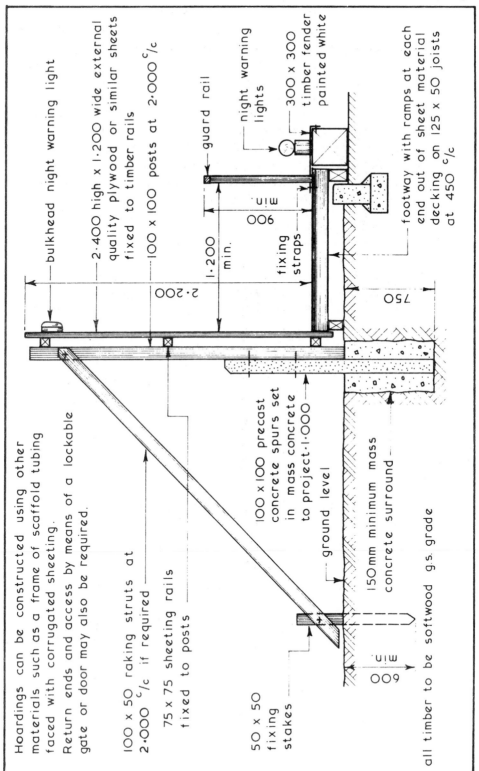

Hoardings can be constructed using other materials such as a frame of scaffold tubing faced with corrugated sheeting.
Return ends and access by means of a lockable gate or door may also be required.

bulkhead night warning light

2·400 high x 1·200 wide external quality plywood or similar sheets fixed to timber rails

100 x 100 posts at 2·000 ᶜ/c

guard rail

night warning lights

300 x 300 timber fender painted white

footway with ramps at each end out of sheet material decking on 125 x 50 joists at 450 ᶜ/c

100 x 50 raking struts at 2·000 ᶜ/c if required

75 x 75 sheeting rails fixed to posts

50 x 50 fixing stakes

100 x 100 precast concrete spurs set in mass concrete to project 1·000

ground level

150mm minimum mass concrete surround

all timber to be softwood g.s. grade

2·200

1·200 min.

fixing straps

900 min.

750

600 min.

Site Lighting ~ this can be used effectively to enable work to continue during periods of inadequate daylight. It can also be used as a deterrent to would-be trespassers. Site lighting can be employed externally to illuminate the storage and circulation areas and internally for general movement and for specific work tasks. The types of lamp available range from simple tungsten filament lamps to tungsten halogen and discharge lamps. The arrangement of site lighting can be static where the lamps are fixed to support poles or mounted on items of fixed plant such as scaffolding and tower cranes. Alternatively the lamps can be sited locally where the work is in progress by being mounted on a movable support or hand held with a trailing lead. Whenever the position of site lighting is such that it can be manhandled it should be run on a reduced voltage of 110 V single phase as opposed to the mains voltage of 240 V.

To plan an adequate system of site lighting the types of activity must be defined and given an illumination target value which is quoted in lux (lx). Recommended minimum target values for building activities are :-

External lighting  –  general circulation  }
                       materials handling   }  10 lx

Internal lighting  –  general circulation       5 lx
                      general working areas     15 lx
                      concreting activities     50 lx
                      carpentry and joinery }
                      bricklaying           }   100 lx
                      plastering            }
                      painting and decorating }
                      site offices            }  200 lx
                      drawing board positions    300 lx

Such target values do not take into account deterioration, dirt or abnormal conditions therefore it is usual to plan for at least twice the recommended target values. Generally the manufacturers will provide guidance as to the best arrangement to use in any particular situation but lamp requirements can be calculated thus :-

$$\text{Total lumens required} = \frac{\text{area to be illuminated (m}^2) \times \text{target value (lx)}}{\text{utilisation factor } 0.23 \text{ [dispersive lights } 0.27]}$$

After choosing lamp type to be used :-

$$\text{Number of lamps required} = \frac{\text{total lumens required}}{\text{lumen output of chosen lamp}}$$

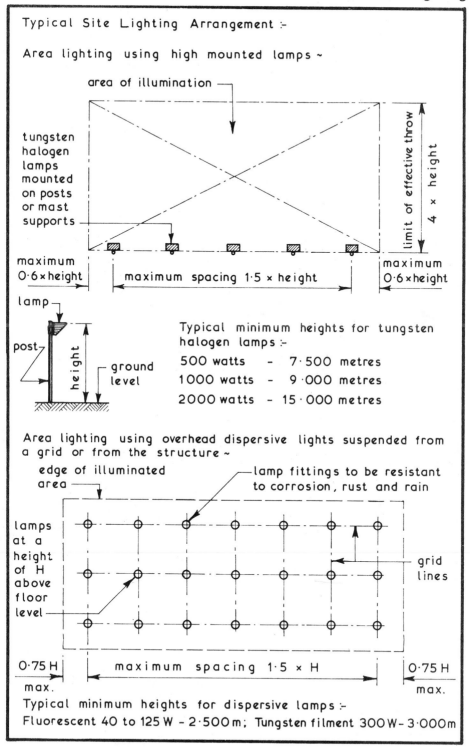

Typical Site Lighting Arrangement :-

Area lighting using high mounted lamps ~

area of illumination

tungsten halogen lamps mounted on posts or mast supports

limit of effective throw

4 × height

maximum 0·6 × height | maximum spacing 1·5 × height | maximum 0·6 × height

lamp

post

height

ground level

Typical minimum heights for tungsten halogen lamps :-

500 watts   -   7·500 metres
1000 watts  -   9·000 metres
2000 watts  -   15·000 metres

Area lighting using overhead dispersive lights suspended from a grid or from the structure ~

edge of illuminated area

lamp fittings to be resistant to corrosion, rust and rain

lamps at a height of H above floor level

grid lines

0·75 H max. | maximum spacing 1·5 × H | 0·75 H max.

Typical minimum heights for dispersive lamps :-
Fluorescent 40 to 125 W – 2·500 m; Tungsten filment 300 W – 3·000 m

Walkway and Local Lighting ~ to illuminate the general circulation routes bulkhead and/or festoon lighting could be used either on a standard mains voltage of 240V or on a reduced voltage of 110V. For local lighting at the place of work hand lamps with trailing leads or lamp fittings on stands can be used and positioned to give the maximum amount of illumination without unacceptable shadow cast.

Typical Walkway and Local Lighting Fittings ~

die-cast aluminium alloy body

supply cable

water and weatherproof front glass

vandal resistant translucent polycarbonate diffuser

white stove enamelled reflector

galvanised steel base

**BULKHEAD LAMP**
(300mm × 8W fluorescent)

**LAMP AND STAND**
(110V tungsten filament)

sealed end to cable

rainproof lampholders

60W tungsten filament bulbs

weather resistant cable

weatherproof fitted plug

glass fibre shades or wire guards if required

**FESTOON LIGHTING**

weatherproof fitted plug

moulded rubber lampholder

fitted wire guard to 110V tungsten filament lamp

weather resistant cable

**HAND HELD LAMP WITH TRAILING LEAD**

Electrical Supply to Building Sites ~ a supply of electricity is usually required at an early stage in the contract to provide light and power to the units of accommodation. As the work progresses power could also be required for site lighting, hand held power tools and large items of plant. The supply of electricity to a building site is the subject of a contract between the contractor and the local area electricity board who will want to know the date when supply is required; site address together with a block plan of the site; final load demand of proposed building and an estimate of the maximum load demand in kilowatts for the construction period. The latter can be estimated by allowing $10W/m^2$ of the total floor area of the proposed building plus an allowance for high load equipment such as cranes. The installation should be undertaken by a competent electrical contractor to ensure that it complies with all the statutory rules and regulations for the supply of electricity to building sites.

Typical Supply and Distribution Equipment ~

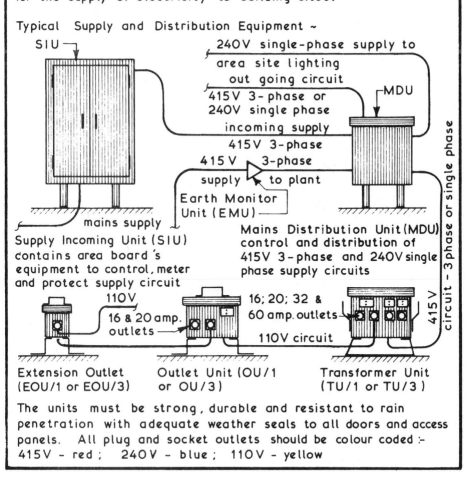

SIU

240V single-phase supply to area site lighting out going circuit

415V 3-phase or 240V single phase incoming supply

MDU

415V 3-phase

415V 3-phase supply to plant

Earth Monitor Unit (EMU)

mains supply

Supply Incoming Unit (SIU) contains area board's equipment to control, meter and protect supply circuit

Mains Distribution Unit (MDU) control and distribution of 415V 3-phase and 240V single phase supply circuits

circuit – 3 phase or single phase

110V

16 & 20 amp. outlets

16; 20; 32 & 60 amp. outlets

110V circuit

415 V

Extension Outlet (EOU/1 or EOU/3)

Outlet Unit (OU/1 or OU/3)

Transformer Unit (TU/1 or TU/3)

The units must be strong, durable and resistant to rain penetration with adequate weather seals to all doors and access panels. All plug and socket outlets should be colour coded :-
415V – red ; 240V – blue ; 110V – yellow

# Site Office Accommodation

Office Accommodation ~ the type of office accommodation to be provided on site is a matter of choice for each individual contractor who can use timber framed huts, prefabricated cabins, mobile offices or even caravans. Generally separate offices would be provided for site agent, clerk of works, administrative staff and site surveyors.

The minimum requirements of such accommodation is governed by the Offices, Shops and Railway Premises Act 1963 unless they are ~

1. Mobile units in use for not more than 6 months.
2. Fixed units in use for not more than 6 weeks.
3 Any type of unit in use for not more than 21 man hours per week.
4. Office for exclusive use of self employed person.
5. Office used by family only staff.

Sizing Example ~

Office for site agent and assistant plus an allowance for 3 visitors.

Assume an internal average height of 2·400.

Allow 3·7 m$^2$ minimum per person and 11·5 m$^3$ minimum per person.

Minimum area = 5 x 3·7 = 18·5 m$^2$

Minimum volume = 5 x 11·5 = 57·5 m$^3$

Assume office width of 3·000 then minimum length required

is = $\dfrac{57·5}{3 \times 2·4}$ = $\dfrac{57·5}{7·2}$ = 7·986   say 8·000

Area check   3 x 8 = 24 m$^2$   which is > 18·5 m$^2$ ∴ satisfactory

Typical Examples ~

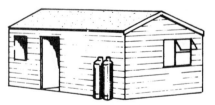

Timber framed site hut with insulated walls and roof. Wall cladding- painted weatherboard. Hut supplied unequipped. Sizes based on 1·500 module.

Portable cabin with four steel tube jacking legs. Shell-timber framing with plywood facing. Cabin insulated and fully equipped. Wide range of sizes available.

The minimum requirements for health and welfare provisions for persons on construction sites are set out in the Construction (Health and Welfare) Regulations 1966 and the Construction (Health and Welfare)(Amended) Regulations 1974.

| Provision | Requirement | No. of persons employed on site |
|---|---|---|
| FIRST AID | Box to be distinctively marked and in charge of responsible person | 5 to 50 – first aid boxes 50 + first aid box and a person trained in first aid |
| AMBULANCES | Stretcher(s) in charge of responsible person | 25 + notify ambulance authority of site details within 24 hours of employing more than 25 persons |
| FIRST AID ROOM | Used only for rest or treatment and in charge of trained person | If more than 250 persons employed on site each employer of more than 40 persons to provide a first aid room |
| SHELTER AND ACCOMMODATION FOR CLOTHING | All persons on site to have shelter and a place for depositing clothes | Up to 5 where possible a means of warming themselves and drying wet clothes 5 + adequate means of warming themselves and drying wet clothing |
| MEALS ROOM | Drinking water, means of boiling water and eating meals for all persons on site | 10 + facilities for heating food if hot meals are not available on site |
| WASHING FACILITIES | Washing facilities to be provided for all persons on site for more than 4 hours | 20 to 100 if work is to last more than 6 weeks - hot and cold or warm water, soap and towel 100 + work lasting more than 12 months – 4 wash places + 1 for every 35 persons over 100 |
| SANITARY FACILITIES | To be maintained, lit and kept clean. Separate facilities for female staff | Up to 100 – 1 convenience for every 25 persons 100 + – 1 convenience for every 35 persons |

Site Storage ~ materials stored on site prior to being used or fixed may require protection for security reasons or against the adverse effects which can be caused by exposure to the elements.

Small and Valuable Items ~ these should be kept in a secure and lockable store. Similar items should be stored together in a rack or bin system and only issued against an authorised requisition.

Large or Bulk Storage Items ~ for security protection these items can be stored within a lockable fenced compound. The form of fencing chosen may give visual security by being of an open nature but these are generally easier to climb than the close boarded type of fence which lacks the visual security property.

Typical Storage Compound Fencing ~

Close boarded fences can be constructed on the same methods **used for hoardings - see pages 60 & 61**

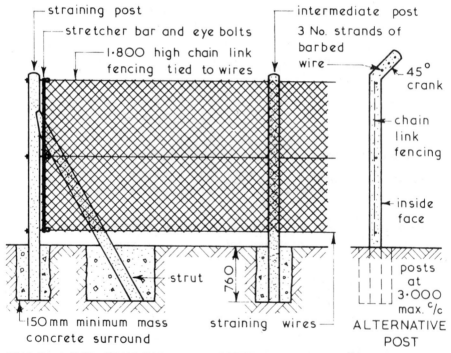

straining post

stretcher bar and eye bolts

1·800 high chain link fencing tied to wires

intermediate post

3 No. strands of barbed wire

45° crank

chain link fencing

inside face

strut

760

150 mm minimum mass concrete surround

straining wires

posts at 3·000 max. c/c

ALTERNATIVE POST

**CHAIN LINK FENCING WITH PRECAST CONCRETE POSTS**

Alternative Fence Types ~ woven wire fence, strained wire fence, cleft chestnut pale fence, wooden palisade fence, wooden post and rail fence and metal fences - see BS1722 for details.

Storage of Materials ~ this can be defined as the provision of adequate space, protection and control for building materials and components held on site during the construction process. The actual requirements for specific items should be familiar to students who have completed studies in construction technology up to and including level 111 but the need for storage and control of materials held on site can be analysed further :-

1. Physical Properties - size, shape, weight and mode of delivery will assist in determining the safe handling and stacking method(s) to be employed on site, which in turn will enable handling and storage costs to be estimated.

2. Organisation - this is the planning process of ensuring that all the materials required are delivered to site at the correct time, in sufficient quantity, of the right quality, the means of unloading is available and that adequate space for storage or stacking has been allocated.

3. Protection - building materials and components can be classified as durable or non-durable, the latter will usually require some form of weather protection to prevent deterioration whilst in store.

4. Security - many building materials have a high resale and/or usage value to persons other than those for whom they were ordered and unless site security is adequate material losses can become unacceptable.

5. Costs - to achieve an economic balance of how much expenditure can be allocated to site storage facilities the following should be taken into account :-
   a. Storage areas, fencing, racks, bins, etc.,
   b. Protection requirements.
   c. Handling, transporting and stacking requirements.
   d. Salaries and wages of staff involved in storage of materials and components.
   e. Heating and/or lighting if required.
   f. Allowance for losses due to wastage, deterioration, vandalism and theft.
   g. Facilities to be provided for sub-contractors.

6. Control - checking quality and quantity of materials at delivery and during storage period, recording delivery and issue of materials and monitoring stock holdings.

# Materials Storage

Site Storage Space ~ the location and size(s) of space to be allocated for any particular material should be planned by calculating the area(s) required and by taking into account all the relevant factors before selecting the most appropriate position on site in terms of handling, storage and convenience. Failure to carry out this simple planning exercise can result in chaos on site or having on site more materials than there is storage space available.

Calculation of Storage Space Requirements ~ each site will present its own problems since a certain amount of site space must be allocated to the units of accommodation, car parking, circulation and working areas, therefore the amount of space available for materials storage may be limited. The size of the materials or component being ordered must be known together with the proposed method of storage and this may vary between different sites of similar building activities. There are therefore no standard solutions for allocating site storage space and each site must be considered separately to suit its own requirements.

Typical Examples ~

Bricks - quantity = 15 200 to be delivered in strapped packs of 380 bricks per pack each being 1100 mm wide x 670 mm long x 850 mm high. Unloading and stacking to be by forklift truck to form 2 rows 2 packs high.

Area required :- number of packs per row = $\dfrac{15\,200}{380 \times 2}$ = 20

length of row = 10 x 670 = 6·700
width of row = 2 x 1100 = 2·200
allowance for forklift approach in front of stack = 5·000
∴ minimum brick storage area = 6·700 long x 7·200 wide

Timber - to be stored in open sided top covered racks constructed of standard scaffold tubes. Maximum length of timber ordered = 5·600. Allow for rack to accept at least 4 No. 300 mm wide timbers placed side by side then minimum width required = 4 x 300 = 1·200
Minimum plan area for timber storage rack = 5·600 x 1·200
Allow for end loading of rack equal to length of rack
∴ minimum timber storage area = 11·200 long x 1·200 wide
Height of rack to be not more than 3 x width = 3·600

Areas for other materials stored on site can be calculated using the basic principles contained in the examples above.

Site Allocation for Materials Storage ~ the area and type of storage required can be determined as shown on pages 69 and 70 but the allocation of an actual position on site will depend on :-

1. Space available after areas for units of accommodation have been allocated.
2. Access facilities on site for delivery vehicles.
3. Relationship of storage area(s) to activity area(s) - the distance between them needs to be kept as short as possible to reduce transportation needs in terms of time and costs to the minimum. Alternatively storage areas and work areas need to be sited within the reach of any static transport plant such as a tower crane.
4. Security - needs to be considered in the context of site operations, vandalism and theft.
5. Stock holding policy - too little storage could result in delays awaiting for materials to be delivered, too much storage can be expensive in terms of weather and security protection requirements apart from the capital used to purchase the materials stored on site.

Typical Example ~

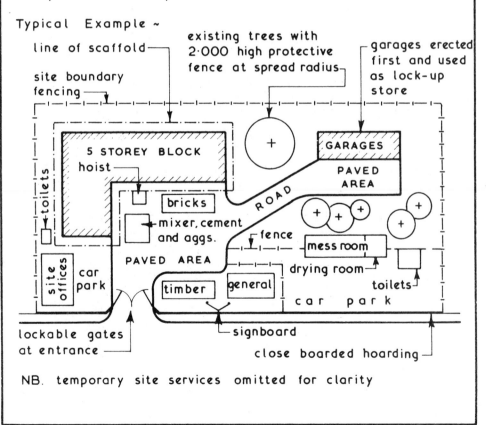

NB. temporary site services omitted for clarity

# Materials Storage

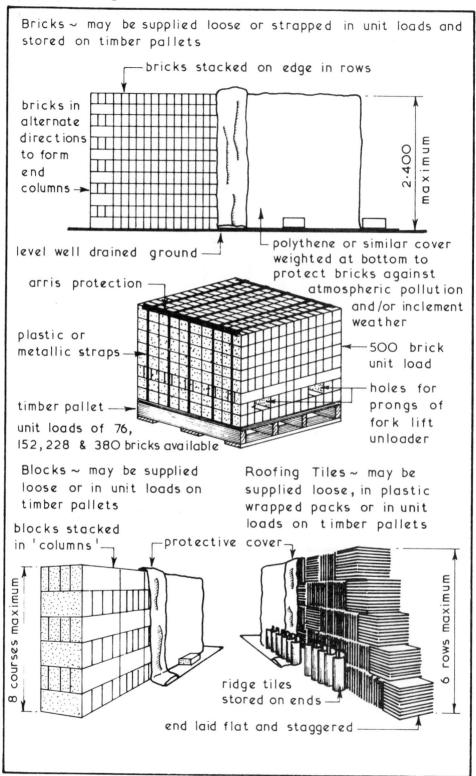

Bricks ~ may be supplied loose or strapped in unit loads and stored on timber pallets

bricks stacked on edge in rows

bricks in alternate directions to form end columns →

2·400 maximum

level well drained ground →

polythene or similar cover weighted at bottom to protect bricks against atmospheric pollution and/or inclement weather

arris protection →

plastic or metallic straps →

500 brick unit load

timber pallet →

holes for prongs of fork lift unloader

unit loads of 76, 152, 228 & 380 bricks available

Blocks ~ may be supplied loose or in unit loads on timber pallets

blocks stacked in 'columns' →

protective cover

8 courses maximum

Roofing Tiles ~ may be supplied loose, in plastic wrapped packs or in unit loads on timber pallets

6 rows maximum

ridge tiles stored on ends →

end laid flat and staggered →

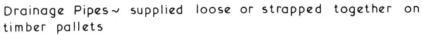

Drainage Pipes~ supplied loose or strapped together on timber pallets

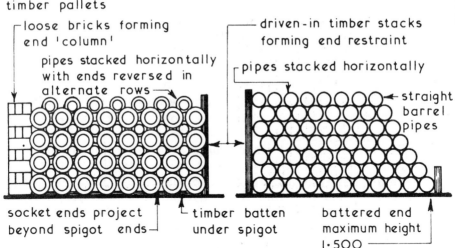

loose bricks forming end 'column'

pipes stacked horizontally with ends reversed in alternate rows

driven-in timber stacks forming end restraint

pipes stacked horizontally

straight barrel pipes

socket ends project beyond spigot ends

timber batten under spigot

battered end maximum height 1·500

gullies etc., should be stored upside down and supported to remain level

Baths~ stacked or nested vertically or horizontally on timber battens

Timber and Joinery Items~ should be stored horizontally and covered but with provison for free air flow

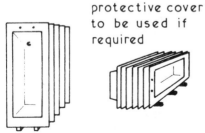

protective cover to be used if required

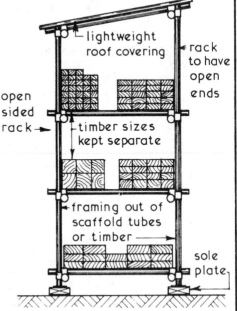

lightweight roof covering

rack to have open ends

open sided rack

timber sizes kept separate

framing out of scaffold tubes or timber

sole plate

Basins~ stored similar to baths but not more than four high if nested one on top of another

Corrugated and Similar Sheet Materials~ stored flat on a level surface and covered with a protective polythene or similar sheet material

Cement, Sand and Aggregates ~ for supply and storage details see pages 212 & 213

Site Tests ~ the majority of materials and components arriving on site will conform to the minimum recommendations of the appropriate British Standard and therefore the only tests which need be applied are those of checking quantity received against amount stated on the delivery note, ensuring quality is as ordered and a visual inspection to reject damaged or broken goods. The latter should be recorded on the delivery note and entered in the site records. Certain site tests can however be carried out on some materials to establish specific data such as the moisture content of timber which can be read direct from a moisture meter. Other simple site tests are given in the various British Standards to ascertain compliance with the recommendations such as the test for compliance with dimensional tolerance given in BS 3921 which covers clay bricks. This test is carried out by measuring a sample of 24 bricks taken at random from a delivered load thus :-

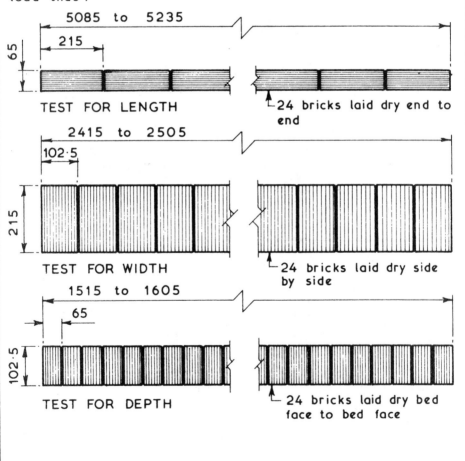

5085 to 5235

215

65

TEST FOR LENGTH

24 bricks laid dry end to end

2415 to 2505

102·5

215

TEST FOR WIDTH

24 bricks laid dry side by side

1515 to 1605

65

102·5

TEST FOR DEPTH

24 bricks laid dry bed face to bed face

Site Tests ~ apart from the test outlined on page 74 site tests on materials which are to be combined to form another material such as concrete can also be tested to establish certain properties which if not known could affect the consistency and/or quality of the final material.

Typical Example – Testing Sand for Bulking ~

this data is required when batching concrete by volume – test made at commencement of mixing and if change in weather

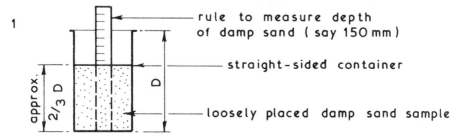

1
rule to measure depth of damp sand ( say 150 mm )

straight-sided container

loosely placed damp sand sample

approx. 2/3 D

D

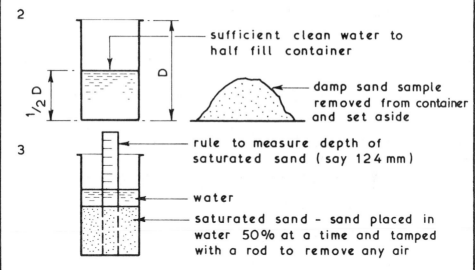

2
sufficient clean water to half fill container

1/2 D

D

damp sand sample removed from container and set aside

3
rule to measure depth of saturated sand ( say 124 mm )

water

saturated sand – sand placed in water 50% at a time and tamped with a rod to remove any air

4    Calculation :-

$$\text{bulking} = \frac{\text{difference in height between damp \& saturated sand}}{\text{depth of saturated sand}}$$

$$\% \text{ bulking} = \frac{150-124}{124} \times 100 = \frac{26}{124} \times 100 = 20 \cdot 96774 \%$$

therefore volume of sand should be increased by 21% over that quoted in the specification

NB. a given weight of saturated sand will occupy the same space as when dry but more space when damp

Silt Test for Sand ~ the object of this test is to ascertain the cleanliness of sand by establishing the percentage of silt present in a natural sand since too much silt will weaken the concrete

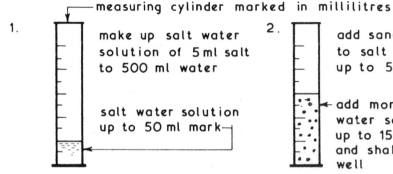

measuring cylinder marked in millilitres

1. make up salt water solution of 5 ml salt to 500 ml water

salt water solution up to 50 ml mark

2. add sand sample to salt water up to 50 ml mark

add more salt water solution up to 150ml mark and shake cylinder well

3. allow mixture to stand for 3 hours and measure height of silt

salt water
silt
sand

4. Height of silt layer should not be more than 6 ml or 6% of height of sand sample

Obtaining Samples for Laboratory Testing ~ these tests may be required for checking aggregate grading by means of a sieve test, checking quality or checking for organic impurities but whatever the reason the sample must be truly representative of the whole :-

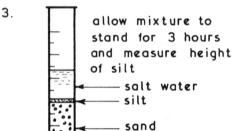

1.

scoop

aggregate pile

samples extracted by means of a scoop from at least ten different positions in the pile

sample required :-

fine aggregate - 50 kg
coarse aggregate - 200 kg

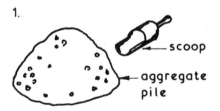

2. well mixed sample divided into four equal parts - opposite quarters are discarded -

remainder of sample remixed and quartered - whole process is repeated until required size of sample is left.

samples required :-

fine aggregate ⟶ 6 mm - 3 kg
⟶ 10mm - 6 kg
coarse aggregate ⟶ 20mm - 25 kg
⟶ 32mm - 50kg

Concrete requires monitoring by means of tests to ensure that subsequent mixes are of the same consistency and this can be carried out on site by means of the slump test and in a laboratory by crushing test cubes to check that the cured concrete has obtained the required designed strength.

Slump Test ~

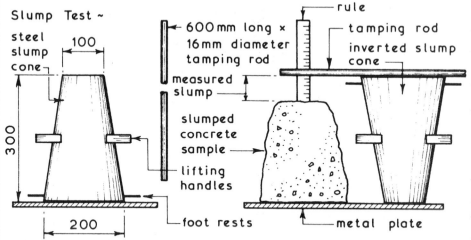

The slump cone is filled to a quarter depth and tamped 25 times – filling and tamping is repeated three more times until the cone is full and the top smoothed off. The cone is removed and the slump measured, for consistent mixes the slump should remain the same for all samples tested. Usual specification 50 mm or 75 mm slump.

Test Cubes – these are required for laboratory strength tests~

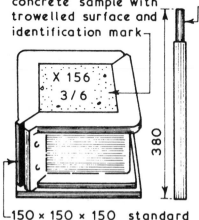

concrete sample with trowelled surface and identification mark

X 156
3/6

150 × 150 × 150 standard steel test cube mould thinly coated inside with mould oil

25 × 25 mm square end tamping bar

1. Sample taken from discharge outlet of mixer or from point of placing using random selection by means of a scoop.

2. Mould filled in three equal layers each layer well tamped with at least 35 strokes from the tamping bar.

3. Sample left in mould for 24 hours and covered with a damp sack or similar at a temperature of 4·4 to 21°C

4. Remove sample from mould and store in water at temperature of 10 to 21°C until required for testing

# Protection Orders for Trees and Structures

Trees~ these are part of our national heritage and are also the source of timber – to maintain this source a control over tree felling has been established under the Forestry Act 1967 which places the control responsibility on the Forestry Commisson. Local planning authorities also have powers under the Town and Country Planning Act 1971 and the Town and Country Amenities Act 1974 to protect trees by making tree preservation orders. Contravention of such an order can lead to a substantial fine and a compulsion to replace any protected tree which has been removed or destroyed. Trees on building sites which are covered by a tree preservation order should be protected by a suitable fence.

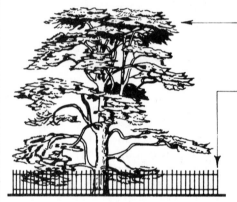

tree covered by a tree preservation order

cleft chestnut or similar fencing at least 1·200 high erected to the full spread and completely encircling the tree

Trees, shrubs, bushes and tree roots which are to be removed from site can usually be grubbed out using hand held tools such as saws, picks and spades. Where whole trees are to be removed for relocation special labour and equipment is required to ensure that the roots, root earth ball and bark are not damaged.

Structures ~ buildings which are considered to be of historic or architectural interest can be protected under the Town and Country Acts provisions. The Department of the Environment lists buildings according to age, architectural, historical and / or intrinsic value. It is an offence to demolish or alter a listed building without first obtaining 'listed building consent' from the local planning authority. Contravention is punishable by a fine and / or imprisonment. It is also an offence to demolish a listed building without giving notice to the Royal Commission on Historic Monuments, this is to enable them to note and record details of the building.

Services which may be encountered on construction sites and the authority responsible are ~

Water — Local Water Company

Electricity - transmission ~ Central Electricity Generating Board in England and Wales.

distribution ~ Area Electricity Board in England and Wales.

South of Scotland Electricity Board and the North of Scotland Hydro-Electric Board.

Gas- Area Gas Board.

Telephones - British Telecom.

Drainage - Local Authority unless a private drain or sewer when owner(s) is responsible.

All the above authorities must be notified of any proposed new services and alterations or terminations to existing services before any work is carried out.

Locating Existing Services on Site ~

Method 1 - By reference to maps and plans prepared and issued by the respective responsible authority

Method 2 - Using visual indicators ~

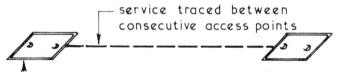

— service traced between consecutive access points

access covers which are sometimes marked as to the service below e.g. soil or surface water

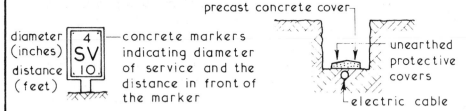

diameter (inches) | 4
SV
distance (feet) | .10

—concrete markers indicating diameter of service and the distance in front of the marker

precast concrete cover

unearthed protective covers

electric cable

Method 3 - Detection-specialist contractor employed to trace all forms of underground service using electronic subsurface survey equipment.

Once located position and type of service can be plotted on a map or plan, marked with special paint on hard surfaces and marked with wood pegs with identification data on earth surfaces.

Setting Out the Building Outline ~ this task is usually undertaken once the site has been cleared of any debris or obstructions and any reduced level excavation work is finished. It is usually the responsibility of the contractor to set out the building(s) using the information provided by the designer or architect. Accurate setting out is of paramount importance and should therefore only be carried out by competent persons and all their work thoroughly checked, preferably by different personnel and by a different method.

The first task in setting out the building is to establish a base line to which all the setting out can be related. The base line very often coincides with the building line which is a line, whose position on site is given by the local authority, in front of which no development is permitted.

Typical Setting Out Example ~

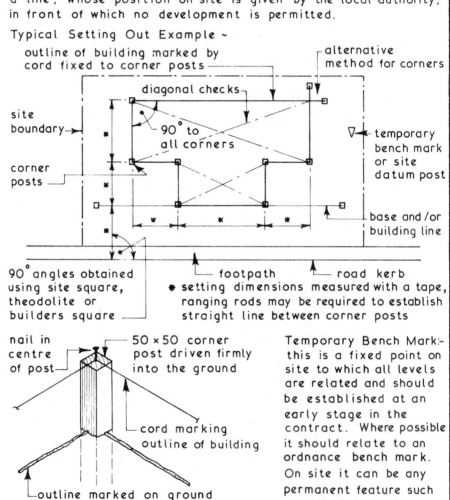

outline of building marked by cord fixed to corner posts

alternative method for corners

diagonal checks

site boundary→

90° to all corners

corner posts

temporary bench mark or site datum post

base and/or building line

90° angles obtained using site square, theodolite or builders square

footpath

road kerb

❋ setting dimensions measured with a tape, ranging rods may be required to establish straight line between corner posts

nail in centre of post

50 × 50 corner post driven firmly into the ground

cord marking outline of building

outline marked on ground with dry lime or similar powder

Temporary Bench Mark:- this is a fixed point on site to which all levels are related and should be established at an early stage in the contract. Where possible it should relate to an ordnance bench mark. On site it can be any permanent feature such as a drain cover or a firmly driven post.

Setting Out Trenches ~ the objective of this task is twofold. Firstly it must establish the excavation size, shape and direction and secondly it must establish the width and position of the walls. The outline of building will have been set out and using this outline profile boards can be set up to control the position, width and possibly the depth of the proposed trenches. Profile boards should be set up at least 2·000 clear of trench positions so they do not obstruct the excavation work. The level of the profile crossboard should related to the site datum and fixed at a convenient height above ground level if a traveller is to be used to control the depth of the trench. Alternatively the trench depth can be controlled using a level and staff related to site datum. The trench width can be marked on the profile with either nails or sawcuts and with a painted band if required for identification.

Typical Details ~

TYPICAL LAYOUT (profiles not to scale)    TYPICAL PROFILE BOARD

NB. corners of walls transferred from intersecting cord lines to mortar spots on concrete foundations using a spirit level

Setting Out a Framed Building ~ framed buildings are usually related to a grid, the intersections of the grid lines being the centre point of an isolated or pad foundation. The grid is usually set out from a base line which does not always form part of the grid. Setting out dimensions for locating the grid can either be given on a drawing or they will have to be accurately scaled off a general layout plan. The grid is established using a theodolite and marking the grid line intersections with stout pegs. Once the grid has been set out offset pegs or profiles can be fixed clear of any subsequent excavation work. Control of excavation depth can be by means of a traveller sighted between sight rails or by level and staff related to site datum.

Typical Details ~

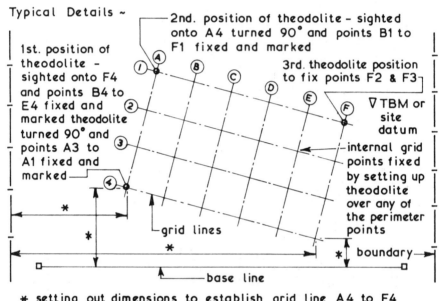

1st. position of theodolite - sighted onto F4 and points B4 to E4 fixed and marked theodolite turned 90° and points A3 to A1 fixed and marked

2nd. position of theodolite - sighted onto A4 turned 90° and points B1 to F1 fixed and marked

3rd. theodolite position to fix points F2 & F3

∇ TBM or site datum

internal grid points fixed by setting up theodolite over any of the perimeter points

grid lines

boundary

base line

* setting out dimensions to establish grid line A4 to F4

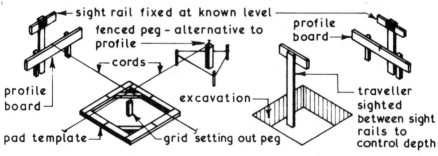

sight rail fixed at known level

fenced peg - alternative to profile

profile board

cords

profile board

excavation

pad template

grid setting out peg

traveller sighted between sight rails to control depth

1. Pad template positioned with cords between profiles and pad outline marked with dry lime or similar powder.

2. Pad pits excavated using traveller sighted between sight rails fixed at a level related to site datum.

Setting Out Reduced Level Excavations ~ the overall outline of the reduced level area can be set out using a theodolite, ranging rods, tape and pegs working from a base line. To control the depth of excavation, sight rails are set up at a convenient height and at positions which will enable a traveller to be used.

Typical Details ~

1. Setting up sight rails :-

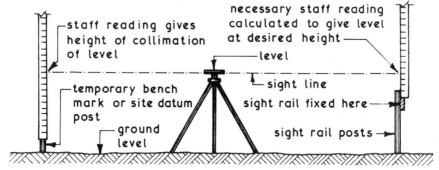

staff reading gives height of collimation of level

necessary staff reading calculated to give level at desired height

level

temporary bench mark or site datum post

sight line

sight rail fixed here

ground level

sight rail posts

2. Controlling excavation depth :-

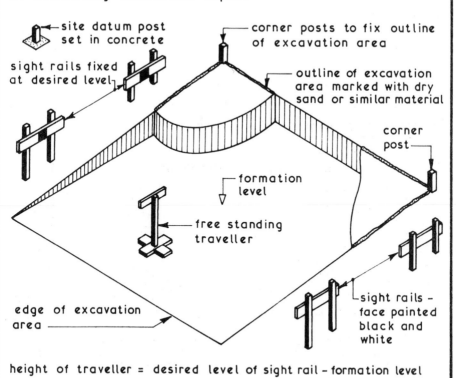

site datum post set in concrete

corner posts to fix outline of excavation area

sight rails fixed at desired level

outline of excavation area marked with dry sand or similar material

corner post

formation level

free standing traveller

edge of excavation area

sight rails - face painted black and white

height of traveller = desired level of sight rail - formation level

Road Construction ~ within the context of building operations roadworks usually consist of the construction of small estate roads, access roads and driveways together with temporary roads laid to define site circulation routes and/or provide a suitable surface for plant movements. The construction of roads can be considered under three headings :-

1. Setting out.

2. **Earthworks ( see page 85 )**

3. **Paving Construction ( see pages 86 & 87 )**

Setting Out Roads ~ this activity is usually carried out after the top soil has been removed using the dimensions given on the layout drawing(s). The layout could include straight lengths, junctions, hammer heads, turning bays and intersecting curves.

Straight Road Lengths - these are usually set out from centre lines which have been established by traditional means.

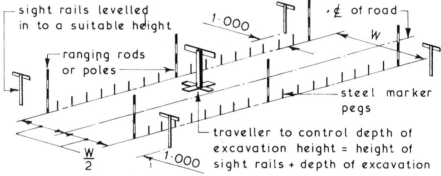

sight rails levelled in to a suitable height

ranging rods or poles

1·000

⅌ of road

W

steel marker pegs

traveller to control depth of excavation height = height of sight rails + depth of excavation

W/2

1·000

NB curve road lengths set out in a similar manner

Junctions and Hammer Heads -

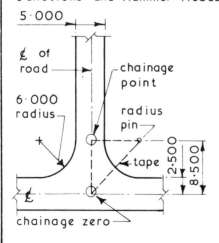

5·000

⅌ of road

chainage point

6·000 radius

radius pin

tape

2·500

8·500

⅌

chainage zero

Centre lines fixed by traditional methods. Tape hooked over pin at chainage zero and passed around chainage point pin at 8·500 then returned to chainage zero with a tape length of 29·021. Radius pin held tape length 17·000 and tape is moved until tight between all pins. Radius pin is driven and a 6·000 tape length is swung from the pin to trace out curve which is marked with pegs or pins.

Tape length = 17 + √2 × 8·5
             = 29·021

Earthworks ~ this will involve the removal of topsoil together with any vegetation, scraping and grading the required area down to formation level plus the formation of any cuttings or embankments. Suitable plant for these operations would be tractor shovels fitted with a 4 in 1 bucket (page 119); graders (page 118) and bulldozers (page 116). The soil immediately below the formation level is called the subgrade whose strength will generally decrease as its moisture content rises therefore if it is to be left exposed for any length of time protection may be required. Subgrade protection may take the form of a covering of medium gauge plastic sheeting with 300 mm laps or alternatively a covering of sprayed bituminous binder with a sand topping applied at a rate of 1 litre per m$^2$ To preserve the strength and durability of the subgrade it may be necessary to install cut off subsoil drains alongside the proposed road (see Road Drainage 1 on page 479)

Paving Construction~ once the subgrade has been prepared and any drainage or other buried services installed the construction of the paving can be undertaken. Paved surfaces can be either flexible or rigid in format. Flexible or bound surfaces are formed of materials applied in layers directly over the subgrade whereas rigid pavings consist of a concrete slab resting on a granular base (see pages 86 & 87)

Typical Flexible Paving Details ~

surfacing = base layer + wearing course

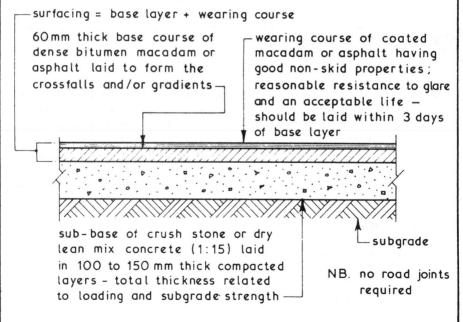

60mm thick base course of dense bitumen macadam or asphalt laid to form the crossfalls and/or gradients

wearing course of coated macadam or asphalt having good non-skid properties; reasonable resistance to glare and an acceptable life — should be laid within 3 days of base layer

sub-base of crush stone or dry lean mix concrete (1:15) laid in 100 to 150 mm thick compacted layers - total thickness related to loading and subgrade strength

subgrade

NB. no road joints required

Rigid Pavings ~ these consist of a reinforced or unreinforced insitu concrete slab laid over a base course of crushed stone or similar material which has been blinded to receive a polythene sheet slip membrane. The primary objective of this membrane is to prevent grout loss from the insitu slab.

Typical Rigid Paving Details ~

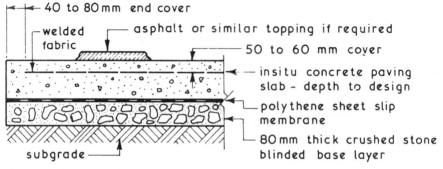

40 to 80 mm end cover
welded fabric
asphalt or similar topping if required
50 to 60 mm cover
insitu concrete paving slab - depth to design
polythene sheet slip membrane
80 mm thick crushed stone blinded base layer
subgrade

The paving can be laid between metal road forms or timber edge formwork. Alternatively the kerb stones could be laid first to act as permanent formwork.

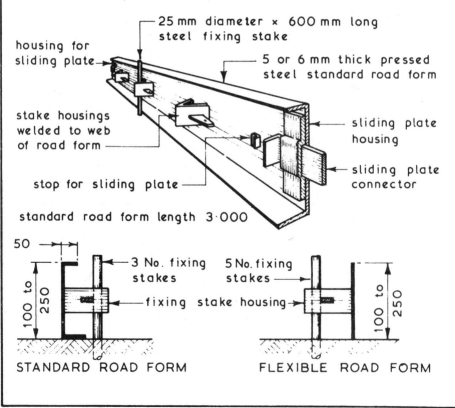

25 mm diameter × 600 mm long steel fixing stake

housing for sliding plate

5 or 6 mm thick pressed steel standard road form

stake housings welded to web of road form

sliding plate housing

sliding plate connector

stop for sliding plate

standard road form length 3·000

50

3 No. fixing stakes    5 No. fixing stakes

100 to 250

fixing stake housing

100 to 250

STANDARD ROAD FORM          FLEXIBLE ROAD FORM

Joints in Rigid Pavings ~ longitudinal and transverse joints are required in rigid pavings to :-

1. Limit size of slab.

2. Limit stresses due to subgrade restraint.

3. Provide for expansion and contraction movements.

The main joints used are classified as expansion, contraction or longitudinal, the latter being the same in detail as the contraction joint differing only in direction. The spacing of road joints is determined by :-

1. Slab thickness.

2. Whether slab is reinforced or unreinforced.

3. Anticipated traffic load and flow rate.

4. Temperature at which concrete is laid.

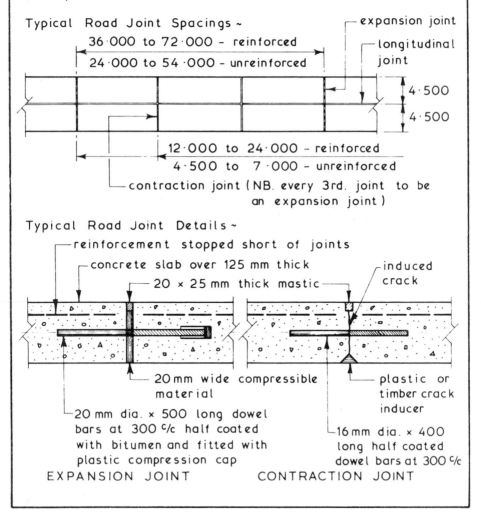

Typical Road Joint Spacings ~

36·000 to 72·000 – reinforced

24·000 to 54·000 – unreinforced

expansion joint

longitudinal joint

4·500

4·500

12·000 to 24·000 – reinforced

4·500 to 7·000 – unreinforced

contraction joint (NB. every 3rd. joint to be an expansion joint)

Typical Road Joint Details ~

reinforcement stopped short of joints

concrete slab over 125 mm thick

20 × 25 mm thick mastic

induced crack

20 mm wide compressible material

20 mm dia. × 500 long dowel bars at 300 c/c half coated with bitumen and fitted with plastic compression cap

EXPANSION JOINT

plastic or timber crack inducer

16 mm dia. × 400 long half coated dowel bars at 300 c/c

CONTRACTION JOINT

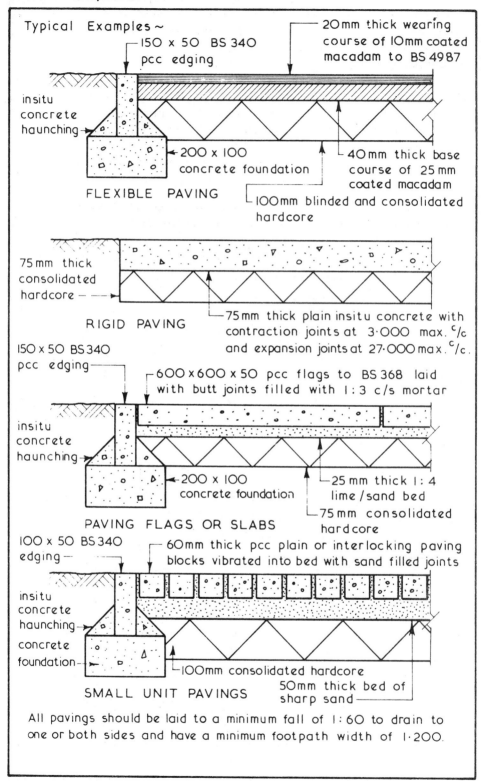

Typical Examples ~

150 x 50 BS 340 pcc edging

20mm thick wearing course of 10mm coated macadam to BS 4987

insitu concrete haunching

200 x 100 concrete foundation

40mm thick base course of 25mm coated macadam

100mm blinded and consolidated hardcore

**FLEXIBLE PAVING**

75mm thick consolidated hardcore

75mm thick plain insitu concrete with contraction joints at 3·000 max. $^c/c$ and expansion joints at 27·000 max. $^c/c$.

**RIGID PAVING**

150 x 50 BS 340 pcc edging

600 x 600 x 50 pcc flags to BS 368 laid with butt joints filled with 1:3 c/s mortar

insitu concrete haunching

200 x 100 concrete foundation

25mm thick 1:4 lime/sand bed

75mm consolidated hardcore

**PAVING FLAGS OR SLABS**

100 x 50 BS 340 edging

60mm thick pcc plain or interlocking paving blocks vibrated into bed with sand filled joints

insitu concrete haunching

concrete foundation

100mm consolidated hardcore

50mm thick bed of sharp sand

**SMALL UNIT PAVINGS**

All pavings should be laid to a minimum fall of 1:60 to drain to one or both sides and have a minimum footpath width of 1·200.

Landscaping ~ in the context of building works this would involve reinstatement of the site as a preparation to the landscaping in the form of lawns, paths, pavings, flower and shrub beds and tree planting. The actual planning, lawn laying and planting activities are normally undertaken by a landscape subcontractor. The main contractor's work would involve clearing away all waste and unwanted materials, breaking up and levelling surface areas, removing all unwanted vegetation, preparing the subsoil for and spreading topsoil to a depth of at least 150 mm.

Services ~ the actual position and laying of services is the responsibility of the various service boards and undertakings. The best method is to use the common trench approach, avoid as far as practicable laying services under the highway.

Typical Common Trench Details ~

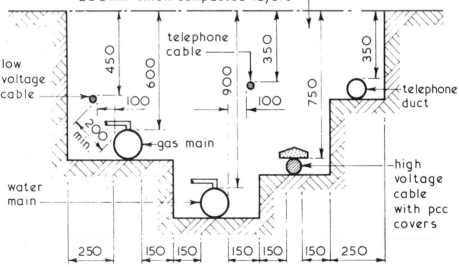

Road Signs ~ these can range from markings painted on roads to define traffic lanes, rights of way and warnings of hazards to signs mounted above the road level to give information, warning or directives, the latter being obligatory.

Typical Examples ~

INFORMATION        WARNING - ROAD WORKS        DIRECTIVE - NO LEFT TURN

# Tubular Scaffolding

Scaffolds ~ these are temporary working platforms erected around the perimeter of a building or structure to provide a safe working place at a convenient height. They are usually required when the working height or level is 1·500 or more above the ground level. All scaffolds must comply with the minimum requirements set out in the Construction (Working Places) Regulations 1966.

Component Parts of a Tubular Scaffold ~

all tubes to comply with BS 1139

standard
transoms or putlogs

transom or putlog

ledger

longitudinal horizontal member usually called a ledger - fixed to standards with double couplers

vertical member usually called a standard spaced at 1·800 to 2·400 centres depending on load to be carried

blade end     standard

transverse horizontal member called a putlog - fixed to ledger with a putlog clip

transom or putlog

standard

putlog clip

ledger

double coupler

ledger

transverse horizontal member called a transom fixed to ledgers

base plate with locating spigot plan size 150 x 150

timber sole plate under base plates on soft or uneven ground

**HORIZONTAL COMPONENTS**

façade brace

cross brace

all bracing fixed with swivel couplers

**VERTICAL COMPONENT**

**SLOPING COMPONENTS**

Putlog Scaffolds ~ these are scaffolds which have an outer row of standards joined together by ledgers which in turn support the transverse putlogs which are built into the bed joints or perpends as the work proceeds, they are therefore only suitable for new work in bricks or blocks.

Typical Details ~

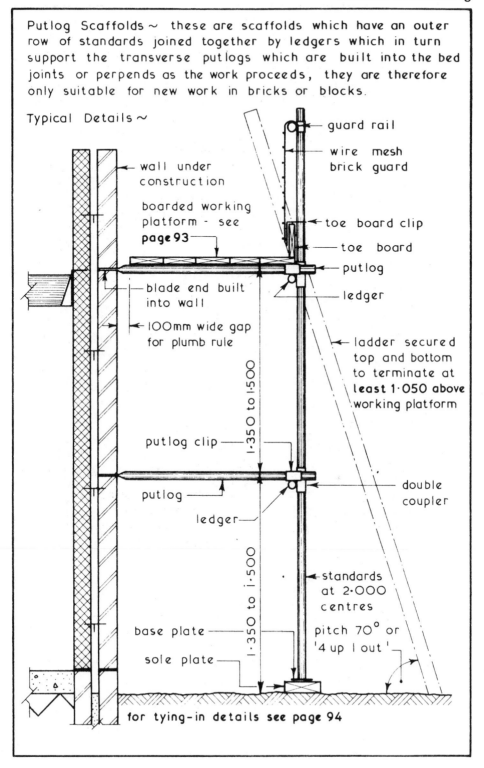

guard rail

wire mesh brick guard

wall under construction

boarded working platform – see **page 93**

toe board clip

toe board

putlog

ledger

blade end built into wall

100mm wide gap for plumb rule

1·350 to 1·500

ladder secured top and bottom to terminate at **least 1·050 above** working platform

putlog clip

putlog

ledger

double coupler

1·350 to 1·500

standards at 2·000 centres

base plate

sole plate

pitch 70° or '4 up 1 out'

for tying-in details see page 94

91

Independent Scaffolds ~ these are scaffolds which have two rows of standards each row joined together with ledgers which in turn support the transverse transoms. The scaffold is erected clear of the existing or proposed building but is tied to the building or structure at suitable intervals – see **page 94**

Typical Details ~

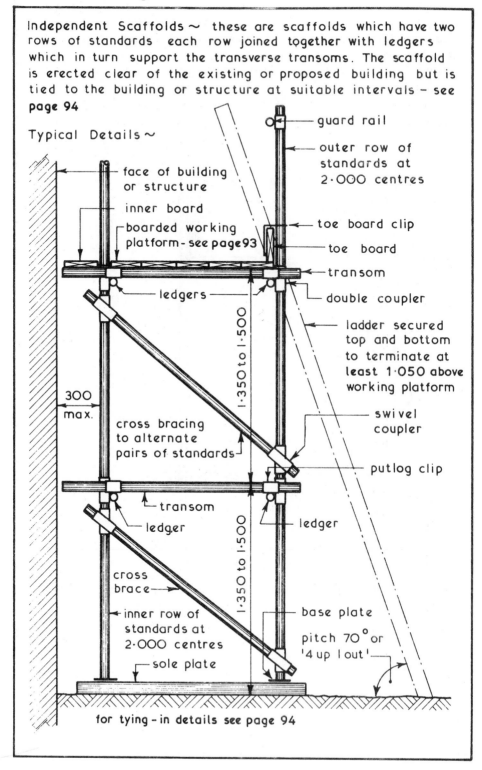

face of building or structure

inner board

boarded working platform- see page 93

ledgers

300 max.

cross bracing to alternate pairs of standards

transom

ledger

cross brace

inner row of standards at 2·000 centres

sole plate

guard rail

outer row of standards at 2·000 centres

toe board clip

toe board

transom

double coupler

ladder secured top and bottom to terminate at least 1·050 above working platform

swivel coupler

putlog clip

1·350 to 1·500

ledger

1·350 to 1·500

base plate

pitch 70° or '4 up 1 out'

for tying-in details see page 94

92

Working Platforms ~ these are close boarded or plated level surfaces at height at which work is being carried out and they must provide a safe working place of sufficient strength to support the imposed loads of operatives and/or materials. All working platforms more than 2·000 above the ground level must be fitted with a toe board and a guard rail.

Typical Details ~

boards to be free of defects

150 mm minimum

225mm wide x 38mm thick x 3·900 long softwood standard scaffold board

25mm wide x 0·9mm thick galvanised hoop iron binding to both ends to prevent splitting

maximum overhang 4 x board thickness

bevelled piece at board overlap

38

1·500 max.   1·500 max.

transom or putlog

boards to be evenly supported on at least 3 supports per board length

## SCAFFOLD BOARDS FOR WORKING PLATFORMS

inner row of standards

300 max.

800 minimum

guard rail

outer row of standards

430 minimum

toe board clip

deposited material

toe board

transom

working platform

ledger

passage way only
600 minimum

NB above dimensions also apply to putlog scaffolds

760 maximum

150 min.

915 to 1150

over 2·000 to ground level

# Tubular Scaffolding

Tying-in ~ all putlog and independent scaffolds should be securely tied to the building or structure at alternate lift heights vertically and at not more than 6·000 centres horizontally. Putlogs should not be classified as ties. Suitable tying-in methods include connecting to tubes fitted between sides of window openings or to internal tubes fitted across window openings, the former method should not be used for more than 50 % of the total number of ties. If there is an insufficient number of window openings for the required number of ties external rakers should be used.

Typical Details ~

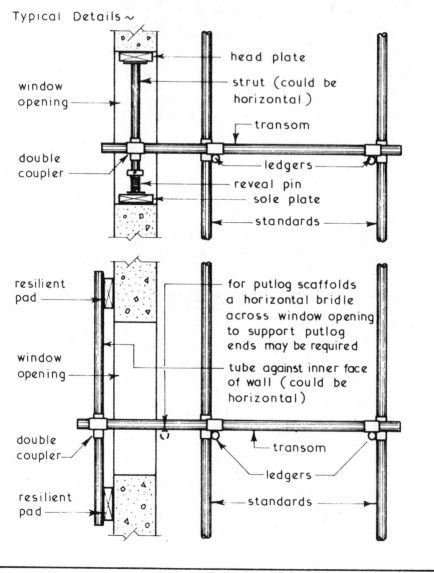

Mobile Scaffolds ~ sometimes called mobile tower scaffolds, are constructed to the basic principles as independent tubular scaffolds and are used to provide access to restricted or small areas and /or where mobility is required.

Typical Details ~

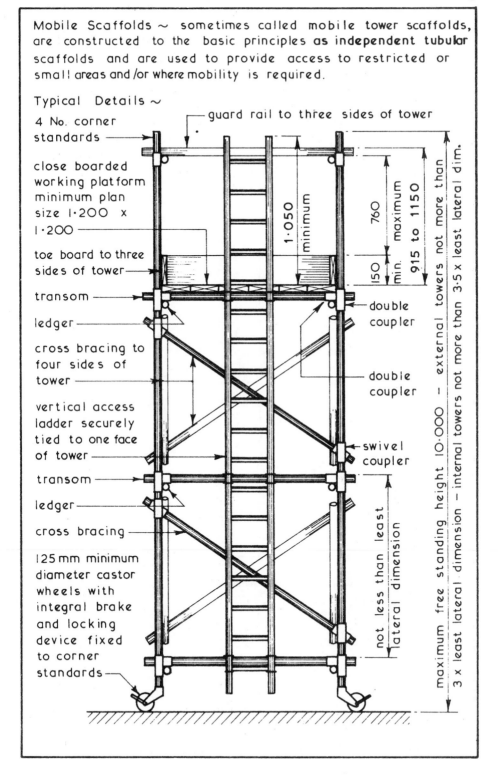

4 No. corner standards

guard rail to three sides of tower

close boarded working platform minimum plan size 1·200 x 1·200

toe board to three sides of tower

transom

ledger

cross bracing to four sides of tower

vertical access ladder securely tied to one face of tower

transom

ledger

cross bracing

125 mm minimum diameter castor wheels with integral brake and locking device fixed to corner standards

double coupler

double coupler

swivel coupler

1·050 minimum

760

150 min.

915 to 1150 min. maximum

maximum free standing height 10·000 – external towers not more than 3 x least lateral dimension – internal towers not more than 3·5 x least lateral dim.

not less than least lateral dimension

Patent Scaffolding ~ these are systems based on an independent scaffold format in which the members are connected together using an integral locking device instead of conventional clips and couplers used with traditional tubular scaffolding. They have the advantages of being easy to assemble and take down using semi-skilled labour and will automatically comply with the majority of the requirements set out in the Construction (Working Places) Regulations 1966. Generally cross bracing is not required with these systems but façade bracing can be fitted if necessary. Although simple in concept patent systems of scaffolding can lack the flexibility of traditional tubular scaffolds in complex layout situations.

Typical Example ~

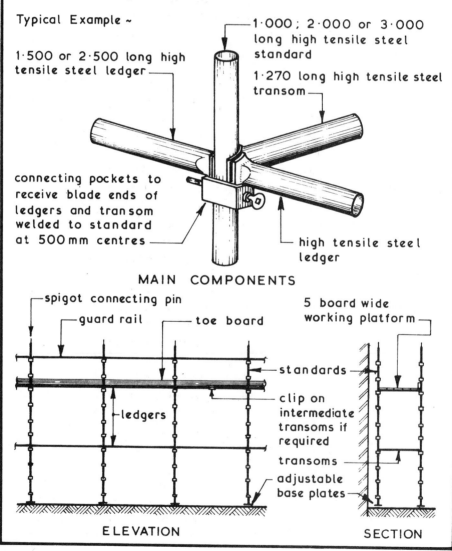

1·500 or 2·500 long high tensile steel ledger

1·000; 2·000 or 3·000 long high tensile steel standard

1·270 long high tensile steel transom

connecting pockets to receive blade ends of ledgers and transom welded to standard at 500mm centres

high tensile steel ledger

**MAIN COMPONENTS**

spigot connecting pin

guard rail

toe board

5 board wide working platform

standards

clip on intermediate transoms if required

ledgers

transoms

adjustable base plates

**ELEVATION**

**SECTION**

Scaffolding Systems ~ these are temporary stagings to provide safe access to and egress from a working platform. The traditional putlog and independent scaffolds have been covered on pages 90 to 95 inclusive. The minimum legal requirements contained in the Construction ( Working Places ) Regulations 1966 applicable to traditional scaffolds apply equally to special scaffolds.   Special scaffolds are designed to fulfil a specific function or to provide access to areas where it is not possible and / or economic to use traditional formats.   They can be constructed from standard tubes or patent systems the latter complying with most regulation requirements, are easy and quick to assemble but lack the complete flexibilty of traditional tubular scaffolds.

Birdcage Scaffolds ~ these are a form of independent scaffold normally used for internal work in large buildings such as public halls and churches to provide access to ceilings and soffits for light maintenance work like painting and cleaning.   They consist of parallel rows of standards connected by ledgers in both directions , the whole arrangement being firmly braced in all directions.   The whole birdcage scaffold assembly is designed to support a single working platform which should be double planked or underlined with polythene or similar sheeting as a means of restricting the amount of dust reaching the floor level.

Slung Scaffolds ~ these are a form of scaffold which is suspended from the main structure by means of wire ropes or steel chains and is not provided with a means of being raised or lowered.   Each working platform of a slung scaffold consists of a supporting framework of ledgers and transoms which should not create a plan size in excess of 2·400 × 2·400 and be held in position by not less than six evenly spaced wire ropes or steel chains securely anchored at both ends.   The working platform should be double planked or underlined with polythene or similar sheeting to restrict the amount of dust reaching the floor level.   Slung scaffolds are an alternative to birdcage scaffolds and although more difficult to erect have the advantage of leaving a clear space beneath the working platform which makes them suitable for cinemas , theatres and high ceiling banking halls.

# Scaffolding Systems

Suspended Scaffolds ~ these consist of a working platform in the form of a cradle which is suspended from cantilever beams or outriggers from the roof of a tall building to give access to the façade for carrying out light maintenace work and cleaning activites. The cradles can have manual or power control and be in single units or grouped together to form a continuous working platform. If grouped together they are connected to one another at their abutment ends with hinges to form a gap of not more than 25 mm wide. Many high rise buildings have a permanent cradle system installed at roof level and this is recommended for all buildings over 30·000 high.

Typical Example ~

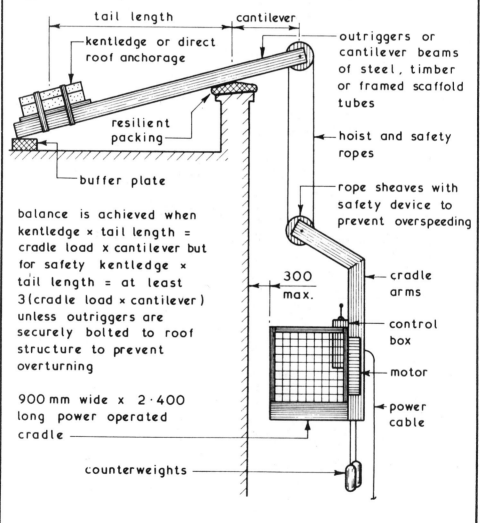

tail length — cantilever

kentledge or direct roof anchorage

outriggers or cantilever beams of steel, timber or framed scaffold tubes

resilient packing

hoist and safety ropes

buffer plate

rope sheaves with safety device to prevent overspeeding

300 max.

cradle arms

balance is achieved when kentledge × tail length = cradle load × cantilever but for safety kentledge × tail length = at least 3 (cradle load × cantilever) unless outriggers are securely bolted to roof structure to prevent overturning

control box

motor

power cable

900 mm wide × 2·400 long power operated cradle

counterweights

Cantilever Scaffolds ~ these are a form of independent
tied scaffold erected on cantilever beams and used where
it is impracticable, undesirable or uneconomic to use a
traditional scaffold raised from ground level. The assembly
of a cantilever scaffold requires special skills and should
therefore always be carried out by trained and experienced
personnel.

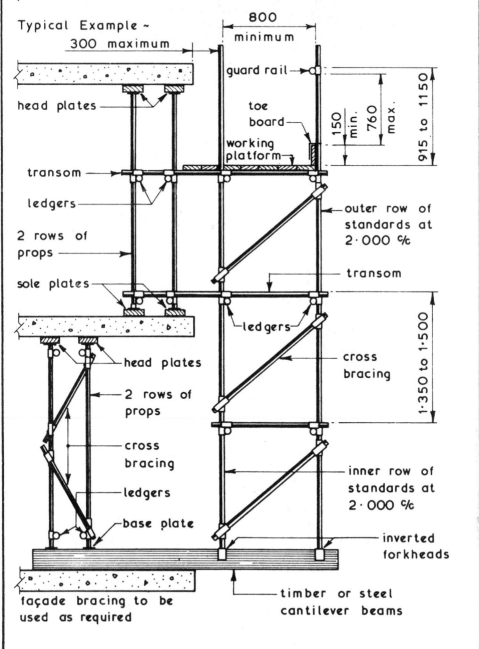

Typical Example ~

façade bracing to be
used as required

99

Truss-out Scaffold ~ this is a form of independent tied scaffold used where it is impracticable, undesirable or uneconomic to build a scaffold from ground level. The supporting scaffold structure is known as the truss-out. The assembly of this form of scaffold requires special skills and should therefore be carried out by trained and experienced personnel.

Typical Example ~

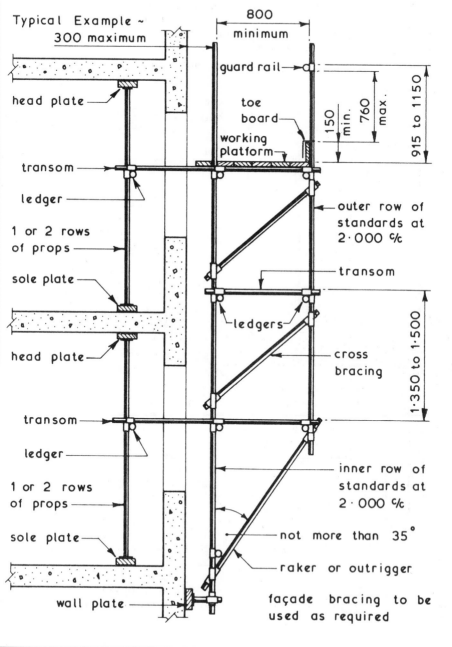

Gantries ~ these are elevated platforms used when the building being maintained or under construction is adjacent to a public footpath. A gantry over a footpath can be used for storage of materials, housing units of accommodation and supporting an independent scaffold. Local authority permission will be required before a gantry can be erected and they have the power to set out the conditions regarding minimum sizes to be used for public walkways and lighting requirements. It may also be necessary to comply with police restrictions regarding the loading and unloading of vehicles at the gantry position. A gantry can be constructed of any suitable structural material and may need to be structurally designed to meet all the necessary safety requirements.

Typical Example ~

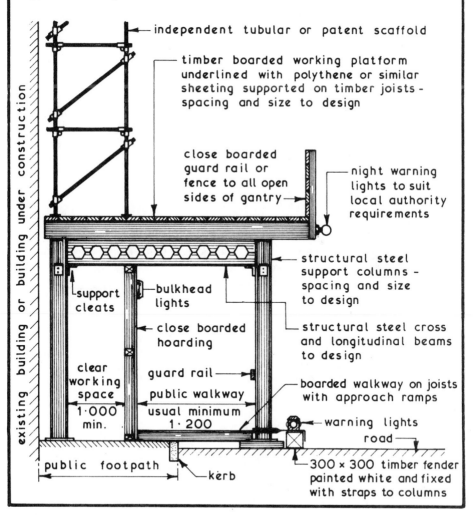

independent tubular or patent scaffold

timber boarded working platform underlined with polythene or similar sheeting supported on timber joists - spacing and size to design

close boarded guard rail or fence to all open sides of gantry

night warning lights to suit local authority requirements

support cleats

bulkhead lights

close boarded hoarding

structural steel support columns - spacing and size to design

structural steel cross and longitudinal beams to design

clear working space 1·000 min.

guard rail

public walkway usual minimum 1·200

boarded walkway on joists with approach ramps

warning lights

road

public footpath

kerb

300 × 300 timber fender painted white and fixed with straps to columns

existing building or building under construction

101

Shoring ~ this is a form of temporary support which can be given to existing buildings with the primary function of providing the necessary precautions to avoid damage to any person from collapse of structure as required by Regulation 50 of the Construction (General Provisions) Regulations 1961.

Shoring Systems ~ there are three basic systems of shoring which can be used separately or in combination with one another to provide the support(s) and these are namely :-

1. Dead Shoring - used primarily to carry vertical loadings.

2. Raking Shoring - used to support a combination of vertical and horizontal loadings.

3. Flying Shoring - an alternative to raking shoring to give a clear working space at ground level.

Typical Shoring Situations ~

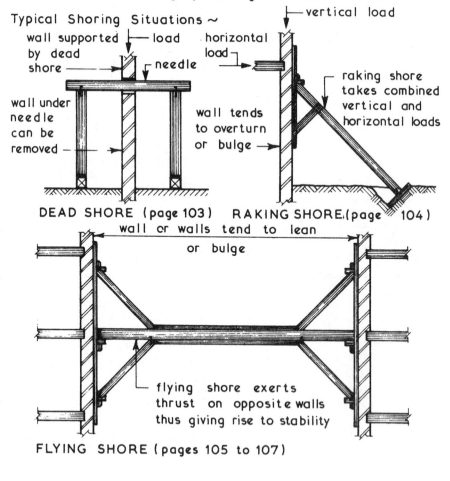

DEAD SHORE (page 103)     RAKING SHORE.(page 104)

FLYING SHORE (pages 105 to 107)

Dead Shores ~ these shores should be placed at approximately 2·000 ᶜ/c and positioned under the piers between the windows, any windows in the vicinity of the shores being strutted to prevent distortion of the openings. A survey should be carried out to establish the location of any underground services so that they can be protected as necessary. The sizes shown in the detail below are typical, actual sizes should be obtained from tables or calculated from first principles. Any suitable structural material such as steel can be substituted for the timber members shown.

Typical Detail ~

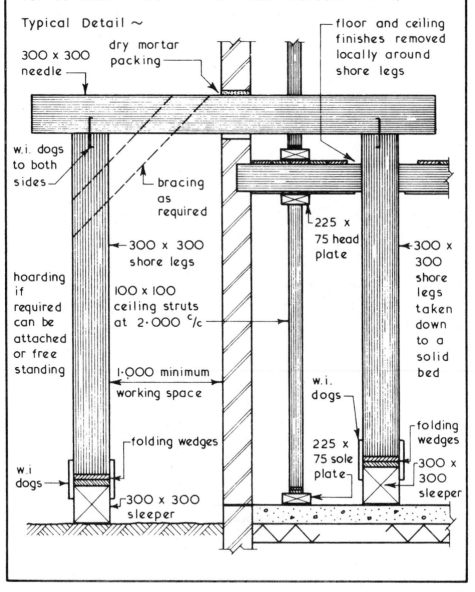

300 x 300 needle

dry mortar packing

floor and ceiling finishes removed locally around shore legs

w.i. dogs to both sides

bracing as required

300 x 300 shore legs

100 x 100 ceiling struts at 2·000 ᶜ/c

hoarding if required can be attached or free standing

1·000 minimum working space

225 x 75 head plate

300 x 300 shore legs taken down to a solid bed

w.i. dogs

225 x 75 sole plate

folding wedges

300 x 300 sleeper

w.i dogs

folding wedges

300 x 300 sleeper

# Shoring

Raking Shoring ~ these are placed at 3·000 to 4·500 $^c/c$ and can be of single, double, triple or multiple raker format. Suitable materials are timber structural steel and framed tubular scaffolding.

Typical Multiple Raking Shore Detail ~

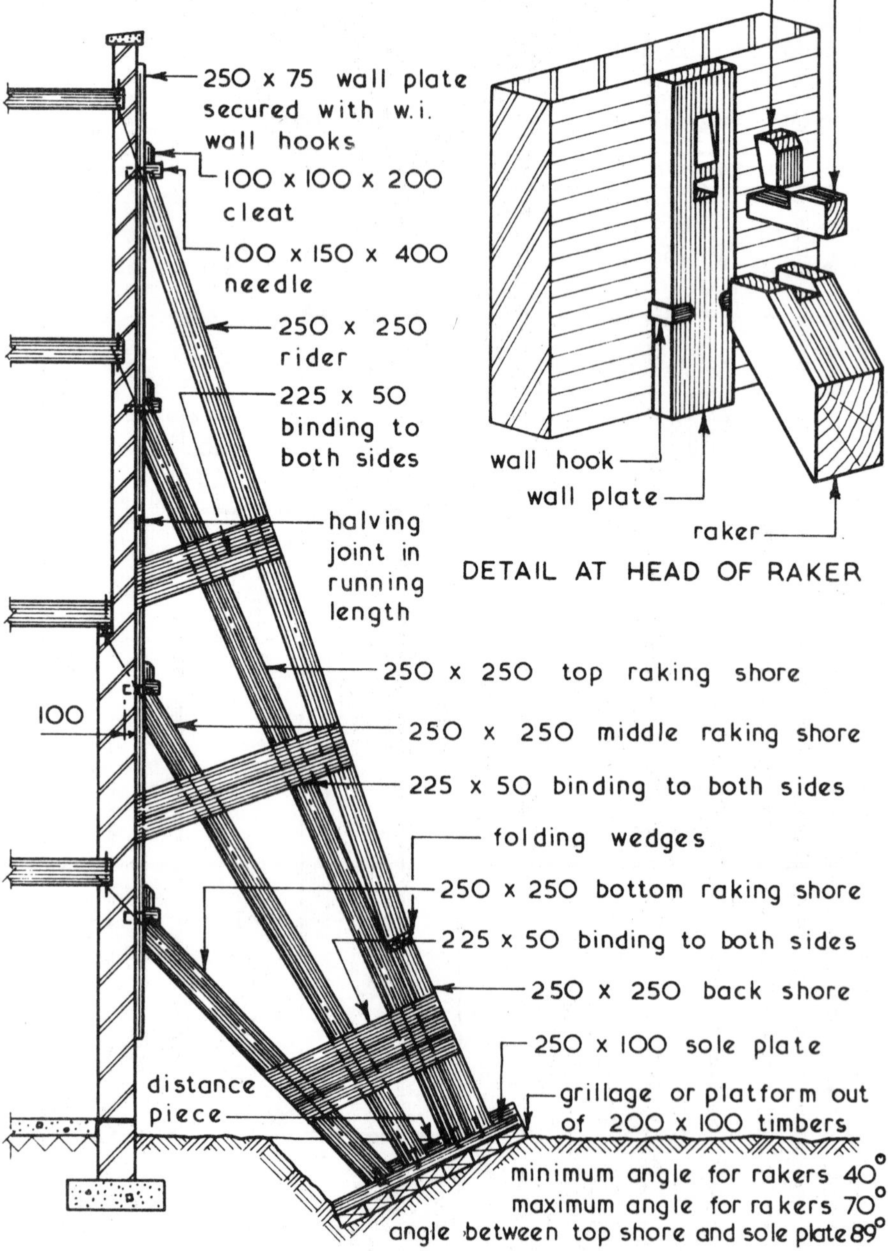

- 250 x 75 wall plate secured with w.i. wall hooks
- 100 x 100 x 200 cleat
- 100 x 150 x 400 needle
- 250 x 250 rider
- 225 x 50 binding to both sides
- halving joint in running length

needle
cleat

DETAIL AT HEAD OF RAKER

wall hook
wall plate
raker

100

- 250 x 250 top raking shore
- 250 x 250 middle raking shore
- 225 x 50 binding to both sides
- folding wedges
- 250 x 250 bottom raking shore
- 225 x 50 binding to both sides
- 250 x 250 back shore
- 250 x 100 sole plate
- grillage or platform out of 200 x 100 timbers

distance piece

minimum angle for rakers 40°
maximum angle for rakers 70°
angle between top shore and sole plate 89°

Flying Shores ~ these are placed at 3·000 to 4·500 °/c and can be of a single or double format. They are designed detailed and constructed to the same basic principles as that shown for raking shores on page 104. Unsymmetrical arrangements are possible providing the basic principles for flying shores are applied - **see page 107.**

Typical Single Flying Shore Detail ~

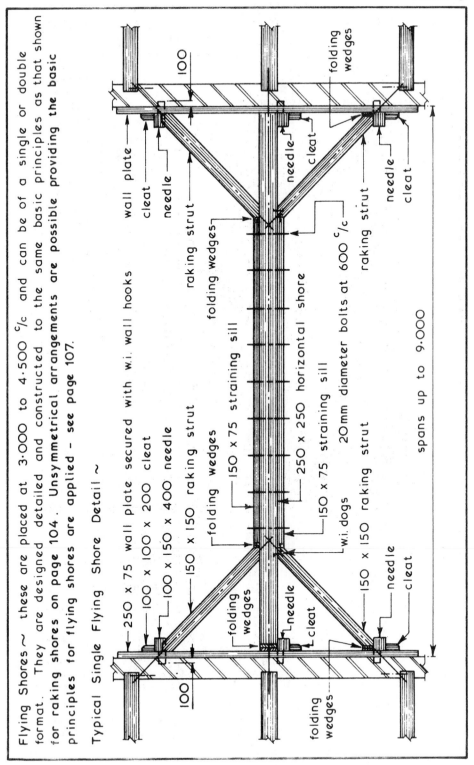

wall plate
cleat
needle

100

folding wedges

needle
cleat

needle
cleat

raking strut

raking strut

folding wedges

150 x 75 straining sill

20mm diameter bolts at 600 °/c

250 x 250 horizontal shore

150 x 75 straining sill

150 x 150 raking strut

w.i. dogs

150 x 150 raking strut

needle
cleat

spans up to 9.000

250 x 75 wall plate secured with w.i. wall hooks

100 x 100 x 200 cleat

100 x 150 x 400 needle

150 x 150 raking strut

folding wedges

folding wedges

needle
cleat

folding wedges

100

105

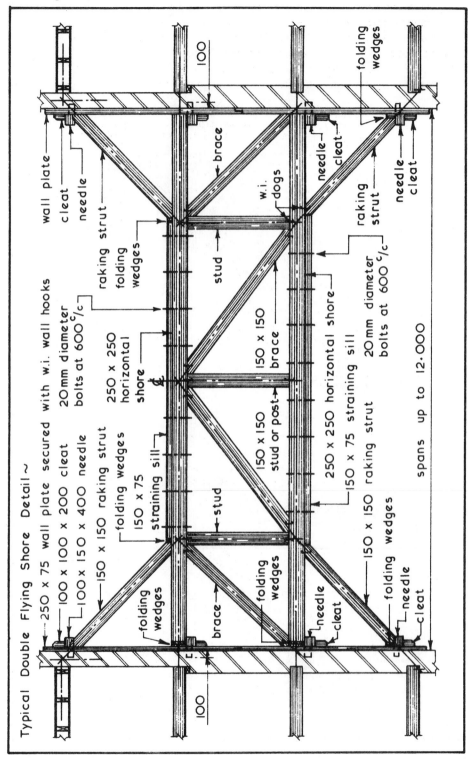

Typical Double Flying Shore Detail ~

- 250 x 75 wall plate secured with w.i. wall hooks
- 100 x 100 x 200 cleat
- 100 x 150 x 400 needle
- 150 x 150 raking strut
- folding wedges
- 150 x 75 straining sill
- wall plate
- cleat
- needle
- raking strut
- folding wedges
- 250 x 250 horizontal shore
- 20mm diameter bolts at 600 c/c
- brace
- stud
- 150 x 150 brace
- 150 x 150 stud or post
- folding wedges
- stud
- folding wedges
- needle cleat
- brace
- w.i. dogs
- needle cleat
- needle cleat
- raking strut
- folding wedges
- 250 x 250 horizontal shore
- 250 x 75 straining sill
- 150 x 150 raking strut
- 20mm diameter bolts at 600 c/c
- folding wedges
- needle cleat
- spans up to 12.000

Unsymmetrical Flying Shores ~ arrangements of flying shores for unsymmetrical situations can be devised if the basic principles for symmetrical shores is applied (see page 105). In some cases the arrangement will consist of a combination of both raking and flying shore principles.

Typical Examples ~

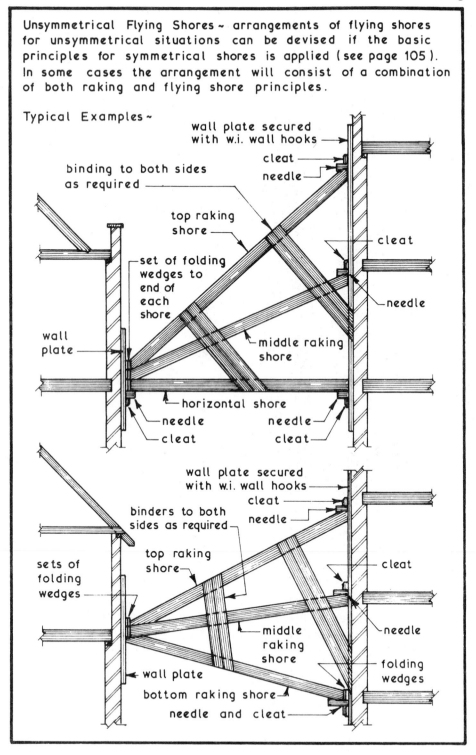

107

Temporary Support Determination ~ the basic sizing of most temporary supports follows the principles of elementary structural design. Readers with this basic knowledge should be able to calculate such support members which are required particularly those used in the context of the maintenance and adaptation of buildings such as a dead shoring system.

Typical Example ~

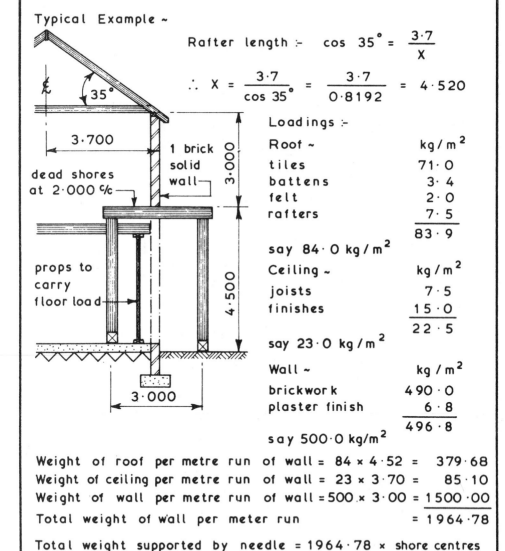

Rafter length :- $\cos 35° = \dfrac{3\cdot7}{X}$

$\therefore X = \dfrac{3\cdot7}{\cos 35°} = \dfrac{3\cdot7}{0\cdot8192} = 4\cdot520$

Loadings :-

| Roof ~ | kg/m² |
|---|---|
| tiles | 71·0 |
| battens | 3·4 |
| felt | 2·0 |
| rafters | 7·5 |
| | 83·9 |

say 84·0 kg/m²

| Ceiling ~ | kg/m² |
|---|---|
| joists | 7·5 |
| finishes | 15·0 |
| | 22·5 |

say 23·0 kg/m²

| Wall ~ | kg/m² |
|---|---|
| brickwork | 490·0 |
| plaster finish | 6·8 |
| | 496·8 |

say 500·0 kg/m²

Weight of roof per metre run of wall = 84 × 4·52 = 379·68
Weight of ceiling per metre run of wall = 23 × 3·70 = 85·10
Weight of wall per metre run of wall = 500 × 3·00 = 1500·00

Total weight of wall per meter run = 1964·78

Total weight supported by needle = 1964·78 × shore centres
= 1964·78 × 2·000
= 3929·56 kg
say 3930 kg

For design calculations see page 109

Design Calculations ~ reference page 108

Needle Design :-

$W$ = 3930 kg
hence force
= 3930 × 9·81 ≙ 39300 N

$R_A$ = $R_B$ = $\dfrac{W}{2}$

= $\dfrac{39300}{2}$

= 19650 N

L = 3·000

$R_A$     $R_B$

$$BM = \frac{WL}{4} = \frac{39300 \times 3000}{4} = 29475000 \ N/mm$$

$$MR = \text{stress} \times \text{section modulus} = fZ = f\frac{bd^2}{6}$$

assume  b = 300 mm  and  f = 7 N/mm$^2$

then  $$29475000 = \frac{7 \times 300 \times d^2}{6}$$

$$d = \sqrt{\frac{29475000 \times 6}{7 \times 300}} = 290 \cdot 2 \ mm$$

use  300 × 300 timber section  or  2 No. 150 × 300 sections bolted together with timber connectors.

Props to Needle Design :-

$$\text{area} = \frac{\text{load}}{\text{stress}} = \frac{19650}{7} = 2807 \cdot 143 \ mm^2$$

∴ minimum timber size = $\sqrt{2807 \cdot 143}$ = 53 × 53

check slenderness ratio assuming  $l$ = 1·0

$$\text{slenderness ratio} = \frac{l}{b} = \frac{4500}{53} = 84 \cdot 9$$

slenderness ratio for medium term load with $K_{18}$ of 1·0 is not more than 17·3 ( from CP 112 )

∴ minimum timber prop size = $\dfrac{l}{sr}$ = $\dfrac{4500}{17 \cdot 3}$ = 260 · 12 mm

for practical reasons use  300 × 300 prop - new sr = 15

Check crushing at point of loading on needle :-

wall loading on needle = 3930 kg = 39300 N = 39·3 kN

area of contact = width of wall × width of needle
= 215 × 300 = 64500 mm$^2$

safe compressive stress perpendicular to grain = 1·72 N/mm$^2$

∴ safe load = $\dfrac{64500 \times 1 \cdot 72}{1000}$ = 110 · 94 kN  which is > 39·3 kN

# 3 BUILDERS PLANT

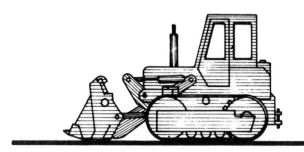

BULLDOZERS

SCRAPERS

GRADERS

TRACTOR SHOVELS

EXCAVATORS

TRANSPORT VEHICLES

HOISTS

CRANES

CONCRETING PLANT

General Considerations ~ items of builders plant ranging from small hand held power tools to larger pieces of plant such as mechanical excavators and tower cranes can be considered for use for one or more of the following reasons :-

1. Increased production.

2. Reduction in overall construction costs.

3. Carry out activities which cannot be carried out by the traditional manual methods in the context of economics.

4. Eliminate heavy manual work thus reducing fatigue and as a consequence increasing productivity.

5. Replacing labour where there is a shortage of personnel with the necessary skills.

6. Maintain the high standards required particularly in the context of structural engineering works.

Economic Considerations ~ the introduction of plant does not always result in economic savings since extra temporary site works such as roadworks, hardstandings, foundations and anchorages may have to be provided at a cost which is in excess of the savings made by using the plant. The site layout and circulation may have to be planned around plant positions and movements rather than around personnel and material movements and accommodation. To be economic plant must be fully utilised and not left standing idle since plant, whether hired or owned, will have to be paid for even if it is non-productive. Full utilisation of plant is usually considered to be in the region of 85% of on site time, thus making an allowance for routine, daily and planned maintenance which needs to be carried out to avoid as far as practicable plant breakdowns which could disrupt the construction programme. Many pieces of plant work in conjunction with other items of plant such as excavators and their attendant haulage vehicles therefore a correct balance of such plant items must be obtained to achieve an economic result.

Maintenance Considerations ~ on large contracts where a a number of plant items are to be used it may be advantageous to employ a skilled mechanic to be on site to carry out all the necessary daily, preventive and planned maintenance tasks together with any running repairs which could be carried out on site.

Plant Costing ~ with the exception of small pieces of plant, which are usually purchased, items of plant can be bought or hired or where there are a number of similar items a combination of buying and hiring could be considered. The choice will be governed by economic factors and the possibility of using the plant on future sites thus enabling the costs to be apportioned over several contracts.

Advantages of Hiring Plant :-

1. Plant can be hired for short periods.

2. Repairs and replacements are usually the responsibility of the hire company.

3. Plant is returned to the hire company after use thus relieving the building contractor of the problem of disposal or finding more work for the plant to justify its purchase or retention.

4. Plant can be hired with the operator, fuel and oil included in the hire rate.

Advantages of Buying Plant :-

1. Plant availability is totally within the control of the contractor.

2. Hourly cost of plant is generally less than hired plant.

3. Owner has choice of costing method used.

Typical Costing Methods ~

1. Straight Line - simple method

Capital Cost = £10 000

Anticipated life = 5 years

Year's working = 1 500 hrs

Resale or scrap value = £ 900

Annual depreciation ~

$$= \frac{10\ 000 - 900}{5} = £1\ 820$$

Hourly depreciation ~

$$= \frac{1\ 820}{1\ 500} = 1 \cdot 210$$

Add  2 % insurance = 0·024
      10 % maintenance = 0·121
      Hourly rate = £1·355

2. Interest on Capital Outlay - widely used more accurate method

Capital Cost = £10 000

C.I. on capital
(8% for 5 yrs ) = 4 693
                        14 693

Deduct resale value     900
                        13 793

+ Insurance at 2% =     200

+ Maintenance at 10% =  1 000
                        14 993

Hourly rate ~

$$= \frac{14\ 993}{5 \times 1\ 500} = £2 \cdot 00$$

NB. add to hourly rate running costs

Output and Cycle Times ~ all items of plant have optimum output and cycle times which can be used as a basis for estimating anticipated productivity taking into account the task involved, task efficiency of the machine, operator's efficiency and in the case of excavators the type of soil. Data for the factors to be taken into consideration can be obtained from timed observations, feedback information or published tables contained in manufacturer's literature or reliable textbooks.

Typical Example ~

Backacter with $1 \, m^3$ capacity bucket engaged in normal trench excavation in a clayey soil and discharging directly into an attendant haulage vehicle.

| | | |
|---|---|---|
| Optimum output | = | 60 bucket loads per hour |
| Task efficiency factor | = | 0·8 (from tables) |
| Operator efficiency factor | = | 75% (typical figure) |
| ∴ Anticipated output | = | 60 × 0·8 × 0·75 |
| | = | 36 bucket loads per hour |
| | = | $36 × 1 = 36 \, m^3$ per hour |

an allowance should be made for the bulking or swell of the solid material due to the introduction of air or voids during the excavation process

∴ Net output allowing for a 30% swell = 36 - (36 × 0·3)
= say $25 \, m^3$ per hr.

If the Bill of Quantities gives a total net excavation of $950 \, m^3$

time required $= \dfrac{950}{25} = $ __38 hours__

or assuming an 8 hour day - $^1/2$ hour maintenance time in

days $= \dfrac{38}{7·5} = $ say __5 days__

Haulage vehicles required $= 1 + \dfrac{\text{round trip time of vehicle}}{\text{loading time of vehicle}}$

if round trip time = 30 minutes and loading time = 10 mins.

number of haulage vehicles required $= 1 + \dfrac{30}{10} = 4$

this gives a vehicle waiting overlap ensuring excavator is fully utilised which is economically desirable.

# Bulldozers

Bulldozers ~ these machines consist of a track or wheel mounted power unit with a mould blade at the front which is usually controlled by hydraulic rams although wire cable operation is preferred on some models. Many bulldozers have the capacity to adjust the mould blade to form an angledozer and the capacity to tilt the mould blade about a central swivel point. Some bulldozers can also be fitted with rear attachments such as rollers and scarifiers.

The main functions of a bulldozer are :-

1. Shallow excavations up to 300 mm deep either on level ground or sidehill cutting.

2. Clearance of shrubs and small trees.

3. Clearance of trees by using raised mould blade as a pusher arm.

4. Acting as a towing tractor.

5. Acting as a pusher to scraper machines (see page 117).

NB. Bulldozers push earth in front of the mould blade with some side spillage whereas angledozers pushes and casts the spoil to one side of the mould blade.

Typical Bulldozer Details ~

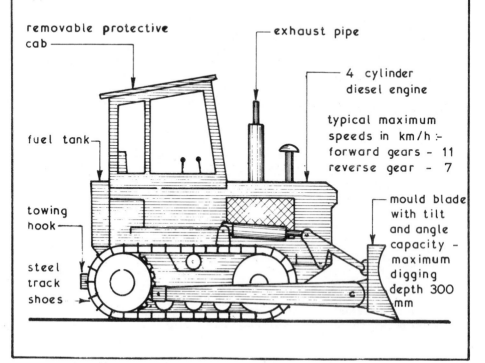

removable protective cab

exhaust pipe

4 cylinder diesel engine

typical maximum speeds in km/h :-
forward gears - 11
reverse gear - 7

fuel tank

towing hook

mould blade with tilt and angle capacity - maximum digging depth 300 mm

steel track shoes

Scrapers ~ these machines consist of a scraper bowl which is lowered to cut and collect soil where site stripping and levelling operations are required involving large volume of earth. When the scraper bowl is full the apron at the cutting edge is closed to retain the earth and the bowl is raised for travelling to the disposal area. On arrival the bowl is lowered, the apron opened and the spoil pushed out by the tailgate as the machine moves forwards. Scrapers are available in three basic formats :-

1. Towed Scrapers - these consist of a four wheeled scraper bowl which is towed behind a power unit such as a crawler tractor. They tend to be slower than other forms of scraper but are useful for small capacities with haul distances up to 300·00.

2. Two Axle Scrapers - these have a two wheeled scraper bowl with an attached two wheeled power unit. They are very manoeuvrable with a low rolling resistance and very good traction.

3. Three Axle Scrapers - these consist of a two wheeled scraper bowl which may have a rear engine to assist the four wheeled traction engine which makes up the complement. Generally these machines have a greater capacity potential than their counterparts, are easier to control and have a faster cycle time.

To obtain maximum efficiency scrapers should operate downhill if possible, have smooth haul roads, hard surfaces broken up before scraping and be assisted over the last few metres by a pushing vehicle such as a bulldozer.

Typical Scraper Details ~

scraper bowl
struck capacity 14 m$^3$
heaped capacity 20m$^3$
width of cut 3·000
depth of cut 450mm max.

8 cylinder diesel engine attached power unit with a top forward speed of 45 km/h

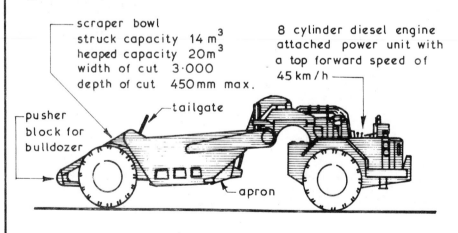

pusher block for bulldozer

tailgate

apron

# Graders

Graders ~ these machines are similar in concept to bulldozers in that they have a long slender adjustable mould blade, which is usually slung under the centre of the machine. A grader's main function is to finish or grade the upper surface of a large area usually as a follow up operation to scraping or bulldozing. They can produce a fine and accurate finish but do not have the power of a bulldozer therefore they are not suitable for oversite excavation work. The mould blade can be adjusted in both the horizontal and vertical planes through an angle of 300° the latter enabling it to be used for grading sloping banks.

Two basic formats of grader are available :-

1. Four Wheeled – all wheels are driven and steered which gives the machine the ability to offset and crab along its direction of travel.

2. Six Wheeled – this machine has 4 wheels in tandem drive at the rear and 2 front tilting idler wheels giving it the ability to counteract side thrust.

Typical Grader Details ~

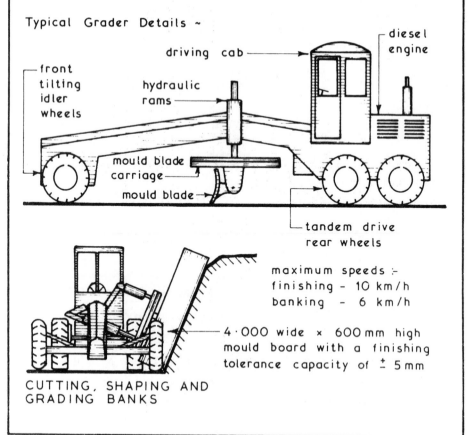

driving cab

diesel engine

front tilting idler wheels

hydraulic rams

mould blade carriage

mould blade

tandem drive rear wheels

maximum speeds :-
finishing – 10 km/h
banking – 6 km/h

4·000 wide × 600 mm high mould board with a finishing tolerance capacity of ± 5 mm

CUTTING, SHAPING AND GRADING BANKS

Tractor Shovels ~ these machines are sometimes called loaders or loader shovels and primary function is to scoop up loose materials in the front mounted bucket, elevate the bucket and manoeuvre into a position to deposit the loose material into an attendant transport vehicle. Tractor shovels are driven towards the pile of loose material with the bucket lowered, the speed and power of the machine will enable the bucket to be filled. Both tracked and wheeled versions are available, the tracked format being more suitable for wet and uneven ground conditions than the wheeled tractor shovel which has greater speed and manoeuvring capabilities. To increase their versatility tractor shovels can be fitted with a 4 in 1 bucket enabling them to carry out bulldozing, excavating, clam lifting and loading activities.

Typical Tractor Shovel Details ~

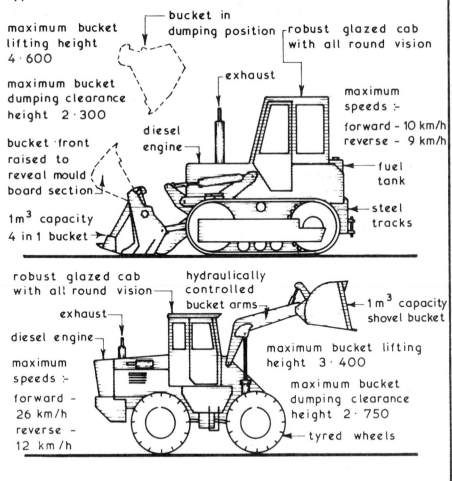

maximum bucket lifting height 4·600

maximum bucket dumping clearance height 2·300

bucket front raised to reveal mould board section

1m³ capacity 4 in 1 bucket

bucket in dumping position

exhaust

diesel engine

robust glazed cab with all round vision

maximum speeds :-
forward – 10 km/h
reverse – 9 km/h

fuel tank

steel tracks

robust glazed cab with all round vision

exhaust

diesel engine

maximum speeds :-
forward – 26 km/h
reverse – 12 km/h

hydraulically controlled bucket arms

1m³ capacity shovel bucket

maximum bucket lifting height 3·400

maximum bucket dumping clearance height 2·750

tyred wheels

# Excavators

Excavating Machines ~ these are one of the major items of builders plant and are used primarily to excavate and load most types of soil. Excavating machines come in a wide variety of designs and sizes but all of them can be placed within one of three categories :-

1. Universal Excavators - this category covers most forms of excavators all of which have a common factor the power unit. The universal power unit is a tracked based machine with a slewing capacity of $360°$ and by altering the boom arrangement and bucket type different excavating functions can be obtained. These machines are selected for high output requirements and are rope controlled.

2. Purpose Designed Excavators - these are machines which have been designed specifically to carry out one mode of excavation and they usually have smaller bucket capacities than universal excavators; they are hydraulically controlled with a shorter cycle time.

3. Multi - purpose Excavators - these machines can perform several excavating functions having both front and rear attachments. They are designed to carry out small excavation operations of low output quickly and efficiently. Multi - purpose excavators can be obtained with a wheeled or tracked base and are ideally suited for a small building firm with low excavation plant utilisation requirements.

Skimmers ~ these excavators are rigged using a universal power unit for surface stripping and shallow excavation work up to 300mm deep where a high degree of accuracy is required. They usually require attendant haulage vehicles to remove the spoil and need to be transported between sites on a low - loader. Because of their limitations and the alternative machines available they are seldom used today.

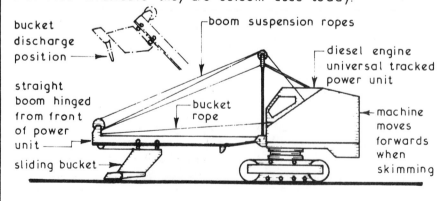

bucket discharge position ─►

boom suspension ropes

diesel engine
universal tracked power unit

straight boom hinged from front of power unit ─┘

bucket rope

◄─ machine moves forwards when skimming

sliding bucket ─►

Face Shovels ~ the primary function of this piece of plant is to excavate above its own track or wheel level. They are available as a universal power unit based machine or as a hydraulic purpose designed unit. These machines can usually excavate any type of soil except rock which needs to be loosened, usually by blasting, prior to excavation. Face shovels generally require attendant haulage vehicles for the removal of spoil and a low loader transport lorry for travel between sites. Most of these machines have a limited capacity of between 300 and 400 mm for excavation below their own track or wheel level.

Typical Face Shovel Details ~

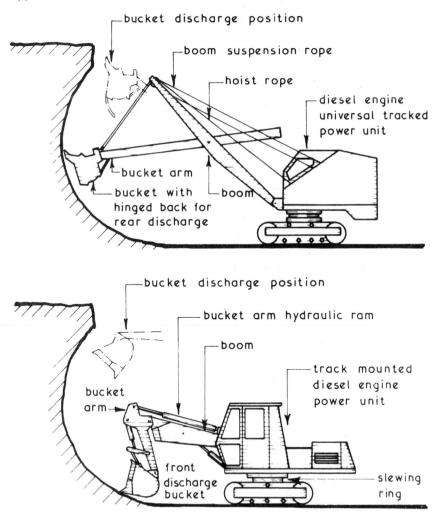

bucket discharge position

boom suspension rope

hoist rope

diesel engine
universal tracked
power unit

bucket arm

bucket with
hinged back for
rear discharge

boom

bucket discharge position

bucket arm hydraulic ram

boom

track mounted
diesel engine
power unit

bucket
arm

front
discharge
bucket

slewing
ring

# Excavators

Backacters ~ these machines are suitable for trench, foundation and basement excavations and are available as a universal power unit base machine or as a purpose designed hydraulic unit. They can be used with or without attendant haulage vehicles since the spoil can be placed alongside the excavation for use in backfilling. These machines will require a low loader transport vehicle for travel between sites. Backacters used in trenching operations with a bucket width equal to the trench width can be very accurate with a high output rating.

Typical Backacter Details ~

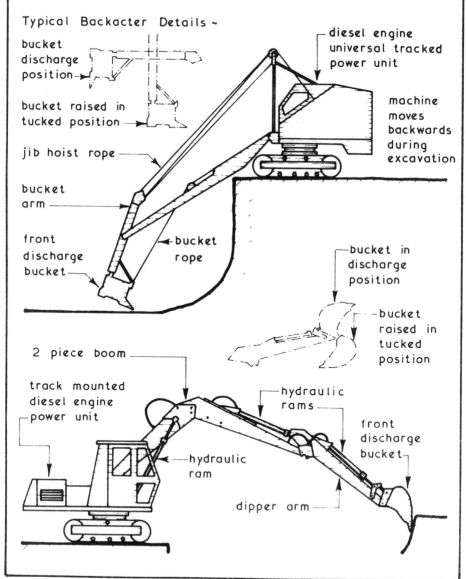

Draglines ~ these machines are based on the universal power unit with basic crane rigging to which is attached a drag bucket. The machine is primarily designed for bulk excavation in loose soils up to 3·000 below its own track level by swinging the bucket out to the excavation position and hauling or dragging it back towards the power unit. Dragline machines can also be fitted with a grab or clamshell bucket for excavating in very loose soils.

Typical Dragline Details ~

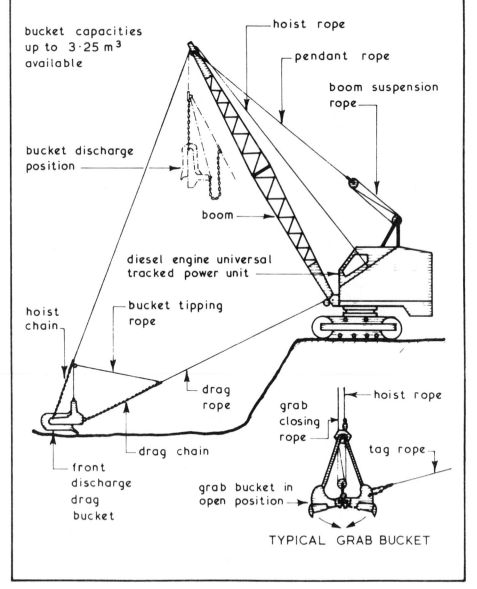

bucket capacities up to 3·25 m³ available

hoist rope

pendant rope

boom suspension rope

bucket discharge position

boom

diesel engine universal tracked power unit

hoist chain

bucket tipping rope

drag rope

drag chain

front discharge drag bucket

grab closing rope

hoist rope

tag rope

grab bucket in open position

TYPICAL GRAB BUCKET

# Excavators

Multi-purpose Excavators ~ these machines are usually based on the agricultural tractor with 2 or 4 wheel drive and are intended mainly for use in conjunction with small excavation works such as those encountered by the small to medium sized building contractor. Most multi-purpose excavators are fitted with a loading/excavating front bucket and a rear backacter bucket both being hydraulically controlled. When in operation using the backacter bucket the machine is raised off its axles by rear mounted hydraulic outriggers or jacks and in some models by placing the front bucket on the ground. Most machines can be fitted with a variety of bucket widths and various attachments such as bulldozer blades, scarifiers, grab buckets and post hole auger borers.

Typical Multi-purpose Excavator Details ~

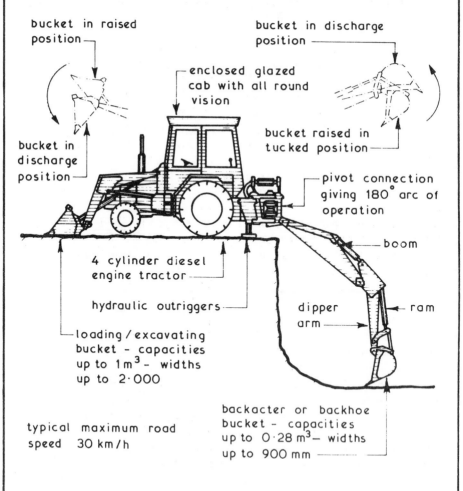

bucket in raised position

bucket in discharge position

enclosed glazed cab with all round vision

bucket raised in tucked position

bucket in discharge position

pivot connection giving 180° arc of operation

boom

4 cylinder diesel engine tractor

hydraulic outriggers

dipper arm

ram

loading/excavating bucket - capacities up to 1 m³ - widths up to 2·000

backacter or backhoe bucket - capacities up to 0·28 m³ - widths up to 900 mm

typical maximum road speed 30 km/h

Transport Vehicles ~ these can be defined as vehicles whose primary function is to convey passengers and /or materials between and around building sites. The types available range from the conventional saloon car to the large low loader lorries designed to transport other items of builders plant between construction sites and the plant yard or depot.

Vans - these transport vehicles range from the small two person plus a limited amount of materials to the large vans with purpose designed bodies such as those built to carry large sheets of glass. Most small vans are usually fitted with a petrol engine and are based on the manufacturer's standard car range whereas the larger vans are purpose designed with either petrol or diesel engines. These basic designs can usually be supplied with an uncovered tipping or non - tipping container mounted behind the passenger cab for use as a 'pick-up' truck.

Passenger Vehicles - these can range from a simple framed cabin which can be placed in the container of a small lorry or 'pick-up' truck to a conventional bus or coach. Vans can also be designed to carry a limited number of seated passengers by having fixed or removable seating together with windows fitted in the van sides thus giving the vehicle a dual function. The number of passengers carried can be limited so that the driver does not have to hold a PSV (public service vehicle) licence.

Lorries - these are sometimes referred to as haul vehicles and are available as road or site only vehicles. Road haulage vehicles have to comply with all the requirements of the Road Traffic Acts which among other requirements limits size and axle loads. The off - highway or site only lorries are not so restricted and can be designed to carry two to three times the axle load allowed on the public highway. Site only lorries are usually specially designed to traverse and withstand the rough terrain encountered on many construction sites. Lorries are available as non - tipping, tipping and special purpose carriers such as those with removable skips and those equipped with self loading and unloading devices. Lorries specifically designed for the transportation of large items of plant are called low loaders and are usually fitted with integral or removable ramps to facilitate loading and some have a winching system to haul the plant onto the carrier platform.

Dumpers ~ these are used for the horizontal transportation of materials on and off construction sites generally by means of an integral tipping skip. Highway dumpers are of a similar but larger design and can be used to carry materials such as excavated spoil along the roads. A wide range of dumpers are available of various carrying capacities and options for gravity or hydraulic discharge control with front tipping, side tipping or elevated tipping facilities. Special format dumpers fitted with flat platforms, rigs to carry materials skips and rigs for concrete skips for crane hoisting are also obtainable. These machines are designed to traverse rough terrain but they are not designed to carry passengers and this misuse is the cause of many accidents involving dumpers.

Typical Dumper Details ~

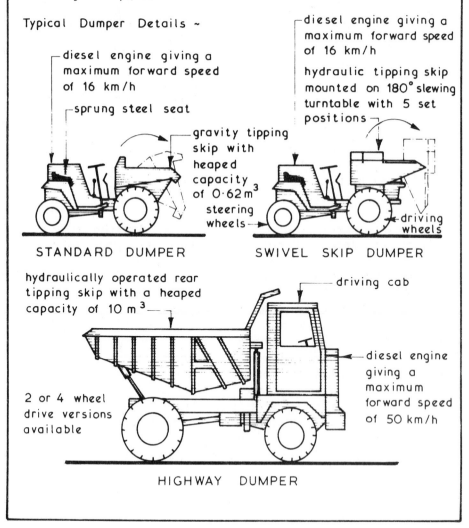

diesel engine giving a maximum forward speed of 16 km/h

sprung steel seat

gravity tipping skip with heaped capacity of $0.62 m^3$

steering wheels

STANDARD DUMPER

diesel engine giving a maximum forward speed of 16 km/h

hydraulic tipping skip mounted on 180° slewing turntable with 5 set positions

driving wheels

SWIVEL SKIP DUMPER

hydraulically operated rear tipping skip with a heaped capacity of $10 m^3$

driving cab

diesel engine giving a maximum forward speed of 50 km/h

2 or 4 wheel drive versions available

HIGHWAY DUMPER

Fork Lift Trucks ~ these are used for the horizontal and limited vertical transportation of materials positioned on pallets or banded together such as brick packs. They are generally suitable for construction sites where the building height does not exceed three storeys. Although designed to negotiate rough terrain site fork lift trucks have a higher productivity on firm and level soils. Three basic fork lift truck formats are available namely straight mast, overhead and telescopic boom with various height, reach and lifting capacities. Scaffolds onto which the load(s) are to be placed should be strengthened locally or a specially constructed loading tower could be built as an attachment to or as an integral part of the main scaffold.

Typical Fork Lift Truck Details ~

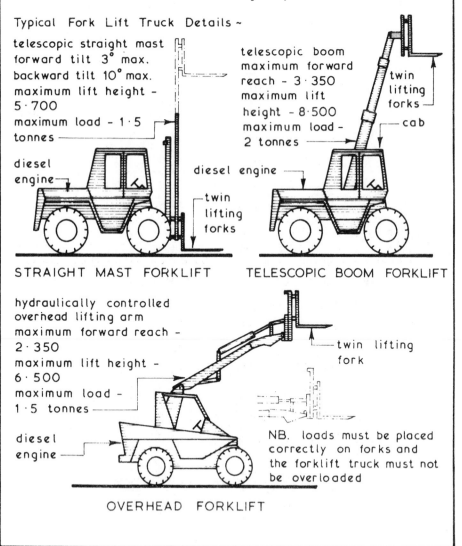

telescopic straight mast
forward tilt 3° max.
backward tilt 10° max.
maximum lift height –
5·700
maximum load – 1·5
tonnes

diesel
engine

twin
lifting
forks

STRAIGHT MAST FORKLIFT

telescopic boom
maximum forward
reach – 3·350
maximum lift
height – 8·500
maximum load –
2 tonnes

twin
lifting
forks

cab

diesel engine

TELESCOPIC BOOM FORKLIFT

hydraulically controlled
overhead lifting arm
maximum forward reach –
2·350
maximum lift height –
6·500
maximum load –
1·5 tonnes

diesel
engine

twin lifting
fork

NB. loads must be placed
correctly on forks and
the forklift truck must not
be overloaded

OVERHEAD FORKLIFT

Hoists ~ these are designed for the vertical transportation of materials, passengers or materials and passengers (see page 129). Materials hoist are designed for one specific use (i.e. the vertical transportation of materials) and under no circumstances should they be used to transport passengers. Most material hoists are of a mobile format which can be dismantled, folded onto the chassis and moved to another position or site under their own power or towed by a haulage vehicle. When in use material hoists need to be stabilised and/or tied to the structure and enclosed with a protective screen.

Typical Materials Hoist Details ~

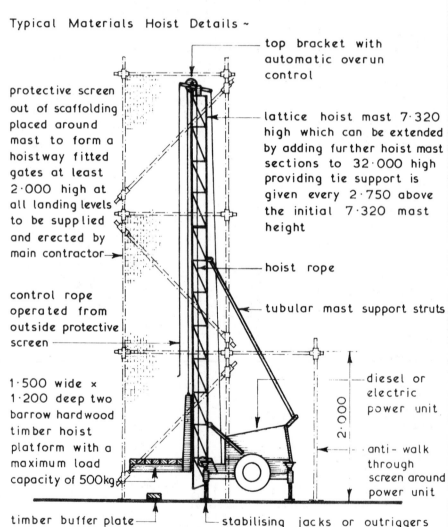

top bracket with automatic overun control

protective screen out of scaffolding placed around mast to form a hoistway fitted gates at least 2·000 high at all landing levels to be supplied and erected by main contractor →

lattice hoist mast 7·320 high which can be extended by adding further hoist mast sections to 32·000 high providing tie support is given every 2·750 above the initial 7·320 mast height

hoist rope

control rope operated from outside protective screen

tubular mast support struts

1·500 wide × 1·200 deep two barrow hardwood timber hoist platform with a maximum load capacity of 500kg

diesel or electric power unit

2·000

anti-walk through screen around power unit

timber buffer plate

stabilising jacks or outriggers

Passenger Hoists ~ these are designed to carry passengers although most are capable of transporting a combined load of materials and passengers within the lifting capacity of the hoist. A wide selection of hoists are available ranging from a single cage with rope suspension to twin cages with rack and pinion operation mounted on two sides of a static tower.

Typical Passenger Hoist Details ~

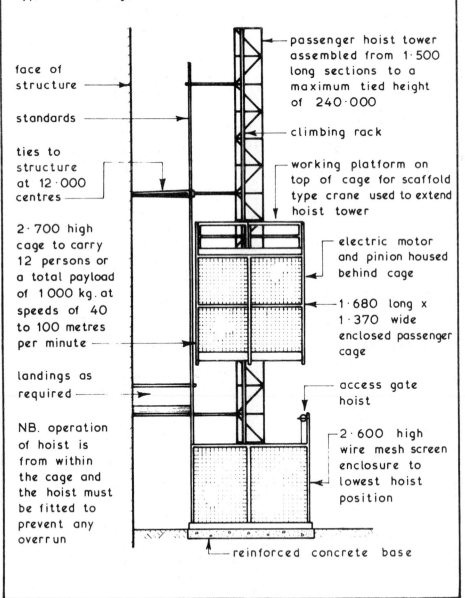

face of structure

standards

ties to structure at 12·000 centres

2·700 high cage to carry 12 persons or a total payload of 1000 kg. at speeds of 40 to 100 metres per minute

landings as required

NB. operation of hoist is from within the cage and the hoist must be fitted to prevent any overrun

passenger hoist tower assembled from 1·500 long sections to a maximum tied height of 240·000

climbing rack

working platform on top of cage for scaffold type crane used to extend hoist tower

electric motor and pinion housed behind cage

1·680 long x 1·370 wide enclosed passenger cage

access gate hoist

2·600 high wire mesh screen enclosure to lowest hoist position

reinforced concrete base

Cranes ~ these are lifting devices designed to raise materials by means of rope operation and move the load horizontally within the limitations of any particular machine. The range of cranes available is very wide.and therefore choice must be based on the loads to be lifted, height and horizontal distance to be covered, time period(s) of lifting operations, utilisation factors and degree of mobility required. Crane types can range from a simple rope and pulley or gin wheel to a complex tower crane but most can be placed within 1 of 3 groups namely mobile, static and tower cranes.

Typical Crane Classifications ~

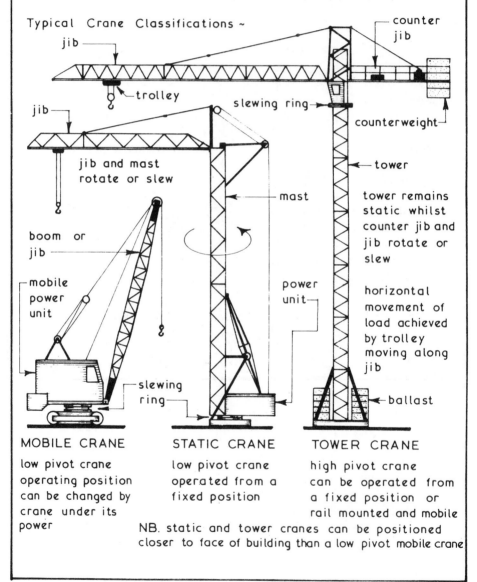

**MOBILE CRANE**

low pivot crane operating position can be changed by crane under its power

**STATIC CRANE**

low pivot crane operated from a fixed position

**TOWER CRANE**

high pivot crane can be operated from a fixed position or rail mounted and mobile

NB. static and tower cranes can be positioned closer to face of building than a low pivot mobile crane

Self Propelled Cranes ~ these are mobile cranes mounted on a wheeled chassis and have only one operator position from which the crane is controlled and the vehicle driven. The road speed of this type of crane is generally low usually not exceeding 30 km.p.h. A variety of self propelled crane formats are available ranging from short height lifting strut booms of fixed length to variable length lattice booms with a fly jib attachment.

Typical Self Propelled Crane Details ~

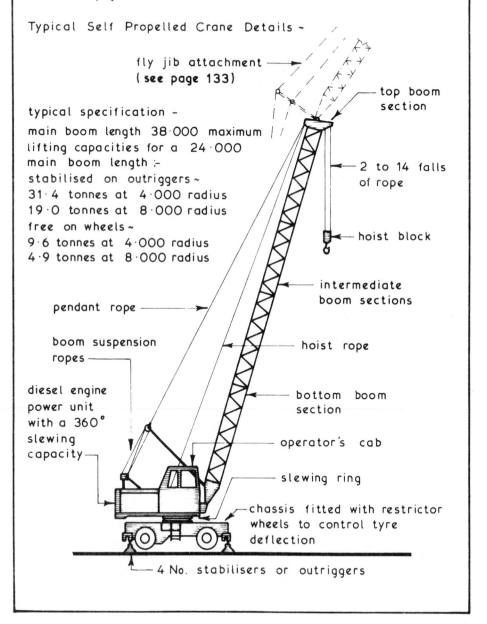

fly jib attachment ⟶
( see page 133 )

top boom section

typical specification –

main boom length 38·000 maximum
lifting capacities for a 24·000
main boom length :-
stabilised on outriggers ~
31·4 tonnes at 4·000 radius
19·0 tonnes at 8·000 radius
free on wheels ~
9·6 tonnes at 4·000 radius
4·9 tonnes at 8·000 radius

2 to 14 falls of rope

hoist block

intermediate boom sections

pendant rope ⟶

boom suspension ropes

hoist rope

diesel engine power unit with a 360° slewing capacity

bottom boom section

operator's cab

slewing ring

chassis fitted with restrictor wheels to control tyre deflection

4 No. stabilisers or outriggers

131

Lorry Mounted Cranes ~ these mobile cranes consist of a lattice or telescopic boom mounted on a specially adapted truck or lorry. They have two operating positions the lorry being driven from a conventional front cab and the crane being controlled from a different location. The lifting capacity of these cranes can be increased by using outrigger stabilising jacks and the approach distance to the face of building decreased by using a fly jib. Lorry mounted telescopic cranes require a firm surface from which to operate and because of their short site preparation time they are ideally suited for short hire periods.

Typical Lorry Mounted Telescopic Crane Details ~

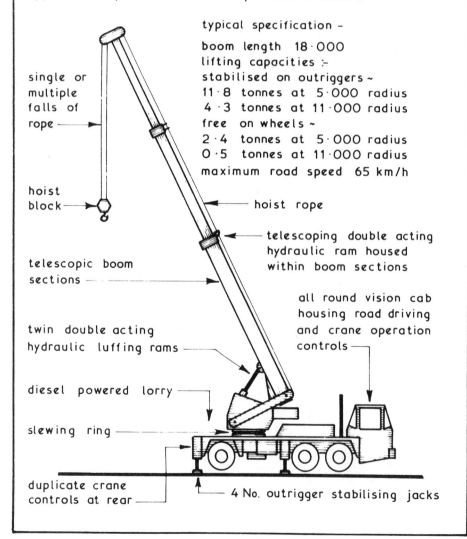

typical specification –
boom length 18·000
lifting capacities :-
stabilised on outriggers ~
11·8 tonnes at 5·000 radius
4·3 tonnes at 11·000 radius
free on wheels ~
2·4 tonnes at 5·000 radius
0·5 tonnes at 11·000 radius
maximum road speed 65 km/h

single or multiple falls of rope

hoist block

hoist rope

telescoping double acting hydraulic ram housed within boom sections

telescopic boom sections

all round vision cab housing road driving and crane operation controls

twin double acting hydraulic luffing rams

diesel powered lorry

slewing ring

duplicate crane controls at rear

4 No. outrigger stabilising jacks

Lorry Mounted Lattice Jib Cranes ~ these cranes follow the same basic principles as the lorry mounted telescopic cranes but they have a lattice boom and are designed as heavy duty cranes with lifting capacities in excess of 100 tonnes. These cranes will require a firm level surface from which to operate and can have a folding or sectional jib which will require the crane to be rigged on site before use.

Typical Lorry Mounted Lattice Jib Crane Details ~

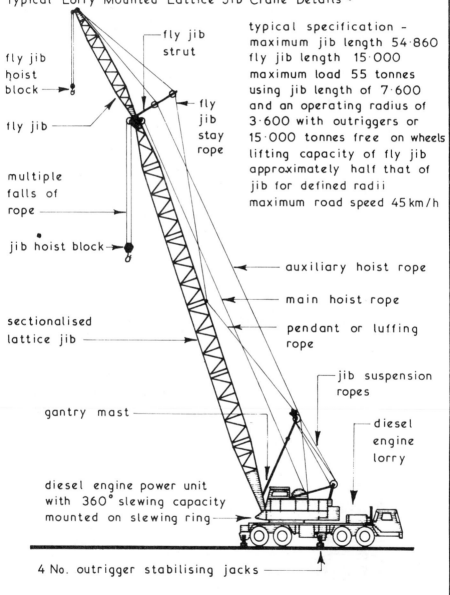

fly jib strut

fly jib hoist block →

fly jib

multiple falls of rope

jib hoist block →

sectionalised lattice jib

gantry mast

diesel engine power unit with 360° slewing capacity mounted on slewing ring

fly jib stay rope

auxiliary hoist rope

main hoist rope

pendant or luffing rope

jib suspension ropes

diesel engine lorry

typical specification –
maximum jib length 54·860
fly jib length 15·000
maximum load 55 tonnes
using jib length of 7·600
and an operating radius of
3·600 with outriggers or
15·000 tonnes free on wheels
lifting capacity of fly jib
approximately half that of
jib for defined radii
maximum road speed 45 km/h

4 No. outrigger stabilising jacks

Track Mounted Cranes ~ these machines can be a universal power unit rigged as a crane (see page 123) or a purpose designed track mounted crane with or without a fly jib attachment. The latter type are usually more powerful with lifting capacities up to 45 tonnes. Track mounted cranes can travel and carry out lifting operations on most sites without the need for special road and hardstand provisons but they have to be rigged on arrival after being transported to site on a low loader lorry.

Typical Track Mounted or Crawler Crane Details ~

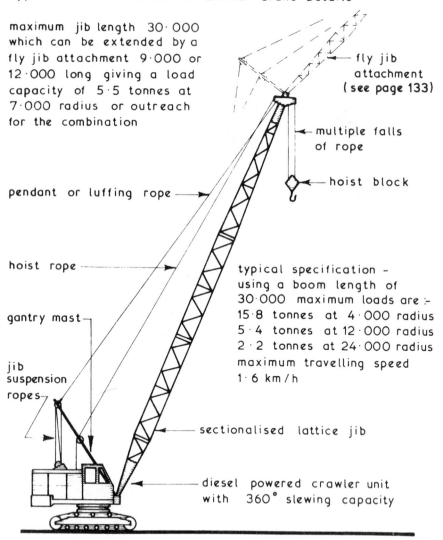

maximum jib length 30·000 which can be extended by a fly jib attachment 9·000 or 12·000 long giving a load capacity of 5·5 tonnes at 7·000 radius or outreach for the combination

fly jib attachment (see page 133)

multiple falls of rope

pendant or luffing rope

hoist block

hoist rope

gantry mast

jib suspension ropes

typical specification – using a boom length of 30·000 maximum loads are :– 15·8 tonnes at 4·000 radius 5·4 tonnes at 12·000 radius 2·2 tonnes at 24·000 radius maximum travelling speed 1·6 km/h

sectionalised lattice jib

diesel powered crawler unit with 360° slewing capacity

Gantry Cranes ~ these are sometimes called portal cranes and consist basically of two 'A' frames joined together with a cross member on which transverses the lifting appliance. In small gantry cranes (up to 10 tonnes lifting capacity) the 'A' frames are usually wheel mounted and manually propelled whereas in the large gantry cranes (up to 100 tonnes lifting capacity) the 'A' frames are mounted on powered bogies running on rail tracks with the driving cab and lifting gear mounted on the cross beam or gantry. Small gantry cranes are used primarily for loading and unloading activities in stock yards whereas the medium and large gantry cranes are used to straddle the work area such as in power station construction or in repetitive low to medium rise developments. All gantry cranes have the advantage of three direction movement –

1. Transverse by moving along the cross beam.

2. Vertical by raising and lowering the hoist block.

3. Horizontal by forward and reverse movements of the whole gantry crane.

Typical Gantry Crane Details ~

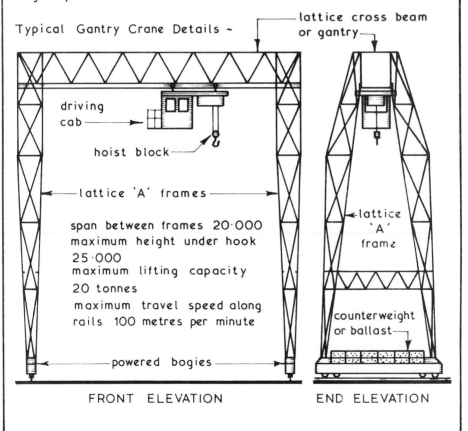

lattice cross beam or gantry

driving cab

hoist block

lattice 'A' frames

span between frames 20·000
maximum height under hook 25·000
maximum lifting capacity 20 tonnes
maximum travel speed along rails 100 metres per minute

powered bogies

FRONT ELEVATION

lattice 'A' frame

counterweight or ballast

END ELEVATION

Mast Cranes ~ these are similar in appearance to the familiar tower cranes but they have one major difference in that the mast or tower is mounted on the slewing ring and thus rotates whereas a tower crane has the slewing ring at the top of the tower and therefore only the jib portion rotates. Mast cranes are often mobile, self erecting, of relatively low lifting capacity and are usually fitted with a luffing jib. A wide variety of models are available and have the advantage over most mobile low pivot cranes of a closer approach to the face of the building.

Typical Mast Crane Details ~

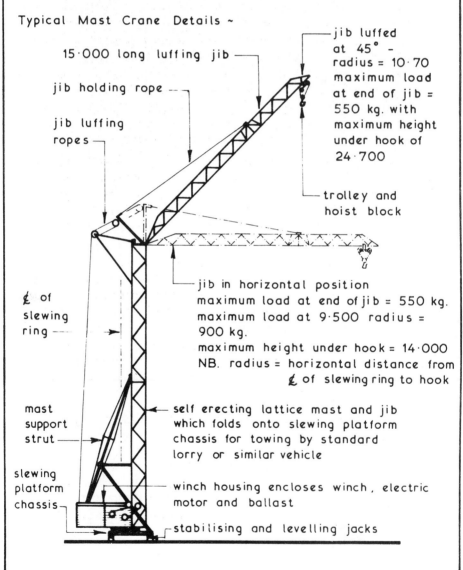

15·000 long luffing jib

jib holding rope

jib luffing ropes

jib luffed at 45° - radius = 10·70 maximum load at end of jib = 550 kg. with maximum height under hook of 24·700

trolley and hoist block

₡ of slewing ring ---

jib in horizontal position
maximum load at end of jib = 550 kg.
maximum load at 9·500 radius = 900 kg.
maximum height under hook = 14·000
NB. radius = horizontal distance from
₡ of slewing ring to hook

mast support strut

self erecting lattice mast and jib which folds onto slewing platform chassis for towing by standard lorry or similar vehicle

slewing platform chassis

winch housing encloses winch, electric motor and ballast

stabilising and levelling jacks

Tower Cranes ~ most tower cranes have to be assembled and erected on site prior to use and can be equipped with a horizontal or luffing jib. The wide range of models available often make it difficult to choose a crane suitable for any particular site but most tower cranes can be classified into one of four basic groups thus :-

1. Self Supporting Static Tower Cranes - high lifting capacity with the mast or tower fixed to a foundation base - they are suitable for confined and open sites. (see page 138)

2. Supported Static Tower Cranes - similar in concept to self supporting cranes and are used where high lifts are required, the mast or tower being tied at suitable intervals to the structure to give extra stability. (see page 139)

3. Travelling Tower Cranes - these are tower cranes mounted on power bogies running on a wide gauge railway track to give greater site coverage - only slight gradients can be accommodated therefore a reasonably level site or specially constructed railway support trestle is required. (see page 140)

4. Climbing Cranes - these are used in conjunction with tall buildings and structures. The climbing mast or tower is housed within the structure and raised as the height of the structure is increased. Upon completion the crane is dismantled into small sections and lowered down the face of the building. (see page 141)

All tower cranes should be left in an 'out of service' condition when unattended and in high wind conditions, the latter varying with different models but generally wind speeds in excess of 60 km.p.h. would require the crane to be placed in an out of service condition thus :-

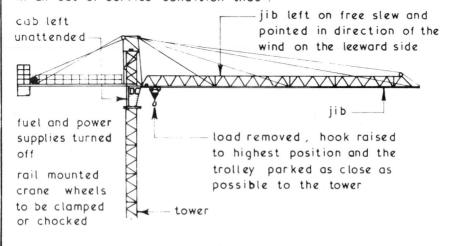

cab left unattended

jib left on free slew and pointed in direction of the wind on the leeward side

jib

fuel and power supplies turned off

rail mounted crane wheels to be clamped or chocked

load removed, hook raised to highest position and the trolley parked as close as possible to the tower

tower

# Cranes

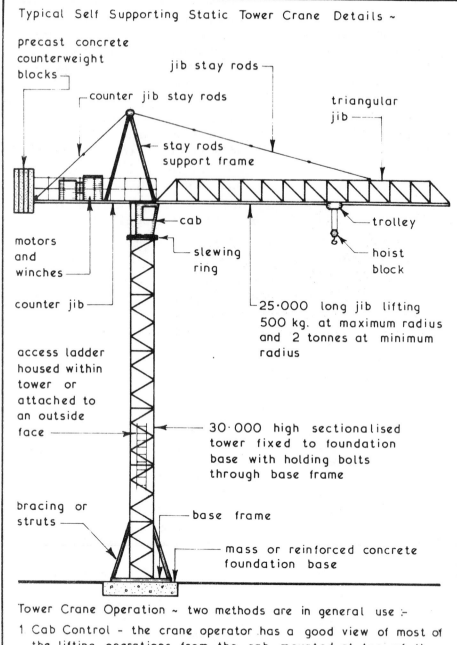

Typical Self Supporting Static Tower Crane Details ~

precast concrete
counterweight
blocks

jib stay rods

counter jib stay rods

triangular
jib

stay rods
support frame

cab

trolley

motors
and
winches

slewing
ring

hoist
block

counter jib

25·000 long jib lifting
500 kg. at maximum radius
and 2 tonnes at minimum
radius

access ladder
housed within
tower or
attached to
an outside
face

30·000 high sectionalised
tower fixed to foundation
base with holding bolts
through base frame

bracing or
struts

base frame

mass or reinforced concrete
foundation base

Tower Crane Operation ~ two methods are in general use :-

1. Cab Control - the crane operator has a good view of most of the lifting operations from the cab mounted at top of the tower, but a second person or banksman is required to give clear signals to the crane operator and to load the crane.

2. Remote Control - the crane operator carries a control box linked by a wandering lead to the crane controls.

Typical Supported Static Tower Crane Details ~

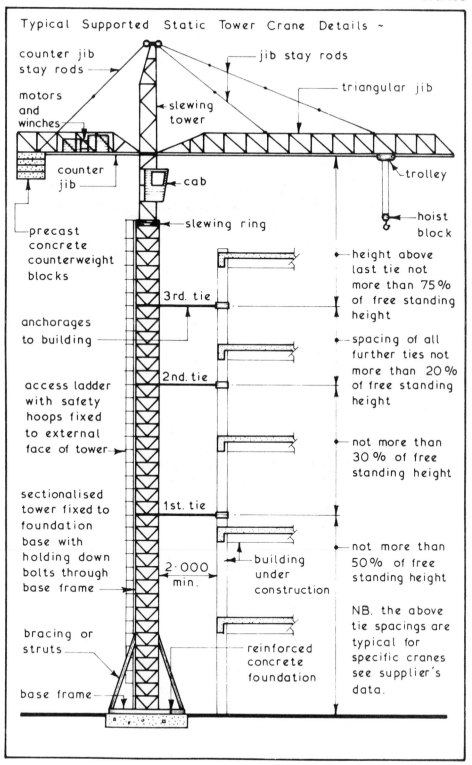

counter jib stay rods

jib stay rods

motors and winches

triangular jib

slewing tower

counter jib

cab

trolley

precast concrete counterweight blocks

slewing ring

hoist block

anchorages to building

3rd. tie

height above last tie not more than 75% of free standing height

access ladder with safety hoops fixed to external face of tower

2nd. tie

spacing of all further ties not more than 20% of free standing height

not more than 30% of free standing height

sectionalised tower fixed to foundation base with holding down bolts through base frame

1st. tie

2·000 min.

building under construction

not more than 50% of free standing height

bracing or struts

reinforced concrete foundation

NB. the above tie spacings are typical for specific cranes see supplier's data.

base frame

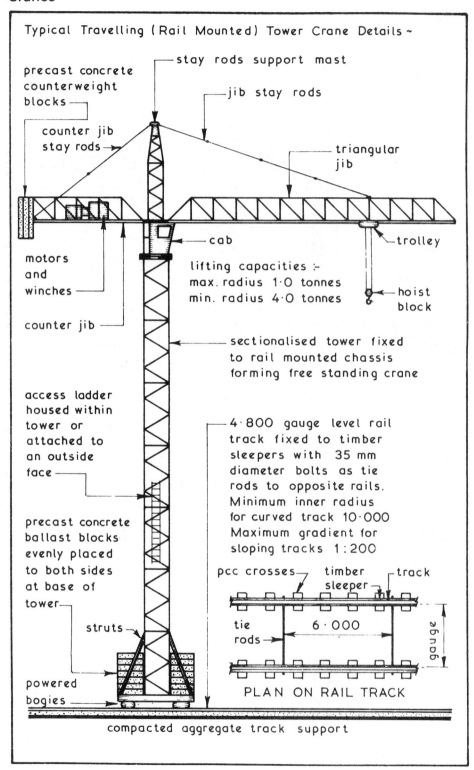

Typical Travelling (Rail Mounted) Tower Crane Details ~

stay rods support mast

precast concrete
counterweight
blocks

jib stay rods

counter jib
stay rods

triangular
jib

cab

trolley

motors
and
winches

lifting capacities :-
max. radius 1·0 tonnes
min. radius 4·0 tonnes

hoist
block

counter jib

sectionalised tower fixed
to rail mounted chassis
forming free standing crane

access ladder
housed within
tower or
attached to
an outside
face

4·800 gauge level rail
track fixed to timber
sleepers with 35 mm
diameter bolts as tie
rods to opposite rails.
Minimum inner radius
for curved track 10·000
Maximum gradient for
sloping tracks 1:200

precast concrete
ballast blocks
evenly placed
to both sides
at base of
tower

pcc crosses    timber
sleeper    track

tie
rods

6·000

gauge

struts

powered
bogies

PLAN ON RAIL TRACK

compacted aggregate track support

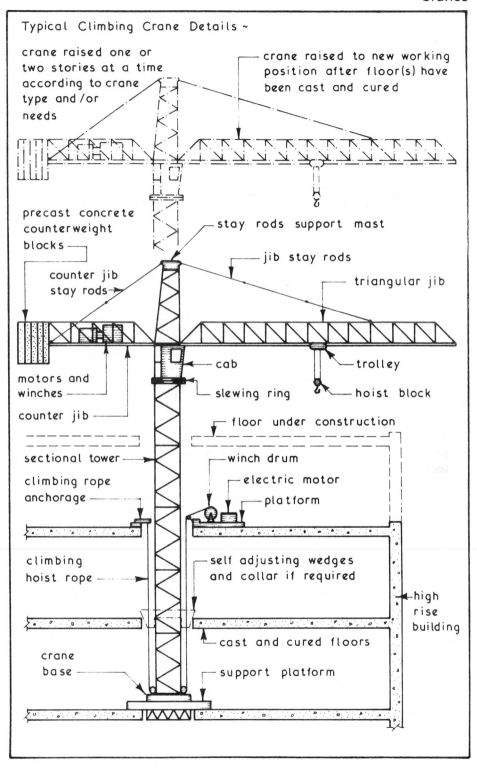

Typical Climbing Crane Details ~

crane raised one or two stories at a time according to crane type and/or needs

crane raised to new working position after floor(s) have been cast and cured

precast concrete counterweight blocks

stay rods support mast

counter jib stay rods

jib stay rods

triangular jib

motors and winches

cab

trolley

counter jib

slewing ring

hoist block

floor under construction

sectional tower

winch drum

climbing rope anchorage

electric motor

platform

climbing hoist rope

self adjusting wedges and collar if required

high rise building

cast and cured floors

crane base

support platform

141

# Concreting Plant

Concreting ~ this site activity consists of four basic procedures :-

1. Material Supply and Storage - this is the receiving on site of the basic materials namely cement, fine aggregate and coarse aggregate and storing them under satisfactory conditions. ( see **Concrete Production - Materials on pages 212 & 213**)

2. Mixing - carried out in small batches this requires only simple hand held tools whereas when demand for increased output is required mixers or ready mixed supplies could **be used. ( see Concrete Production on pages 214 to 216 and Concreting Plant on pages 143 to 145)**

3. Transporting - this can range from a simple bucket to barrows and dumpers for small amounts. For larger loads, especially those required at high level, crane skips could be used :-

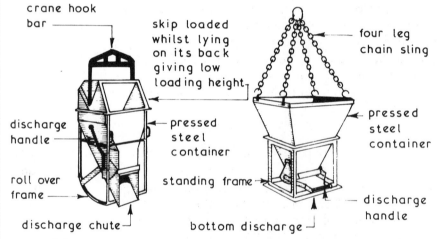

crane hook bar

skip loaded whilst lying on its back giving low loading height

four leg chain sling

discharge handle

pressed steel container

pressed steel container

roll over frame

standing frame

discharge handle

discharge chute

bottom discharge

**ROLL OVER SKIP**
capacities - 0·4 to 2·3 m$^3$

**STANDING SKIP**
capacities - 0·4 to 6·0 m$^3$

For the transportation of large volumes of concrete over a **limited distance concrete pumps could be used. (see page 146)**

4. Placing Concrete - this activity involves placing the wet concrete in the excavation, formwork or mould ; working the concrete between and around any reinforcement; vibrating and / or tamping and curing in accordance with **the recommendations of BS 8110 which also covers the striking or removal of the formwork. ( see Concreting Plant on page 147 and Formwork on page 297)**

Concrete Mixers ~ apart from the very large output mixers most concrete mixers in general use have a rotating drum designed to produce a concrete without segregation of the mix.

Concreting Plant ~ the selection of concreting plant can be considered under three activity headings —
1. Mixing.   2. Transporting.   3. Placing.

Choice of Mixer ~ the factors to be taken into consideration when selecting the type of concrete mixer required are -
1. Maximum output required ($m^3$/hour).
2. Total output required ($m^3$).
3. Type or method of transporting the mixed concrete.
4. Discharge height of mixer (compatibility with transporting method).

Concrete mixer types are generally related to their designed output performance, therefore when the answer to the question 'How much concrete can be placed in a given time period?' or alternatively 'What mixing and placing methods are to be employed to mix and place a certain amount of concrete in a given time period?' has been found the actual mixer can be selected. Generally a batch mixing time of 5 minutes per cycle or 12 batches per hour can be assumed as a reasonable basis for assessing mixer output.

Small Batch Mixers ~ these mixers have outputs of up to 200 litres per batch with wheelbarrow transportation an hourly placing rate of 2 to 3$m^3$ can be achieved. Most small batch mixers are of the tilting drum type. Generally these mixers are hand loaded which makes the quality control of successive mixes difficult to regulate.

Typical Example ~

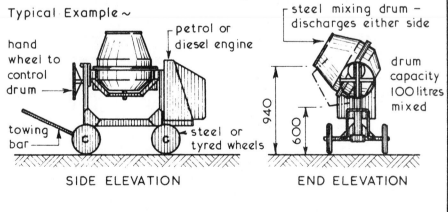

hand wheel to control drum →

┌ petrol or diesel engine

┌ steel mixing drum – discharges either side

drum capacity 100 litres mixed

940

600

towing bar

steel or tyred wheels

SIDE ELEVATION          END ELEVATION

Medium Batch Mixers ~ outputs of these mixers range from 200 to 750 litres and can be obtained at the lower end of the range as a tilting drum mixer or over the complete range as a non-tilting drum mixer with either reversing drum or chute discharge the latter usually having a lower discharge height.

These mixers usually have integral weigh batching loading hoppers, scraper shovels and water tanks thus giving better quality control than the small batch mixers. Generally they are unsuitable for wheelbarrow transportation because of their high output.

Typical Examples ~

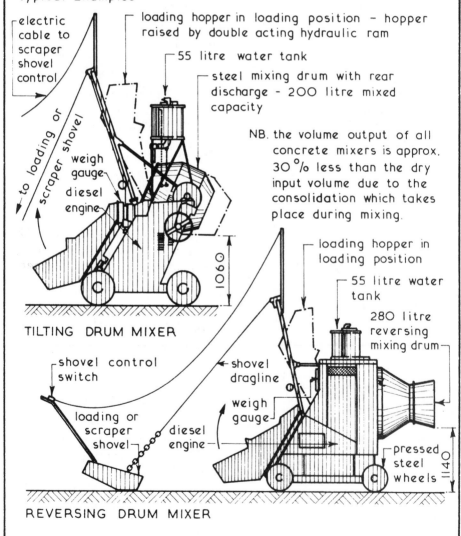

electric cable to scraper shovel control

loading hopper in loading position – hopper raised by double acting hydraulic ram

55 litre water tank

steel mixing drum with rear discharge – 200 litre mixed capacity

to loading or scraper shovel

weigh gauge

diesel engine

NB. the volume output of all concrete mixers is approx. 30% less than the dry input volume due to the consolidation which takes place during mixing.

1060

TILTING DRUM MIXER

loading hopper in loading position

55 litre water tank

280 litre reversing mixing drum

shovel control switch

shovel dragline

loading or scraper shovel

weigh gauge

diesel engine

pressed steel wheels

1140

REVERSING DRUM MIXER

Transporting Concrete ~ the usual means of transporting mixed concrete produced in a small capacity mixer is by wheelbarrow. The run between the mixing and placing positions should be kept to a minimum and as smooth as possible by using planks or similar materials to prevent segregation of the mix within the wheelbarrow.

Dumpers ~ these can be used for transporting mixed concrete from mixers up to 600 litre capacity when fitted with an integral skip and for lower capacities when designed to take a crane skip.

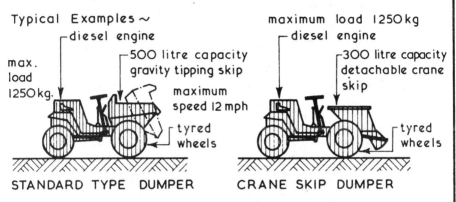

Typical Examples ~

diesel engine

max. load 1250 kg.

500 litre capacity gravity tipping skip

maximum speed 12 mph

tyred wheels

STANDARD TYPE DUMPER

maximum load 1250 kg

diesel engine

300 litre capacity detachable crane skip

tyred wheels

CRANE SKIP DUMPER

Ready Mixed Concrete Trucks ~ these are used to transport mixed concrete from a mixing plant or depot to the site. Usual capacity range of ready mixed concrete trucks is 4 to 6 m$^3$. Discharge can be direct into placing position via a chute or into some form of site transport such as a dumper, crane skip or concrete pump.

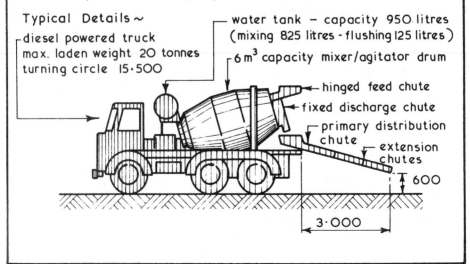

Typical Details ~

diesel powered truck
max. laden weight 20 tonnes
turning circle 15·500

water tank – capacity 950 litres
(mixing 825 litres - flushing 125 litres)

6 m$^3$ capacity mixer/agitator drum

hinged feed chute

fixed discharge chute

primary distribution chute

extension chutes

600

3·000

Concrete Pumps ~ these are used to transport large volumes of concrete in a short time period ( up to 100 m³ per hour ) in both the vertical and horizontal directions from the pump position to the point of placing. Concrete pumps can be trailer or lorry mounted and are usually of a twin cylinder hydraulically driven format with a small bore pipeline (100 mm diameter) with pumping ranges of up to 85·000 vertically and 200·000 horizontally depending on the pump model and the combination of vertical and horizontal distances. It generally requires about 45 minutes to set up a concrete pump on site including coating the bore of the pipeline with a cement grout prior to pumping the special concrete mix The pump is supplied with pumpable concrete by means of a constant flow of ready mixed concrete lorries throughout the pumping period after which the pipeline is cleared and cleaned. Usually a concrete pump and its operator(s) are hired for the period required.

Typical Concrete Pump Details ~

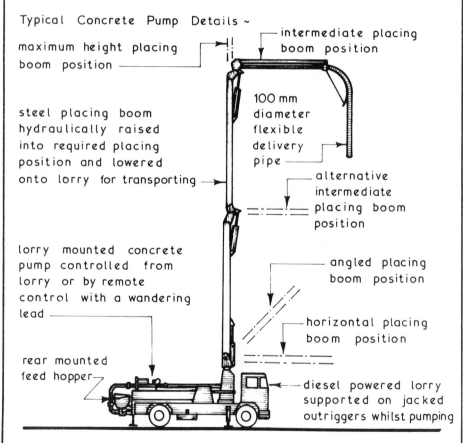

maximum height placing boom position

intermediate placing boom position

steel placing boom hydraulically raised into required placing position and lowered onto lorry for transporting ⟶

100 mm diameter flexible delivery pipe

alternative intermediate placing boom position

lorry mounted concrete pump controlled from lorry or by remote control with a wandering lead

angled placing boom position

horizontal placing boom position

rear mounted feed hopper

diesel powered lorry supported on jacked outriggers whilst pumping

Placing Concrete ~ this activity is usually carried out by hand with the objectives of filling the mould, formwork or excavated area to the correct depth, working the concrete around any inserts or reinforcement and finally compacting the concrete to the required consolidation.   The compaction of concrete can be carried out using simple tamping rods or boards or alternatively it can be carried out with the aid of plant such as vibrators.

Poker Vibrators ~ these consist of a hollow steel tube casing in which is a rotating impellor which generates vibrations as its head comes into contact with the casing –

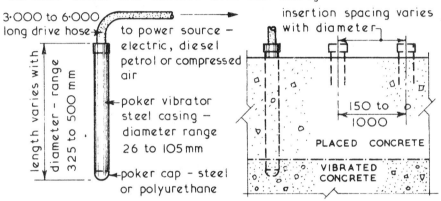

3·000 to 6·000 long drive hose

insertion spacing varies with diameter

to power source – electric, diesel petrol or compressed air

poker vibrator steel casing – diameter range 26 to 105 mm

poker cap – steel or polyurethane

length varies with diameter – range 325 to 500 mm

150 to 1000

PLACED CONCRETE

VIBRATED CONCRETE

Poker vibrators should be inserted vertically and allowed to penetrate 75 mm into any previously vibrated concrete.

Clamp or Tamping Board Vibrators ~ clamp vibrators are powered either by compressed air or electricity whereas tamping board vibrators are usually petrol driven –

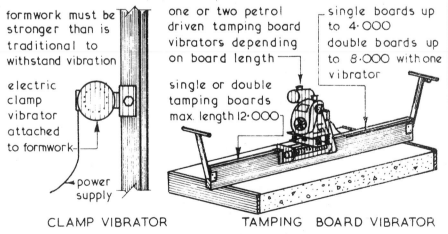

formwork must be stronger than is traditional to withstand vibration

electric clamp vibrator attached to formwork

power supply

one or two petrol driven tamping board vibrators depending on board length

single or double tamping boards max. length 12·000

single boards up to 4·000 double boards up to 8·000 with one vibrator

CLAMP VIBRATOR

TAMPING BOARD VIBRATOR

147

# 4 SUBSTRUCTURE

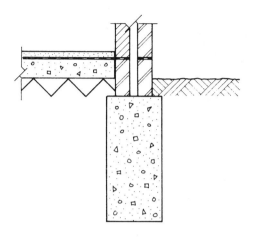

FOUNDATIONS - FUNCTION, MATERIALS AND SIZING

FOUNDATION BEDS

SHORT BORED PILE FOUNDATIONS

FOUNDATION TYPES AND SELECTION

PILED FOUNDATIONS

RETAINING WALLS

BASEMENT CONSTRUCTION

WATERPROOFING BASEMENTS

EXCAVATIONS

CONCRETE PRODUCTION

COFFERDAMS

CAISSONS

UNDERPINNING

GROUND WATER CONTROL

SOIL STABILISATION AND IMPROVEMENT

Foundations~ the function of any foundation is to safely sustain and transmit to the ground on which it rests the combined dead, imposed and wind loads in such a manner as not to cause any settlement or other movement which would impair the stability or cause damage to any part of the building.

Example ~

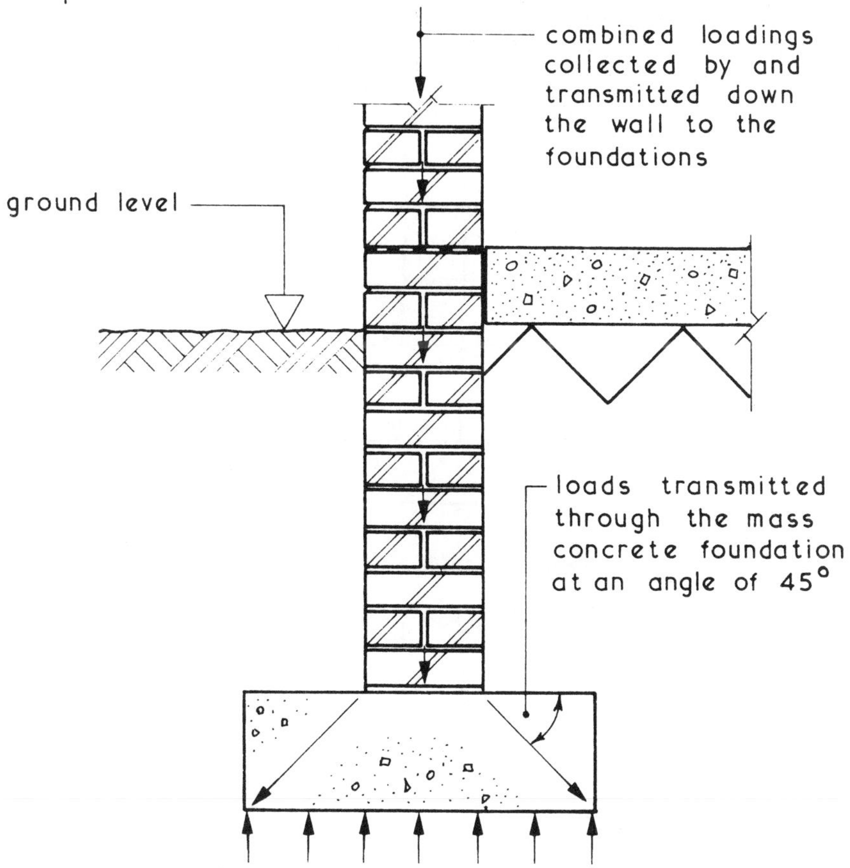

combined loadings collected by and transmitted down the wall to the foundations

ground level

loads transmitted through the mass concrete foundation at an angle of 45°

subsoil beneath foundation is compressed and reacts by exerting an upward pressure to resist foundation loading. If foundation load exceeds maximum passive pressure of ground (i.e. bearing capacity) a downward movement of the foundation could occur. Remedy is to increase plan size of foundation to reduce the load per unit area or alternatively reduce the loadings being carried by the foundations.

Subsoil Movements ~ these are due primarily to changes in volume when the subsoil becomes wet or dry and occurs near the upper surface of the soil. Compact granular soils such as gravel suffer very little movement whereas cohesive soils such as clay do suffer volume changes near the upper surface. Similar volume changes can occur due to water held in the subsoil freezing and expanding – this is called Frost Heave.

Typical Examples ~

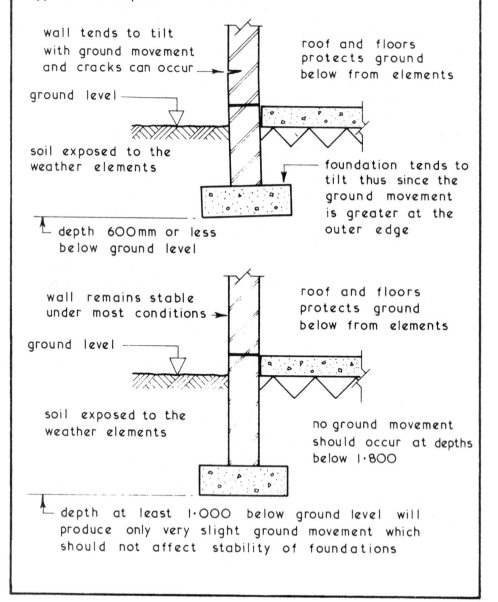

wall tends to tilt with ground movement and cracks can occur

roof and floors protects ground below from elements

ground level

soil exposed to the weather elements

foundation tends to tilt thus since the ground movement is greater at the outer edge

depth 600mm or less below ground level

wall remains stable under most conditions

roof and floors protects ground below from elements

ground level

soil exposed to the weather elements

no ground movement should occur at depths below 1·800

depth at least 1·000 below ground level will produce only very slight ground movement which should not affect stability of foundations

**Foundation Materials ~** from page 151 one of the functions of a foundation can be seen to be the ability to spread its load evenly over the ground on which it rests. It must of course be constructed of a durable material of adequate strength. Experience has shown that the most suitable material is concrete.

Concrete is a mixture of cement + aggregates + water in controlled proportions.

## CEMENT

Manufactured from clay and chalk and is the matrix or binder of the concrete mix. Cement powder can be supplied in bags or bulk —

Bags ~

50 kg.

air-tight sealed bags requiring a dry damp free store.

Bulk ~

12 to 50 tonne

delivered by tanker and pumped into storage silo.

## AGGREGATES

Coarse aggregate is defined as a material which is retained on a 5mm sieve.

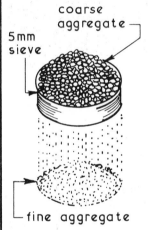

5mm sieve

coarse aggregate

fine aggregate

Fine aggregate is defined as a material which passes a 5mm sieve.

Aggregates can be either natural rock which has disintegrated or crushed stone or gravel.

## WATER

Must be of a quality fit for drinking.

## MIXES

These are expressed as a ratio thus:-
1 : 3 : 6 / 20mm
which means —
1 part cement.
3 parts of fine aggregate.
6 parts of coarse aggregate
20mm — maximum size of coarse aggregate for the mix.

Water is added to start the chemical reaction and to give the mix workability ~ the amount used is called the — Water/Cement Ratio and is usually about 0.4 to 0.5.

Too much water will produce a weak concrete of low strength whereas too little water will produce a concrete mix of low and inadequate work-ability.

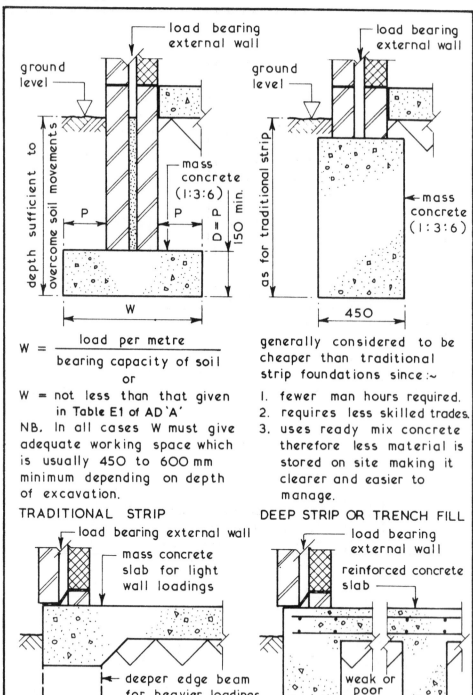

$$W = \frac{\text{load per metre}}{\text{bearing capacity of soil}}$$

or

W = not less than that given in **Table E1 of AD 'A'**

NB. In all cases W must give adequate working space which is usually 450 to 600 mm minimum depending on depth of excavation.

generally considered to be cheaper than traditional strip foundations since :~

1. fewer man hours required.
2. requires less skilled trades.
3. uses ready mix concrete therefore less material is stored on site making it clearer and easier to manage.

TRADITIONAL STRIP

DEEP STRIP OR TRENCH FILL

SOLID SLAB RAFT

BEAM AND SLAB RAFT

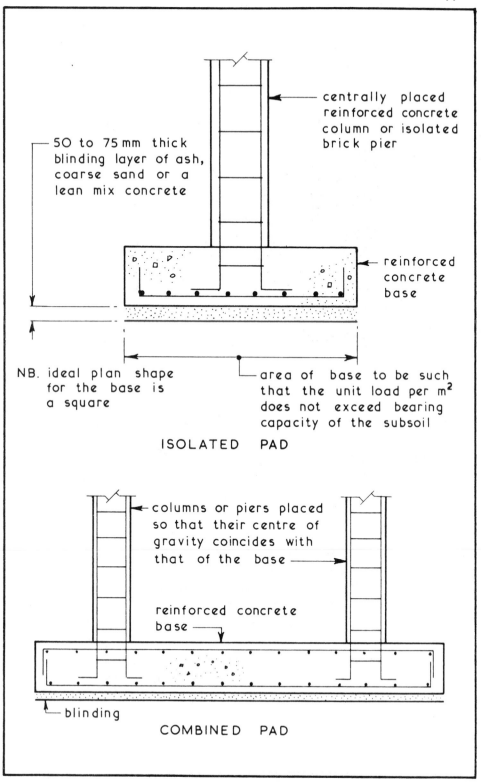

centrally placed
reinforced concrete
column or isolated
brick pier

50 to 75 mm thick
blinding layer of ash,
coarse sand or a
lean mix concrete

reinforced
concrete
base

NB. ideal plan shape
for the base is
a square

area of base to be such
that the unit load per m$^2$
does not exceed bearing
capacity of the subsoil

ISOLATED PAD

columns or piers placed
so that their centre of
gravity coincides with
that of the base

reinforced concrete
base

blinding

COMBINED PAD

155

Bed ~ a concrete slab resting on and supported by the subsoil, usually forming the ground floor surface. Beds (sometimes called oversite concrete) are usually cast on a layer of hardcore which is used to make up the reduced level excavation and thus raise the level of the concrete bed to a position above ground level.

Typical Example ~

mass concrete bed (1 : 3 : 6 / 20 mm mix) thickness for domestic work is usually 100 to 150 mm and the bed is constructed so as to prevent the passage of moisture from the ground to the upper surface of the floor — this is usually achieved by incorporating into the design a damp-proof membrane ~ **see page 409** for details.

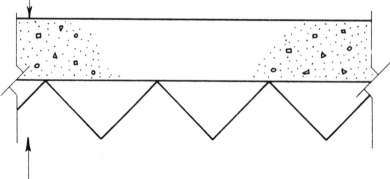

100 to 150 mm layer of hardcore ~ material used should be inert and not affected by water. Suitable materials~ gravel ; crushed rock ; quarry waste ; concrete rubble ; brick or tile rubble ; blast furnace slag and pulverised fuel ash. The hardcore material should be laid evenly and well compacted with upper surface blinded with fine grade material as required.

Basic Sizing ~ the size of a foundation is basically dependent on two factors -

1. Load being transmitted.
2. Bearing capacity of subsoil under proposed foundation.

Bearing capacities for different types of subsoils may be **obtained from tables such as those** in CP 101 **or from soil** investigation results.

Typical Examples ~

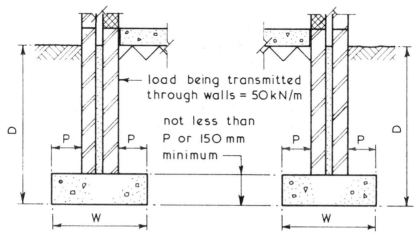

load being transmitted through walls = 50 kN/m

not less than P or 150 mm minimum

safe bearing capacity of compact gravel subsoil = 100 kN/m²

$$W = \frac{\text{load}}{\text{bearing capacity}} = \frac{50}{100}$$

= 500 mm minimum

safe bearing capacity of clay subsoil = 80 kN/m²

$$W = \frac{\text{load}}{\text{bearing capacity}} = \frac{50}{80}$$

= 625 mm minimum

The above widths may not provide adequate working space within the excavation and can be increased to give required space. Minimum width for a limited range of strip foundations can be taken direct from **Table E1 in Approved Document A.**

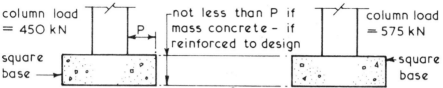

column load = 450 kN

not less than P if mass concrete - if reinforced to design

square base

column load = 575 kN

square base

bearing capacity of subsoil 150 kN/m²

$$\text{area of base} = \frac{\text{load}}{\text{b.c.}} = \frac{450}{150}$$

$= 3\,m^2 \therefore \text{side} = \sqrt{3} = 1 \cdot 732 \cdot \text{min.}$

bearing capacity of subsoil 85 kN/m²

$$\text{area of base} = \frac{\text{load}}{\text{b.c.}} = \frac{575}{85}$$

$= 6 \cdot 765\,m^2 \therefore \text{side} = \sqrt{6 \cdot 765} = 2 \cdot 6\,m.$

Stepped Foundations ~ these are usually considered in the context of strip foundations and are used mainly on sloping sites to reduce the amount of excavation and materials required to produce an adequate foundation.

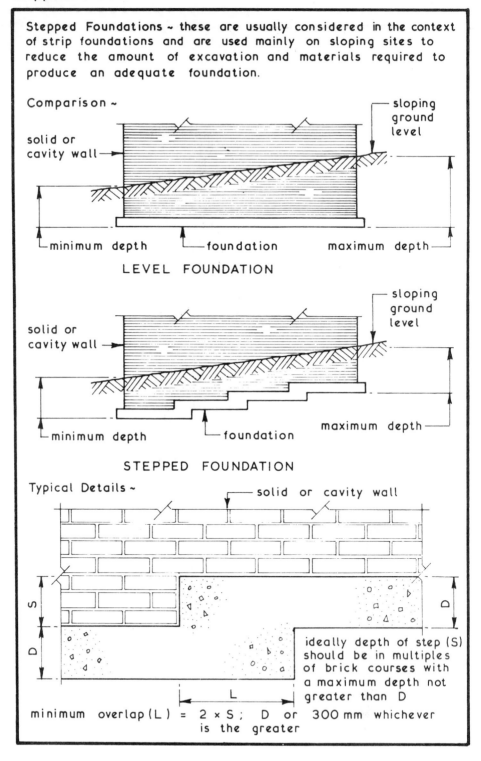

Comparison ~

solid or cavity wall →

sloping ground level

minimum depth — foundation — maximum depth

## LEVEL FOUNDATION

solid or cavity wall →

sloping ground level

minimum depth — foundation — maximum depth

## STEPPED FOUNDATION

Typical Details ~

solid or cavity wall

S

D

D

L

ideally depth of step (S) should be in multiples of brick courses with a maximum depth not greater than D

minimum overlap (L) = 2 × S; D or 300 mm whichever is the greater

Concrete Foundations ~ concrete is a material which is strong in compression but weak in tension. If its tensile strength is exceeded cracks will occur resulting in a weak and unsuitable foundation. One method of providing tensile resistance is to include in the concrete foundation bars of steel as a form of reinforcement to resist all the tensile forces induced into the foundation. Steel is a material which is readily available and has high tensile strength.

Comparisons ~

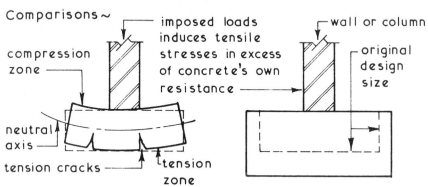

imposed loads induces tensile stresses in excess of concrete's own resistance

compression zone

neutral axis

tension cracks — tension zone

wall or column

original design size

foundation tends to bend, the upper fibres being compressed and the lower fibres being stretched and put in tension - remedies increase size of base or design as a reinforced concrete foundation

size of foundation increased to provide the resistance against the induced tensile stresses - generally not economic due to the extra excavation and materials required

Typical RC Foundation

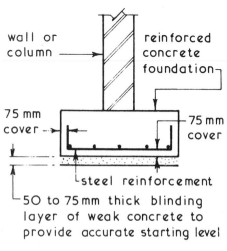

wall or column

reinforced concrete foundation

75 mm cover

75 mm cover

steel reinforcement

50 to 75 mm thick blinding layer of weak concrete to provide accurate starting level

Reinforcement Patterns

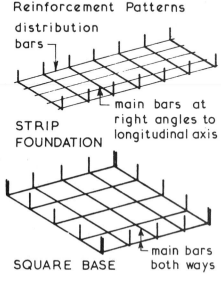

distribution bars

main bars at right angles to longitudinal axis

STRIP FOUNDATION

main bars both ways

SQUARE BASE

## Short Board Pile Foundations

Short Bored Piles ~ these are a form of foundation which are suitable for domestic loadings and clay subsoils where ground movements can occur below the 900 to 1000 mm depth associated with traditional strip and trench fill foundations. They can be used where trees are planted close to the new building since the trees may eventually cause damaging ground movements due to extracting water from the subsoil and root growth. Conversely where trees have been removed this may lead to ground swelling.

Typical Details ~

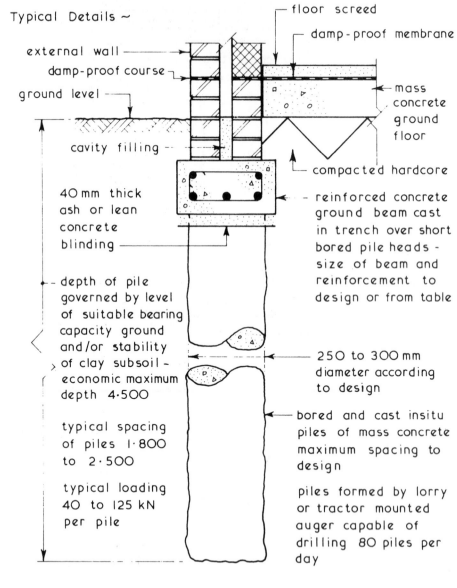

floor screed

damp-proof membrane

external wall

damp-proof course

ground level

mass concrete ground floor

cavity filling

compacted hardcore

40 mm thick ash or lean concrete blinding

reinforced concrete ground beam cast in trench over short bored pile heads - size of beam and reinforcement to design or from table

depth of pile governed by level of suitable bearing capacity ground and/or stability of clay subsoil - economic maximum depth 4·500

250 to 300 mm diameter according to design

bored and cast insitu piles of mass concrete maximum spacing to design

typical spacing of piles 1·800 to 2·500

typical loading 40 to 125 kN per pile

piles formed by lorry or tractor mounted auger capable of drilling 80 piles per day

Simple Raft Foundations ~ these can be used for lightly loaded building on poor soils or where the top 450 to 600mm of soil is overlaying a poor quality substrata.

Typical Details ~

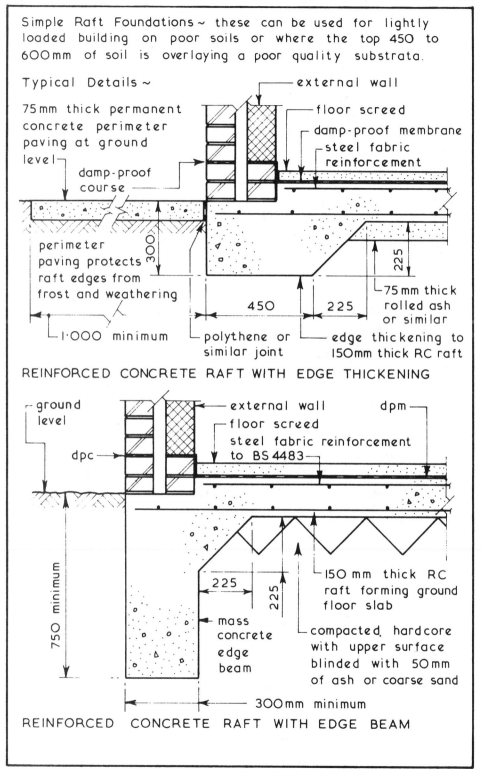

75mm thick permanent concrete perimeter paving at ground level

damp-proof course

external wall

floor screed

damp-proof membrane

steel fabric reinforcement

perimeter paving protects raft edges from frost and weathering

300

225

1·000 minimum

polythene or similar joint

edge thickening to 150mm thick RC raft

75mm thick rolled ash or similar

450  225

**REINFORCED CONCRETE RAFT WITH EDGE THICKENING**

ground level

dpc

external wall

floor screed

steel fabric reinforcement to BS 4483

dpm

750 minimum

225

225

mass concrete edge beam

150 mm thick RC raft forming ground floor slab

compacted hardcore with upper surface blinded with 50mm of ash or coarse sand

300mm minimum

**REINFORCED CONCRETE RAFT WITH EDGE BEAM**

# Foundations Types and Selection

Foundation Design Principles ~ the main objectives of foundation design are to ensure that the structural loads are transmitted to the subsoil(s) safely, economically and without any unacceptable movement during the construction period and throughout the anticipated life of the building or structure.

Basic Design Procedure ~ this can be considered as a series of steps or stages -

1. Assessment of site conditions in the context of the site and soil investigation report.

2. Calculation of anticipated structural loading(s).

3. Choosing the foundation type taking into consideration -
   a. Soil conditions;
   b. Type of structure;
   c. Structural loading(s);
   d. Economic factors;
   e. Time factors relative to the proposed contract period;
   f. Construction problems.

4. Sizing the chosen foundation in the context of loading(s), ground bearing capacity and any likely future movements of the building or structure.

Foundation Types ~ apart from simple domestic foundations most foundation types are constructed in reinforced concrete and may be considered as being shallow or deep. Most shallow types of foundation are constructed within 2·000 of the ground level but in some circumstances it may be necessary to take the whole or part of the foundations down to a depth of 2·000 to 5·000 as in the case of a deep basement where the structural elements of the basement are to carry the superstructure loads. Generally foundations which need to be taken below 5·000 deep are cheaper when designed and constructed as piled foundations and such foundations are classified as deep foundations.
(For piled foundation details see pages 167 to 183)

Foundations are usually classified by their type such as strips, pads, rafts and piles. It is also possible to combine foundation types such as strip foundations connected by beams to and working in conjunction with pad foundations.

Strip Foundations ~ these are suitable for most subsoils and light structural loadings such as those encountered in low to medium rise domestic dwelling where mass concrete can be used. Reinforced concrete is usually required for all other situations.

Typical Strip Foundation Types ~

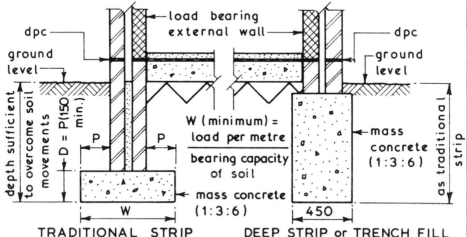

$$W \text{ (minimum)} = \frac{\text{load per metre}}{\text{bearing capacity of soil}}$$

TRADITIONAL STRIP
low rise domestic dwellings or similar buildings

DEEP STRIP or TRENCH FILL
alternative to traditional strip

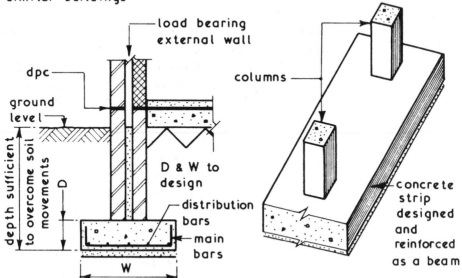

REINFORCED CONCRETE STRIP
used where induced tension exceeds concrete's own tensile resistance

CONTINUOUS COLUMN
used for closely spaced or close to boundary columns

Pad Foundations ~ suitable for most subsoils except loose sands, loose gravels and filled areas. Pad foundations are usually constructed of reinforced concrete and where possible are square in plan.

Typical Pad Foundation Types ~

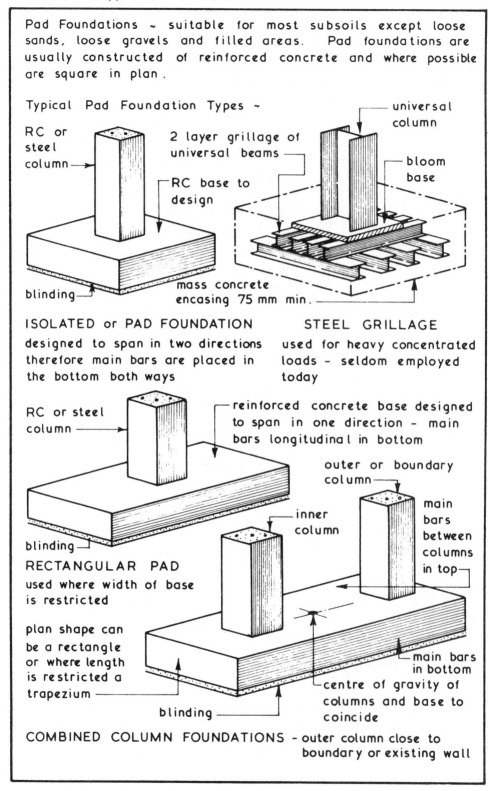

RC or steel column→

2 layer grillage of universal beams ┐

RC base to design

blinding┐

mass concrete encasing 75 mm min.

universal column

bloom base

**ISOLATED or PAD FOUNDATION**
designed to span in two directions therefore main bars are placed in the bottom both ways

**STEEL GRILLAGE**
used for heavy concentrated loads – seldom employed today

RC or steel column →

reinforced concrete base designed to span in one direction – main bars longitudinal in bottom

blinding┐

**RECTANGULAR PAD**
used where width of base is restricted

plan shape can be a rectangle or where length is restricted a trapezium

outer or boundary column →

inner column

main bars between columns in top┐

main bars in bottom

centre of gravity of columns and base to coincide

blinding ┘

**COMBINED COLUMN FOUNDATIONS** - outer column close to boundary or existing wall

Raft Foundations ~ these are used to spread the load of the superstructure over a large base to reduce the load per unit area being imposed on the ground and this is particularly useful where low bearing capacity soils are encountered and where individual column loads are heavy.

Typical Raft Foundation Types ~

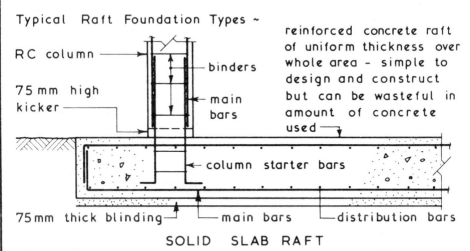

reinforced concrete raft of uniform thickness over whole area - simple to design and construct but can be wasteful in amount of concrete used

RC column

75 mm high kicker

binders

main bars

column starter bars

75 mm thick blinding — main bars — distribution bars

SOLID SLAB RAFT

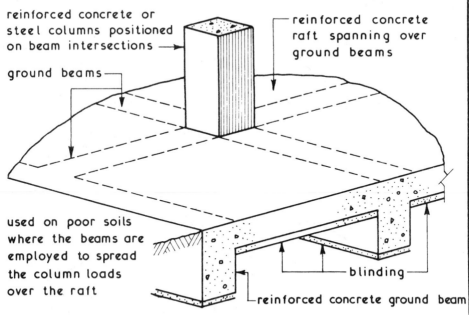

reinforced concrete or steel columns positioned on beam intersections

reinforced concrete raft spanning over ground beams

ground beams

used on poor soils where the beams are employed to spread the column loads over the raft

blinding

reinforced concrete ground beam

NB. Ground beams can be designed as upstand beams with a precast concrete suspended floor at ground level thus creating a void space between raft and ground floor.

BEAM AND SLAB RAFT

165

Cantilever Foundations ~ these can be used where it is necessary to avoid imposing any pressure on an adjacent foundation or underground service.

Typical Cantilever Foundation Types ~

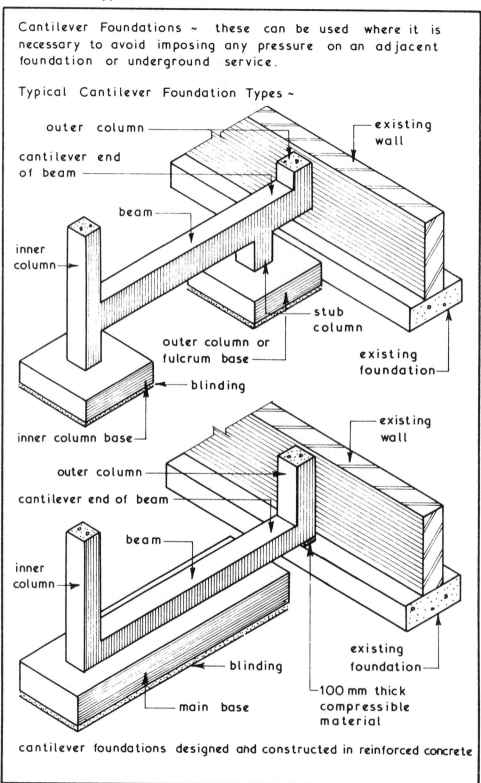

cantilever foundations designed and constructed in reinforced concrete

Piled Foundations ~ these can be defined as a series of columns constructed or inserted into the ground to transmit the load(s) of a structure to a lower level of subsoil. Piled foundations can be used when suitable foundation conditions are not present at or near ground level making the use of deep traditional foundations uneconomic. The lack of suitable foundation conditions may be caused by :-

1. Natural low bearing capacity of subsoil.
2. High water table - giving rise to high permanent dewatering costs.
3. Presence of layers of highly compressible subsoils such as peat and recently placed filling materials which have not sufficiently consolidated.
4. Subsoils which may be subject to moisture movement or plastic failure.

Classification of Piles ~ piles may be classified by their basic design function or by their method of construction :-

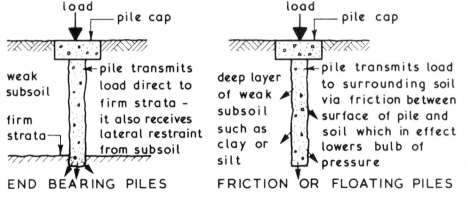

END BEARING PILES

weak subsoil

firm strata

←pile transmits load direct to firm strata - it also receives lateral restraint from subsoil

FRICTION OR FLOATING PILES

deep layer of weak subsoil such as clay or silt

←pile transmits load to surrounding soil via friction between surface of pile and soil which in effect lowers bulb of pressure

NB. Piles can work in a combination of the above design functions

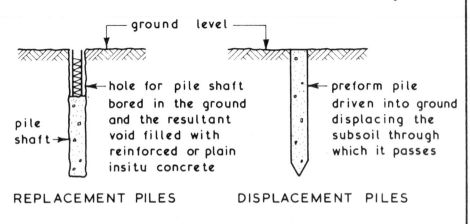

REPLACEMENT PILES

pile shaft→

←hole for pile shaft bored in the ground and the resultant void filled with reinforced or plain insitu concrete

DISPLACEMENT PILES

←preform pile driven into ground displacing the subsoil through which it passes

Replacement Piles ~ these are often called bored piles since the removal of the spoil to form the hole for the pile is always carried out by a boring technique. They are used primarily in cohesive subsoils for the formation of friction piles and when forming pile foundations close to existing buildings where the allowable amount of noise and / or vibration is limited.

Replacement Pile Types ~

## PERCUSSION BORED

small or medium size contracts with up to 300 piles

load range - 300 to 1300 kN

length range - up to 24·000

diameter range - 300 to 900

may have to be formed as a pressure pile in waterlogged subsoils - see page 169

## FLUSH BORED

large projects - these are basically a rotary bored pile using bentonite as a drilling fluid

load range - 1000 to 5000 kN

length range - up to 30·000

diameter range - 600 to 1500

see page 170

## ROTARY BORED

### Small Diameter - < 600 mm

light loadings - can also be used in groups or clusters with a common pile cap to receive heavy loads

load range - 50 to 400 kN

length range - up to 15·000

diameter range - 240 to 600

see page 171

### Large Diameter - > 600 mm

heavy concentrated loadings - may have an underreamed or belled toe

load range - 800 to 15000 kN

length range - up to 60·000

diameter range - 600 to 2400

see page 172

NB. The above given data depicts typical economic ranges. More than one pile type can be used on a single contract.

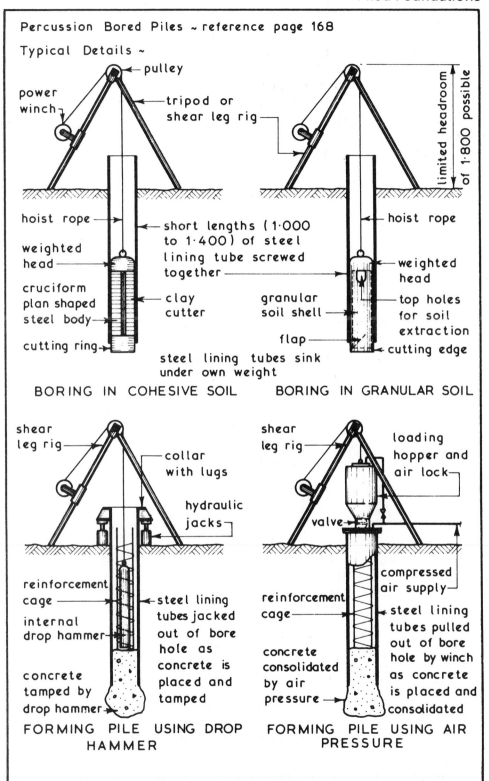

Percussion Bored Piles ~ reference page 168

Typical Details ~

pulley

power winch

tripod or shear leg rig

limited headroom of 1·800 possible

hoist rope

weighted head

short lengths (1·000 to 1·400) of steel lining tube screwed together

cruciform plan shaped steel body

clay cutter

cutting ring

hoist rope

weighted head

granular soil shell

top holes for soil extraction

flap

cutting edge

steel lining tubes sink under own weight

BORING IN COHESIVE SOIL

BORING IN GRANULAR SOIL

shear leg rig

collar with lugs

hydraulic jacks

reinforcement cage

internal drop hammer

steel lining tubes jacked out of bore hole as concrete is placed and tamped

concrete tamped by drop hammer

FORMING PILE USING DROP HAMMER

shear leg rig

loading hopper and air lock

valve

reinforcement cage

compressed air supply

steel lining tubes pulled out of bore hole by winch as concrete is placed and consolidated

concrete consolidated by air pressure

FORMING PILE USING AIR PRESSURE

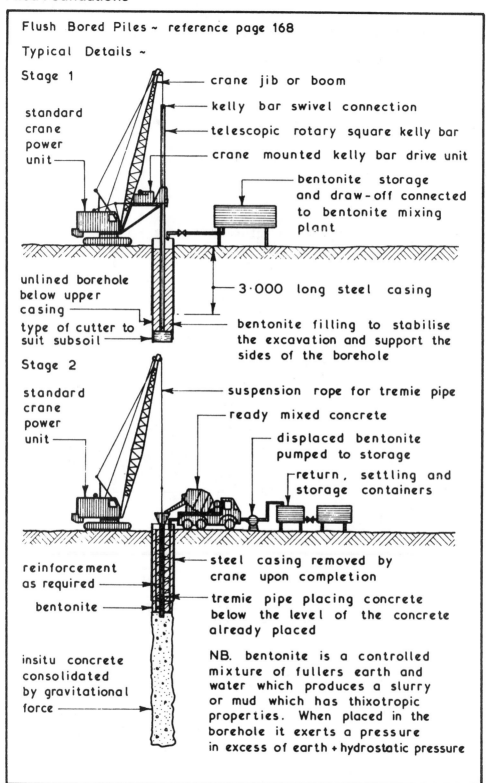

Flush Bored Piles ~ reference page 168

Typical Details ~

Stage 1

standard crane power unit

crane jib or boom

kelly bar swivel connection

telescopic rotary square kelly bar

crane mounted kelly bar drive unit

bentonite storage and draw-off connected to bentonite mixing plant

unlined borehole below upper casing

type of cutter to suit subsoil

3·000 long steel casing

bentonite filling to stabilise the excavation and support the sides of the borehole

Stage 2

standard crane power unit

suspension rope for tremie pipe

ready mixed concrete

displaced bentonite pumped to storage

return, settling and storage containers

reinforcement as required

bentonite

insitu concrete consolidated by gravitational force

steel casing removed by crane upon completion

tremie pipe placing concrete below the level of the concrete already placed

NB. bentonite is a controlled mixture of fullers earth and water which produces a slurry or mud which has thixotropic properties. When placed in the borehole it exerts a pressure in excess of earth + hydrostatic pressure

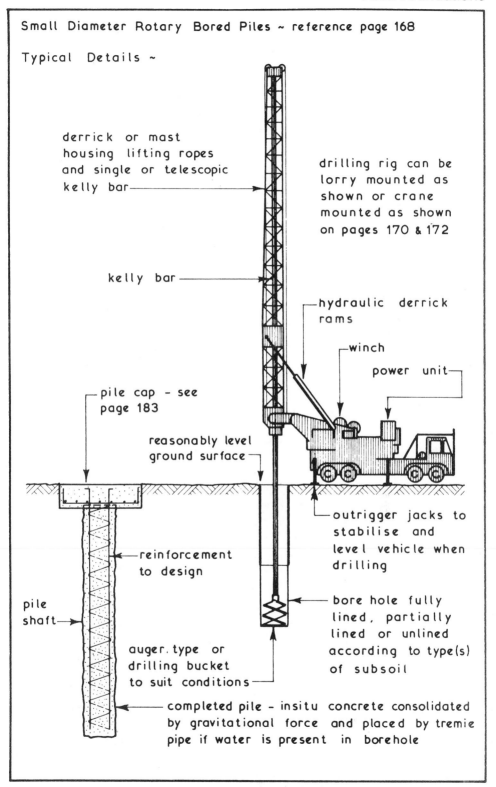

Small Diameter Rotary Bored Piles ~ reference page 168

Typical Details ~

derrick or mast housing lifting ropes and single or telescopic kelly bar

drilling rig can be lorry mounted as shown or crane mounted as shown on pages 170 & 172

kelly bar

hydraulic derrick rams

winch

power unit

pile cap - see page 183

reasonably level ground surface

outrigger jacks to stabilise and level vehicle when drilling

reinforcement to design

pile shaft

bore hole fully lined, partially lined or unlined according to type(s) of subsoil

auger type or drilling bucket to suit conditions

completed pile - insitu concrete consolidated by gravitational force and placed by tremie pipe if water is present in borehole

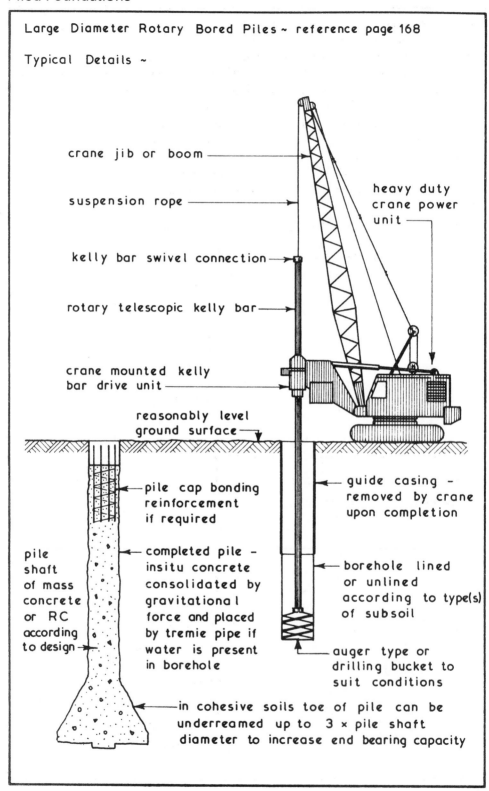

Large Diameter Rotary Bored Piles ~ reference page 168

Typical Details ~

crane jib or boom

suspension rope

kelly bar swivel connection

rotary telescopic kelly bar

crane mounted kelly bar drive unit

heavy duty crane power unit

reasonably level ground surface

pile cap bonding reinforcement if required

pile shaft of mass concrete or RC according to design

completed pile – insitu concrete consolidated by gravitational force and placed by tremie pipe if water is present in borehole

guide casing – removed by crane upon completion

borehole lined or unlined according to type(s) of subsoil

auger type or drilling bucket to suit conditions

in cohesive soils toe of pile can be underreamed up to 3 × pile shaft diameter to increase end bearing capacity

Displacement Piles ~ these are often called driven piles since they are usually driven into the ground displacing the earth around the pile shaft. These piles can be either preformed or partially preformed if they are not cast insitu and are available in a wide variety of types and materials. The pile or forming tube is driven into the required position to a predetermined depth or to the required 'set' which is a measure of the subsoils resistance to the penetration of the pile and hence its bearing capacity by noting the amount of penetration obtained by a fixed number of hammer blows.

Displacement Pile Types ~

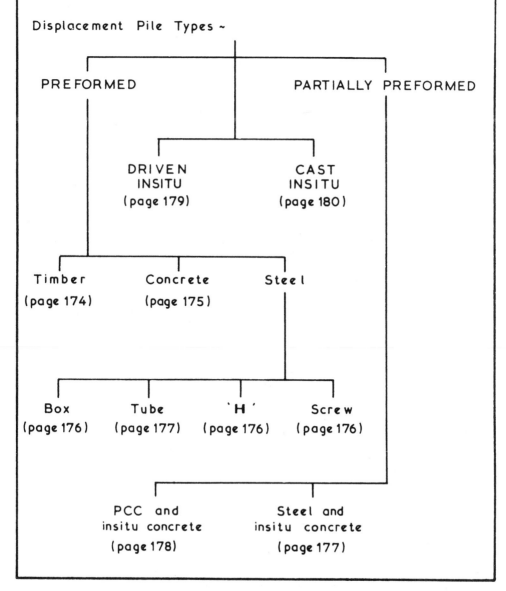

Timber Piles ~ these are usually square sawn and can be used for small contracts on sites with shallow alluvial deposits overlying a suitable bearing strata (e.g. river banks and estuaries.) Timber piles are percussion driven.

Typical Example ~

mild
steel
band

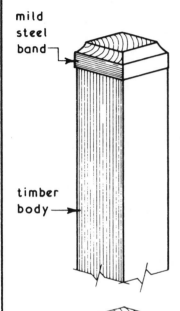

timber
body

Typical Data :-

load range - 50 to 350 kN

length range - up to 12·000
                    without splicing

size range -    225 × 225
                300 × 300 *
                350 × 350 *
                400 × 400 *
                450 × 450
                600 × 600

* common sizes

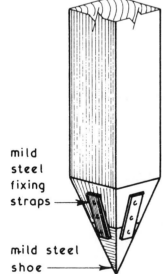

mild
steel
fixing
straps

mild steel
shoe

NB. timber piles are not easy
    to splice and are liable to
    attack by marine borers
    when set in water therefore
    such piles should always
    be treated with a suitable
    preservative before being
    driven.

Preformed Concrete Piles ~ variety of types available which are generally used on medium to large contracts of not less than one hundred piles where soft soil deposits overlie a firmer strata. These piles are percussion driven using a drop or single acting hammer.

Typical Example [West's Hardrive Precast Modular Pile] ~

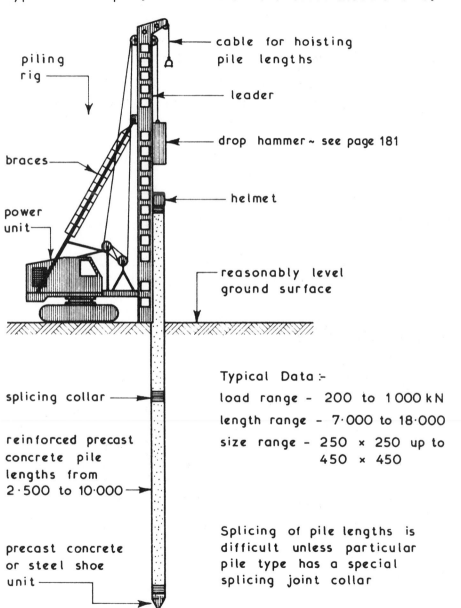

cable for hoisting pile lengths

piling rig

leader

braces

drop hammer ~ see page 181

power unit

helmet

reasonably level ground surface

splicing collar

reinforced precast concrete pile lengths from 2·500 to 10·000

precast concrete or steel shoe unit

Typical Data :-

load range - 200 to 1 000 kN

length range - 7·000 to 18·000

size range - 250 × 250 up to 450 × 450

Splicing of pile lengths is difficult unless particular pile type has a special splicing joint collar

Steel Box and `H` Sections ~ standard steel sheet pile sections can be used to form box section piles whereas the `H` section piles are cut from standard rolled sections. These piles are percussion driven and are used mainly in connection with marine structures.

Typical Examples ~

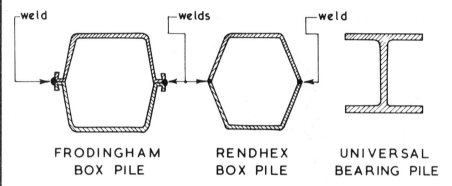

FRODINGHAM
BOX PILE

RENDHEX
BOX PILE

UNIVERSAL
BEARING PILE

Typical Data :-

load range - box piles 300 to 1500 kN
bearing piles 300 to 1700 kN

length range - all types up to 36·000

size range - wide and various range available

Steel Screw Piles ~ rotary driven and used for dock and jetty works where support at shallow depths in soft silts and sands is required.

Typical Example ~

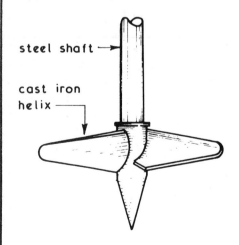

steel shaft

cast iron
helix

Typical Data :-

load range - 400 to 3000 kN

length range - up to 24·000

size range - shafts 150 to
350 mm dia.
overall blades
600 to 1200

**Fig 37   Piled foundations — 10**

Steel Tube Piles ~ used on small to medium size contracts for marine structures and foundations in soft subsoils over a suitable bearing strata. Tube piles are usually bottom driven with an internal drop hammer. The loading can be carried by the tube alone but it is usual to fill the tube with mass concrete to form a composite pile. Reinforcement, except for pile cap bonding bars is not normally required.

Typical Example [BSP Cased Pile] ~

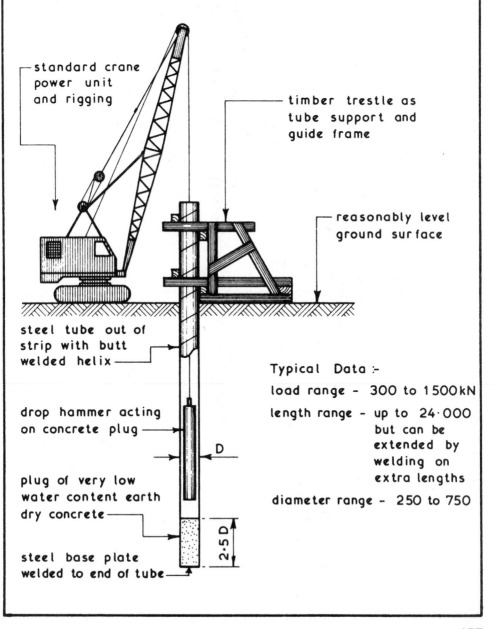

standard crane power unit and rigging

timber trestle as tube support and guide frame

reasonably level ground surface

steel tube out of strip with butt welded helix

drop hammer acting on concrete plug

plug of very low water content earth dry concrete

steel base plate welded to end of tube

D

2·5 D

Typical Data :-

load range - 300 to 1500kN

length range - up to 24·000 but can be extended by welding on extra lengths

diameter range - 250 to 750

Partially Preformed Piles ~ these are composite piles of precast concrete and insitu concrete or steel and insitu concrete (see page 177). These percussion driven piles are used on medium to large contracts where bored piles would not be suitable owing to running water or very loose soils.

Typical Example [ West's Shell Pile ] ~

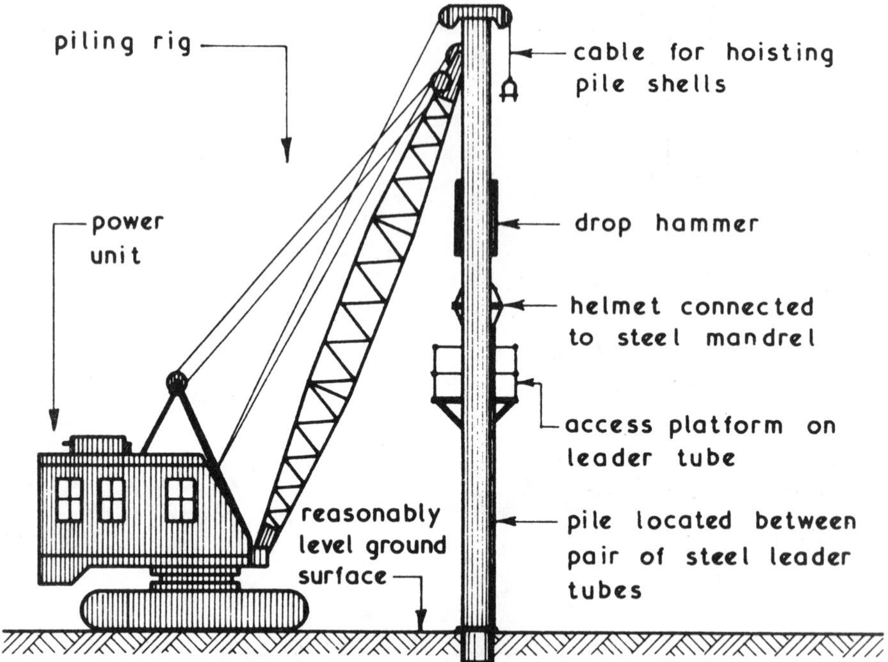

piling rig

power unit

cable for hoisting pile shells

drop hammer

helmet connected to steel mandrel

access platform on leader tube

reasonably level ground surface

pile located between pair of steel leader tubes

Typical Pile Details ~

polypropylene reinforced concrete shells

steel jointing band

reinforcement to design

insitu concrete filling to core

steel jointing band

900

900

pile shells located on mandrel while being driven

steel mandrel inside shell pile core

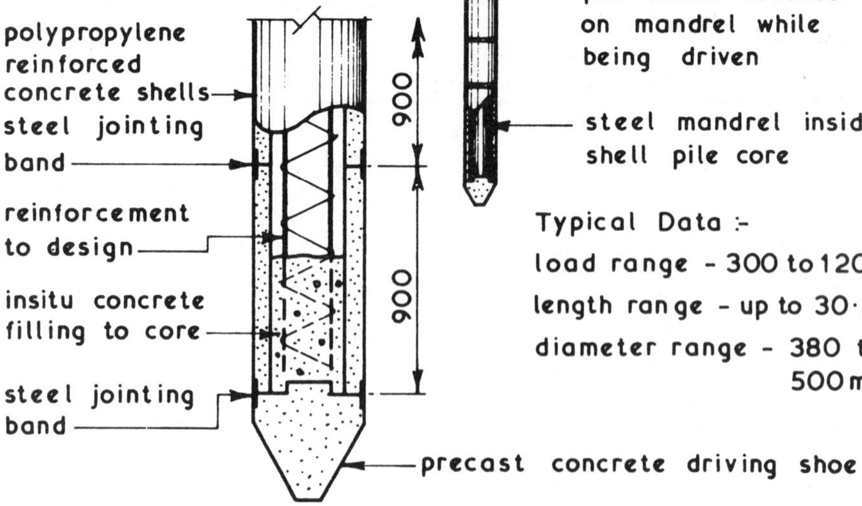

Typical Data :-
load range - 300 to 1200 kN
length range - up to 30·000
diameter range - 380 to 500 mm

precast concrete driving shoe

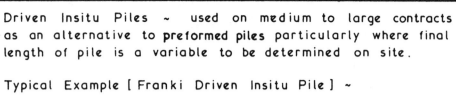

Driven Insitu Piles ~ used on medium to large contracts as an alternative to preformed piles particularly where final length of pile is a variable to be determined on site.

Typical Example [ Franki Driven Insitu Pile ] ~

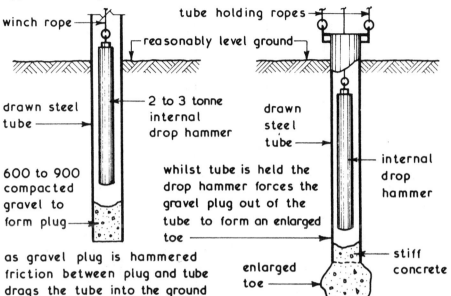

winch rope

tube holding ropes

reasonably level ground

drawn steel tube

2 to 3 tonne internal drop hammer

drawn steel tube

internal drop hammer

600 to 900 compacted gravel to form plug

whilst tube is held the drop hammer forces the gravel plug out of the tube to form an enlarged toe

stiff concrete

enlarged toe

as gravel plug is hammered friction between plug and tube drags the tube into the ground

1. DRIVING TUBE

2. FORMING ENLARGED TOE

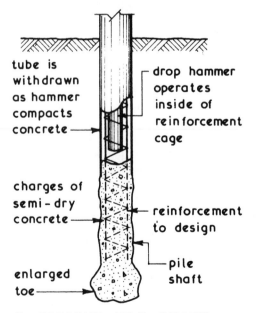

tube is withdrawn as hammer compacts concrete

drop hammer operates inside of reinforcement cage

charges of semi-dry concrete

reinforcement to design

enlarged toe

pile shaft

3. FORMING PILE SHAFT

The drawn steel tube is supported in the leaders of a piling rig or frame during the formation of the pile

Typical Data :-
load range - 300 to 1300 kN
length range - up to 18·000
diameter range - 300 to 600 mm

# Piled Foundations

Cast Insitu Piles ~ an alternative to the driven insitu piles
(see page 179)

Typical Example [ Vibro Cast Insitu Pile] ~

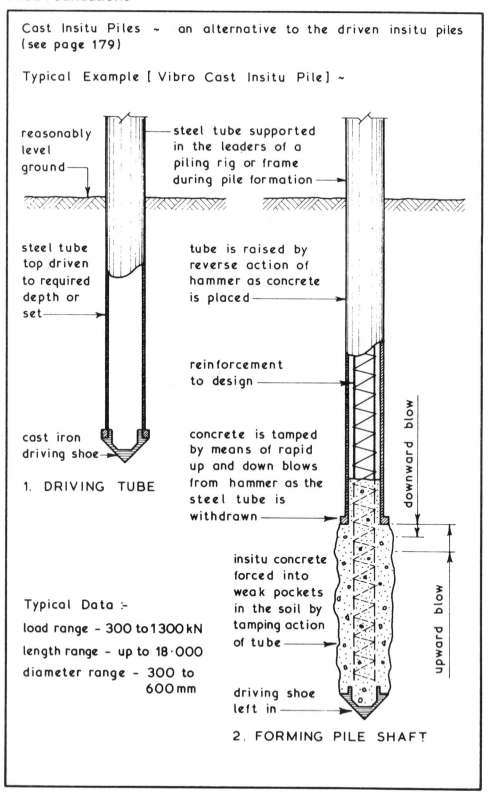

reasonably
level
ground

steel tube supported
in the leaders of a
piling rig or frame
during pile formation

steel tube
top driven
to required
depth or
set

tube is raised by
reverse action of
hammer as concrete
is placed

reinforcement
to design

cast iron
driving shoe

concrete is tamped
by means of rapid
up and down blows
from hammer as the
steel tube is
withdrawn

**1. DRIVING TUBE**

downward blow

upward blow

insitu concrete
forced into
weak pockets
in the soil by
tamping action
of tube

Typical Data :-

load range - 300 to 1300 kN

length range - up to 18·000

diameter range - 300 to
600 mm

driving shoe
left in

**2. FORMING PILE SHAFT**

Piling Hammers ~ these are designed to deliver an impact blow to the top of the pile to be driven. The hammer weight and drop height is chosen to suit the pile type and nature of subsoil(s) through which it will be driven. The head of the pile being driven is protected against damage with a steel helmet which is padded with a sand bed or similar material and is cushioned with a plastic or hardwood block called a dolly.

Drop Hammers ~ these are blocks of iron with a rear lug(s) which locate in the piling rig guides or leaders and have a top eye for attachment of the winch rope. The number of blows which can be delivered with a free fall of 1·200 to 1·500 ranges from 10 to 20 per minute. The weight of the hammer should be not less than 50% of the concrete or steel pile weight and 1 to 1·5 times the weight of a timber pile.

Single Acting Hammers ~ these consist of a heavy falling cylinder raised by steam or compressed air sliding up and down a fixed piston. Guide lugs or rollers are located in the piling frame leaders to maintain the hammer position relative to the pile head. The number of blows delivered ranges from 36 to 75 per minute with a total hammer weight range of 2 to 15 tonnes.

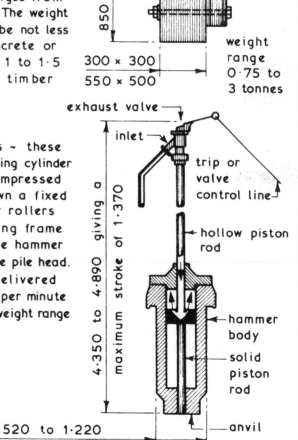

Double Acting Hammers ~ these consist of a cast iron cylinder which remains stationary on the pile head whilst a ram powered by steam or compressed air for both up and down strokes delivers a series of rapid blows which tends to keep the pile on the move during driving. The blow delivered is a smaller force than that from a drop or single acting hammer. The number of blows delivered ranges from 95 to 300 per minute with a total hammer weight range of 0·7 to 6·5 tonnes. Diesel powered double acting hammers are also available.

Diesel Hammers ~ these are self contained hammers which are located in the leaders of a piling rig and rest on the head of the pile. The driving action is started by raising the ram within the cylinder which activates the injection of a measured amount of fuel. The free falling ram compresses the fuel above the anvil causing the fuel to explode and expand resulting in a downward force on the anvil and upward force which raises the ram to recommence the cycle which is repeated until the fuel is cut off. The number of blows delivered ranges from 40 to 60 per minute with a total hammer weight range of 1·0 to 4·5 tonnes.

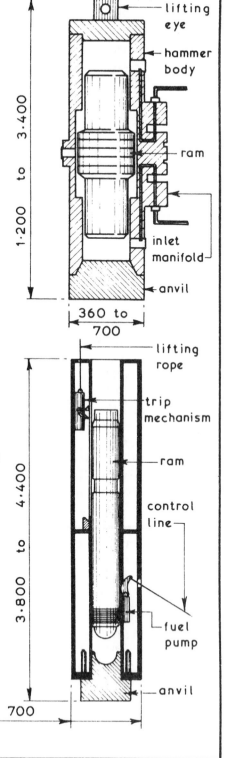

Pile Caps ~ piles can be used singly to support the load but often it is more economical to use piles in groups or clusters linked together with a reinforced concrete cap. The pile caps can also be linked together with reinforced concrete ground beams.

The usual minimum spacing for piles is :-

1. Friction Piles - 1·100 or not less than 3 × pile diameter whichever is the greater.

2. End Bearing Piles - 750 mm or not less than 2 x pile diameter whichever is the greater.

Typical Examples ~

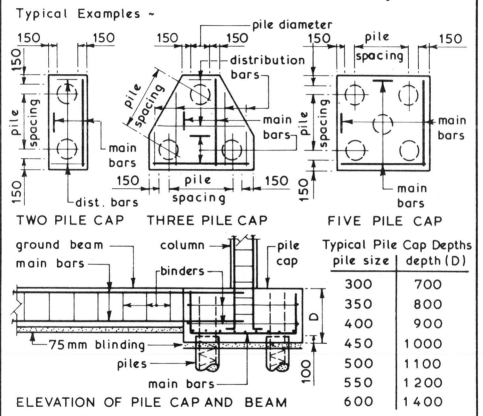

TWO PILE CAP   THREE PILE CAP   FIVE PILE CAP

ELEVATION OF PILE CAP AND BEAM

| Typical Pile Cap Depths | |
|---|---|
| pile size | depth (D) |
| 300 | 700 |
| 350 | 800 |
| 400 | 900 |
| 450 | 1000 |
| 500 | 1100 |
| 550 | 1200 |
| 600 | 1400 |

Pile Testing ~ it is advisable to test load at least one pile per scheme. The test pile should be overloaded by at least 50% of its working load and this load should be held for 24 hours. The test pile should not form part of the actual foundations. Suitable testing methods are :-

1. Jacking against kentledge placed over test pile.

2. Jacking against a beam fixed to anchor piles driven in on two sides of the test pile.

Retaining Walls ~ the major function of any retaining wall is to act as an earth retaining structure for the whole or part of its height on one face, the other being exposed to the elements. Most small height retaining walls are built entirely of brickwork or a combination of brick facing and blockwork or mass concrete backing. To reduce hydrostatic pressure on the wall from ground water an adequate drainage system in the form of weep holes should be used, alternatively subsoil drainage behind the wall could be employed.

Typical Example of Combination Retaining Wall ~

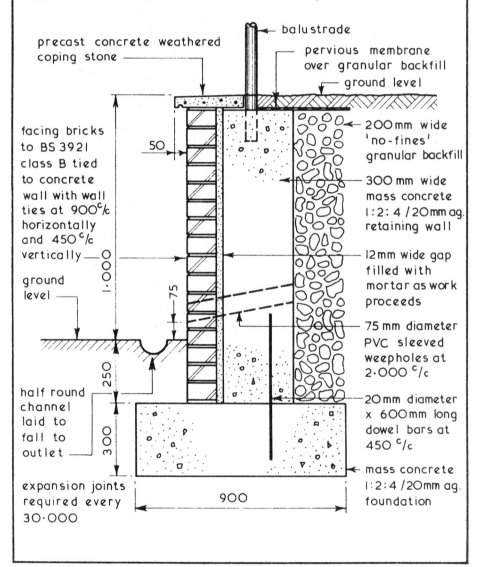

Small Height Retaining Walls~ retaining walls must be stable and the usual rule of thumb for small height brick retaining walls is for the height to lie between 2 and 4 times the wall thickness. Stability can be checked by applying the middle third rule —

middle third

$^1/_3$ W    $^1/_3$ W

W

P = earth pressure

L = self weight of wall

reaction under wall to fall within middle third

R = reaction

P

L

R

Triangle of Forces

P and L drawn to scale and direction – closing line of triangle gives magnitude and direction of R

Typical Example of Brick Retaining Wall~

pcc weathered coping stone

50

pervious membrane over granular backfill

ground level

free standing height

retaining height not greater than 1·000

retaining wall of BS 3921 class B bricks laid to English bond in cm.mt 1 : 3

drainage channel

75

200mm wide 25mm nominal diameter 'no-fines' granular backfill

75mm diameter PVC sleeved weepholes at 2·000 $^c/c$

mass concrete 1:2:4 / 20 mm agg. foundation

225

225

20mm wide flexcell or similar expansion joints at 30·000 $^c/c$

900

# Medium Height Retaining Walls

Retaining Walls up to 6·000 High ~ these can be classified as medium height retaining walls and have the primary function of retaining soils at an angle in excess of the soil's natural angle of repose. Walls within this height range are designed to provide the necessary resistance by either their own mass or by the principles of leverage.

Design ~ the actual design calculations are usually carried out by a structural engineer who endeavours to ensure that :-

1. Overturning of the wall does not occur.
2. Forward sliding of the wall does not occur.
3. Materials used are suitable and not overstressed.
4. The subsoil is not overloaded.
5. In clay subsoils slip circle failure does not occur.

The factors which the designer will have to take into account :-

1. Nature and characteristics of the subsoil(s).
2. Height of water table - the presence of water can create hydrostatic pressure on the rear face of the wall, it can also affect the bearing capacity of the subsoil together with its shear strength, reduce the frictional resistance between the underside of the foundation and the subsoil and reduce the passive pressure in front of the toe of the wall.
3. Type of wall.
4. Material(s) to be used in the construction of the wall.

Retaining Wall Terminology ~

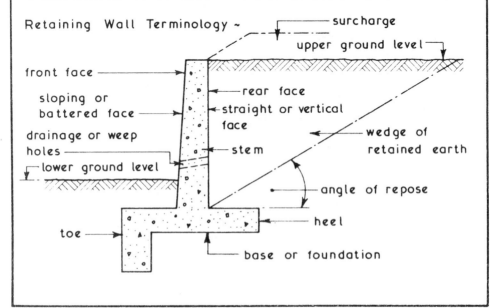

Earth Pressures ~ these can take one of two forms namely :-

1. Active Earth Pressures - these are those pressures which tend to move the wall at all times and consists of the wedge of earth retained plus any hydrostatic pressure. The latter can be reduced by including a subsoil drainage system behind and/or through the wall.

2. Passive Earth Pressures - these are a reaction of an equal and opposite force to any imposed pressure thus giving stability by resisting movement.

Typical Examples ~

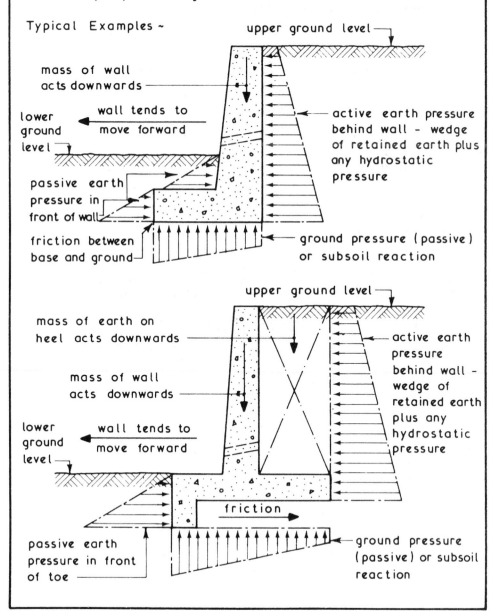

Mass Retaining Walls ~ these walls rely mainly on their own mass to overcome the tendency to slide forwards. Mass retaining walls are not generally considered to be economic over a height of 1·800 when constructed of brick or concrete and 1·000 high in the case of natural stonework. Any mass retaining wall can be faced with another material but generally any applied facing will not increase the strength of the wall and is therefore only used for aesthetic reasons.

Typical Brick Mass Retaining Wall Details ~

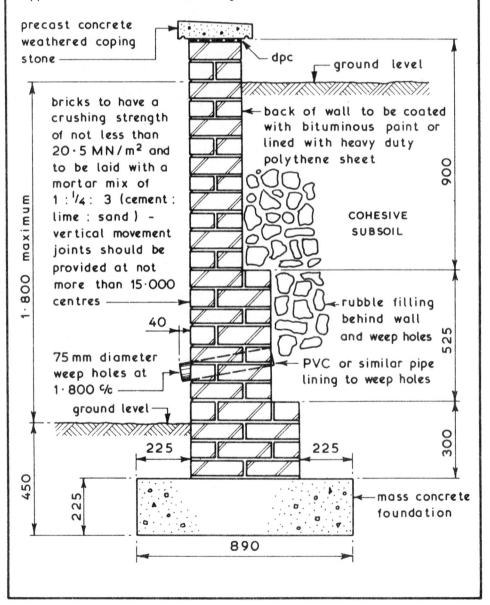

precast concrete weathered coping stone

dpc

ground level

bricks to have a crushing strength of not less than 20·5 MN/m² and to be laid with a mortar mix of 1 : ¹/₄ : 3 (cement : lime : sand) – vertical movement joints should be provided at not more than 15·000 centres

back of wall to be coated with bituminous paint or lined with heavy duty polythene sheet

COHESIVE SUBSOIL

rubble filling behind wall and weep holes

40

75 mm diameter weep holes at 1·800 ℅

PVC or similar pipe lining to weep holes

ground level

225

225

mass concrete foundation

890

450

225

1·800 maximum

900

525

300

Typical Brick Faced Mass Concrete Retaining Wall Detail ~

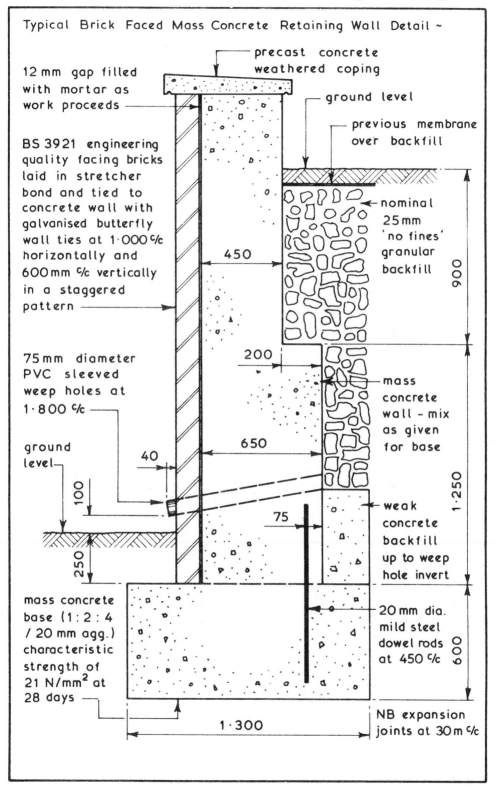

precast concrete
weathered coping

12 mm gap filled
with mortar as
work proceeds

ground level

previous membrane
over backfill

BS 3921 engineering
quality facing bricks
laid in stretcher
bond and tied to
concrete wall with
galvanised butterfly
wall ties at 1·000 c/c
horizontally and
600 mm c/c vertically
in a staggered
pattern

nominal
25 mm
'no fines'
granular
backfill

450

900

75 mm diameter
PVC sleeved
weep holes at
1·800 c/c

200

mass
concrete
wall – mix
as given
for base

ground
level

40

650

1·250

weak
concrete
backfill
up to weep
hole invert

100

75

250

mass concrete
base (1:2:4
/ 20 mm agg.)
characteristic
strength of
21 N/mm² at
28 days

20 mm dia.
mild steel
dowel rods
at 450 c/c

600

1·300

NB expansion
joints at 30 m c/c

189

Cantilever Retaining Walls ~ these are constructed of reinforced concrete with an economic height range of 1·200 to 6·000. They work on the principles of leverage where the stem is designed as a cantilever fixed at the base and base is designed as a cantilever fixed at the stem. Several formats are possible and in most cases a beam is placed below the base to increase the total passive resistance to sliding. Facing materials can be used in a similar manner to that shown on page 189.

Typical Formats ~

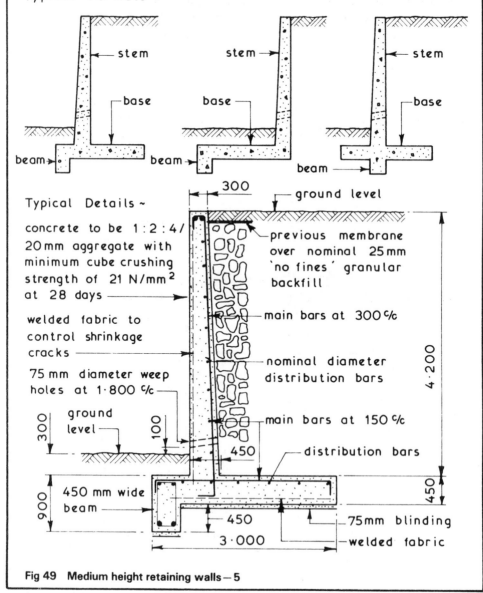

Typical Details ~

concrete to be 1:2:4/ 20mm aggregate with minimum cube crushing strength of 21 N/mm² at 28 days

welded fabric to control shrinkage cracks

75 mm diameter weep holes at 1·800 ℅

ground level

450 mm wide beam

300

ground level

previous membrane over nominal 25mm 'no fines' granular backfill

main bars at 300 ℅

nominal diameter distribution bars

main bars at 150 ℅

distribution bars

4·200

450

450

900

100

300

75mm blinding

welded fabric

450

3·000

**Fig 49   Medium height retaining walls — 5**

Formwork ~ concrete retaining walls can be cast in one of three ways - full height ; climbing (page 192) or against earth face (page 193).

Full Height Casting ~ this can be carried out if the wall is to be cast as a freestanding wall and allowed to cure and gain strength before the earth to be retained is backfilled behind the wall. Considerations are the height of the wall, anticipated pressure of wet concrete, any strutting requirements and the availability of suitable materials to fabricate the formwork. As with all types of formwork a traditional timber format or a patent system using steel forms could be used.

Typical Details ~

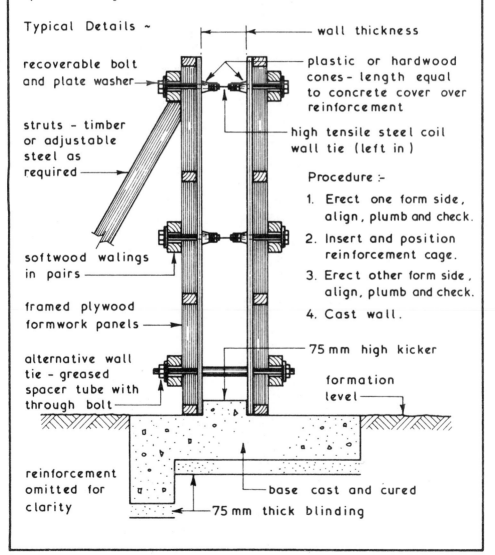

recoverable bolt and plate washer

wall thickness

plastic or hardwood cones - length equal to concrete cover over reinforcement

struts - timber or adjustable steel as required

high tensile steel coil wall tie (left in )

Procedure :-

1. Erect one form side, align, plumb and check.

2. Insert and position reinforcement cage.

3. Erect other form side, align, plumb and check.

4. Cast wall.

softwood walings in pairs

framed plywood formwork panels

75 mm high kicker

alternative wall tie - greased spacer tube with through bolt

formation level

reinforcement omitted for clarity

base cast and cured

75 mm thick blinding

Climbing Formwork or Lift Casting ~ this method can be employed on long walls, high walls or where the amount of concrete which can be placed in a shift is limited.

Typical Details ~

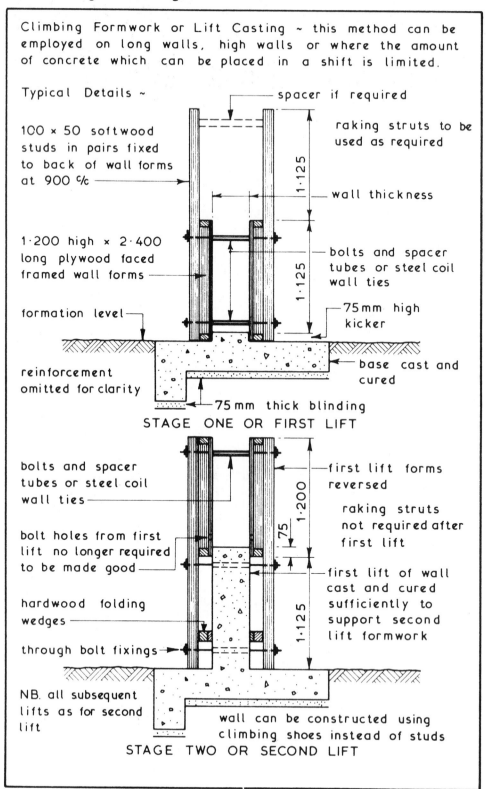

spacer if required

100 × 50 softwood studs in pairs fixed to back of wall forms at 900 c/c

raking struts to be used as required

wall thickness

1·125

1·200 high × 2·400 long plywood faced framed wall forms

1·125

bolts and spacer tubes or steel coil wall ties

formation level

75 mm high kicker

reinforcement omitted for clarity

base cast and cured

75 mm thick blinding

STAGE ONE OR FIRST LIFT

bolts and spacer tubes or steel coil wall ties

first lift forms reversed

1·200

raking struts not required after first lift

bolt holes from first lift no longer required to be made good

75

first lift of wall cast and cured sufficiently to support second lift formwork

hardwood folding wedges

1·125

through bolt fixings

NB. all subsequent lifts as for second lift

wall can be constructed using climbing shoes instead of studs

STAGE TWO OR SECOND LIFT

Casting Against Earth Face ~ this method can be an adaptation of the full height or climbing formwork systems. The latter uses a steel wire loop tie fixing to provide the support for the second and subsequent lifts.

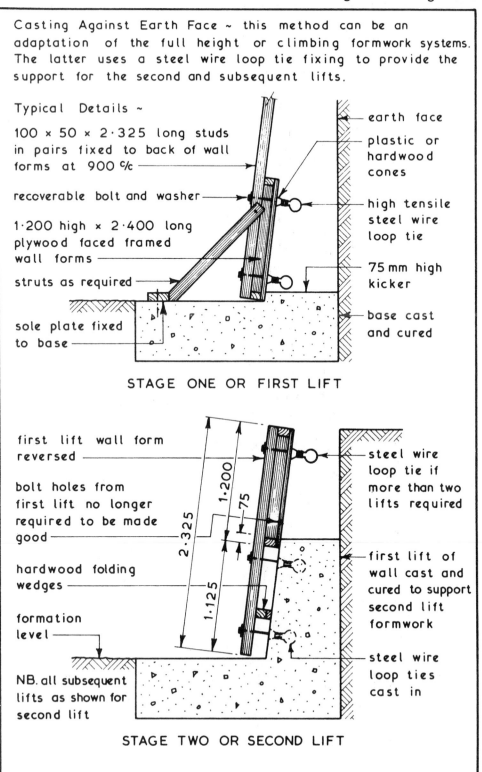

Typical Details ~

100 × 50 × 2·325 long studs in pairs fixed to back of wall forms at 900 ℅

recoverable bolt and washer

1·200 high × 2·400 long plywood faced framed wall forms

struts as required

sole plate fixed to base

earth face

plastic or hardwood cones

high tensile steel wire loop tie

75 mm high kicker

base cast and cured

### STAGE ONE OR FIRST LIFT

first lift wall form reversed

bolt holes from first lift no longer required to be made good

hardwood folding wedges

formation level

NB. all subsequent lifts as shown for second lift

steel wire loop tie if more than two lifts required

first lift of wall cast and cured to support second lift formwork

steel wire loop ties cast in

### STAGE TWO OR SECOND LIFT

## Basement Excavations

Open Excavations ~ one of the main problems which can be encountered with basement excavations is the need to provide temporary support or timbering to the sides of the excavation. This can be intrusive when the actual construction of the basement floor and walls is being carried out. One method is to use battered excavation sides cut back to a safe angle of repose thus eliminating the need for temporary support works to the sides of the excavation.

Typical Example of Open Basement Excavations ~

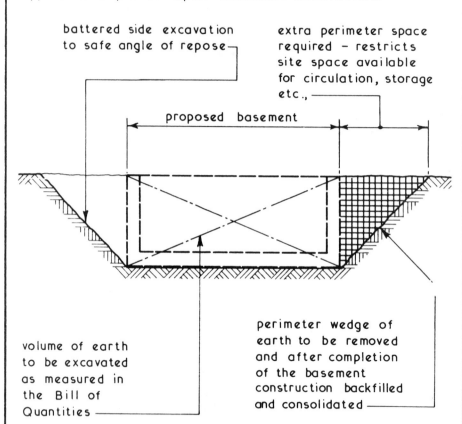

battered side excavation to safe angle of repose

extra perimeter space required – restricts site space available for circulation, storage etc.,

proposed basement

perimeter wedge of earth to be removed and after completion of the basement construction backfilled and consolidated

volume of earth to be excavated as measured in the Bill of Quantities

In economic terms the costs of plant and manpower to cover the extra excavation, backfilling and consolidating must be offset by the savings made by omitting the temporary support works to the sides of the excavation. The main disadvantage of this method is the large amount of free site space required.

Perimeter Trench Excavations ~ in this method a trench wide enough for the basement walls to be constructed is excavated and supported with timbering as required. It may be necessary for runners or steel sheet piling to be driven ahead of the excavation work. This method can be used where weak subsoils are encountered so that the basement walls act as permanent timbering whilst the mound or dumpling is excavated and the base slab cast. Perimeter trench excavations can also be employed in firm subsoils when the mechanical plant required for excavating the dumpling is not available at the right time.

Typical Details ~

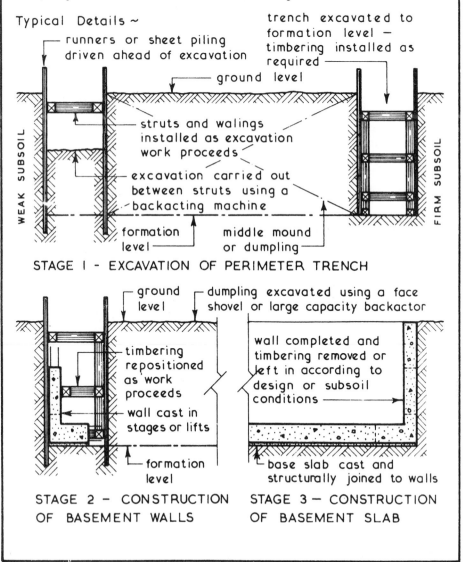

STAGE 1 - EXCAVATION OF PERIMETER TRENCH

STAGE 2 — CONSTRUCTION OF BASEMENT WALLS

STAGE 3 — CONSTRUCTION OF BASEMENT SLAB

# Basement Excavations

Complete Excavation ~ this method can be used in firm subsoils where the centre of the proposed basement can be excavated first to enable the basement slab to be cast thus giving protection to the subsoil at formation level. The sides of excavation to the perimeter of the basement can be supported from the formation level using raking struts or by using raking struts pitched from the edge of the basement slab.

Typical Details ~

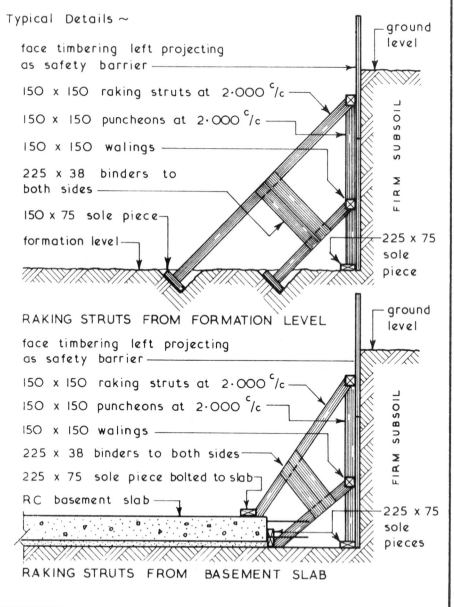

face timbering left projecting as safety barrier

150 x 150 raking struts at 2·000 $^c$/c

150 x 150 puncheons at 2·000 $^c$/c

150 x 150 walings

225 x 38 binders to both sides

150 x 75 sole piece

formation level

ground level

FIRM SUBSOIL

225 x 75 sole piece

**RAKING STRUTS FROM FORMATION LEVEL**

face timbering left projecting as safety barrier

150 x 150 raking struts at 2·000 $^c$/c

150 x 150 puncheons at 2·000 $^c$/c

150 x 150 walings

225 x 38 binders to both sides

225 x 75 sole piece bolted to slab

RC basement slab

ground level

FIRM SUBSOIL

225 x 75 sole pieces

**RAKING STRUTS FROM BASEMENT SLAB**

Excavating Plant ~ the choice of actual pieces of plant to be used in any construction activity is a complex matter taking into account many factors. Specific details of various types of excavators are given on pages 120 to 124. At this stage it is only necessary to consider basic types for particular operations. In the context of basement excavation two forms of excavator could be considered.

1. Backactors — these machines are available as cable rigged or hydraulic excavators suitable for trench and bulk excavating. Cable rigged backactors are usually available with larger bucket sizes and deeper digging capacities than the hydraulic machines but these have a more positive control and digging operation and are also easier to operate.

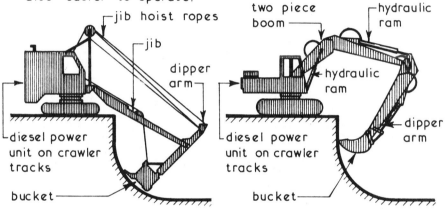

2. Face Shovels — these are robust machines designed to excavate above their own wheel or track level and are suitable for bulk excavation work. In basement work they will require a ramp approach unless they are to be lifted out of the excavation area by means of a crane. Like backactors face shovels are available as cable rigged or hydraulic machines.

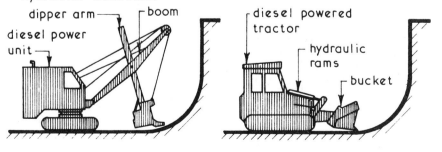

197

## Basement Construction

Basement Construction ~ in the general context of buildings a basement can be defined as a storey which is below the ground storey and is therefore constructed below ground level. Most basements can be classified into one of three groups :-

1. Retaining Wall and Raft Basements - this is the general format for basement construction and consists of a slab raft foundation which forms the basement floor and helps to distribute the structural loads transmitted down the retaining walls.

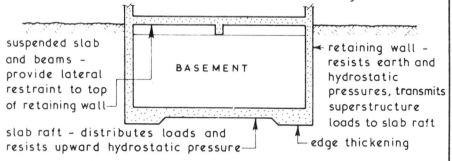

suspended slab and beams - provide lateral restraint to top of retaining wall

retaining wall - resists earth and hydrostatic pressures, transmits superstructure loads to slab raft

slab raft - distributes loads and resists upward hydrostatic pressure

edge thickening

2. Box and Cellular Raft Basements - similar method to above except that internal walls are used to transmit and spread loads over raft as well as dividing basement into cells.

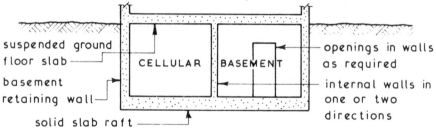

suspended ground floor slab

basement retaining wall

solid slab raft

openings in walls as required

internal walls in one or two directions

3. Piled Basements - the main superstructure loads are carried to the basement floor level by columns where they are finally transmitted to the ground via pile caps and bearing piles. This method can be used where low bearing capacity soils are found at basement floor level.

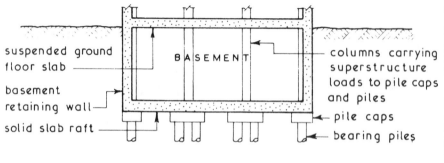

suspended ground floor slab

basement retaining wall

solid slab raft

columns carrying superstructure loads to pile caps and piles

pile caps

bearing piles

Deep Basement Construction ~ basements can be constructed within a cofferdam or other temporary supported excavation (**see Basement Excavations on pages 194 to 196**) up to the point when these methods become uneconomic, unacceptable or both due to the amount of necessary temporary support work. Deep basements can be constructed by installing diaphragm walls within a trench and providing permanent support with ground anchors or by using the permanent lateral support given by the internal floor during the excavation period (see **page 200**). Temporary lateral support during the excavation period can be provided by lattice beams spanning between the diaphragm walls (**see page 200**).

Typical Ground Anchor Support Details~

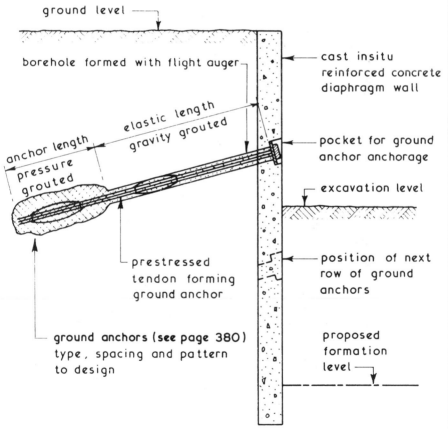

NB   vertical ground anchors installed through the lowest floor can be used to overcome any tendency to floatation during the construction period

# Basement Construction

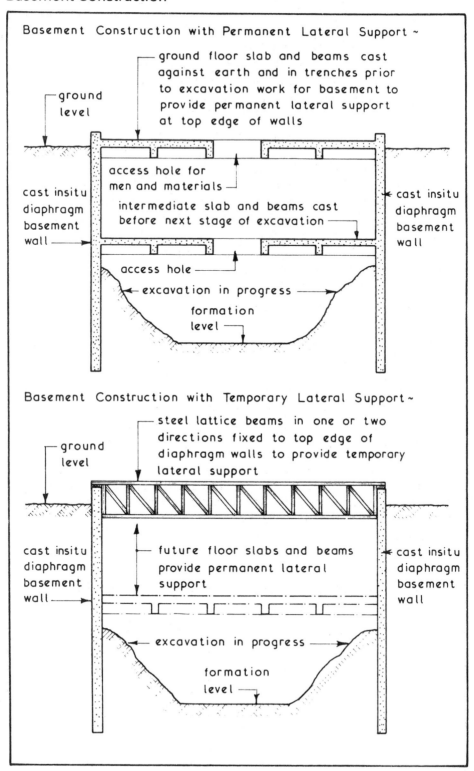

Basement Construction with Permanent Lateral Support ~

ground floor slab and beams cast against earth and in trenches prior to excavation work for basement to provide permanent lateral support at top edge of walls

ground level

access hole for men and materials

cast insitu diaphragm basement wall

intermediate slab and beams cast before next stage of excavation

cast insitu diaphragm basement wall

access hole

excavation in progress

formation level

Basement Construction with Temporary Lateral Support ~

steel lattice beams in one or two directions fixed to top edge of diaphragm walls to provide temporary lateral support

ground level

cast insitu diaphragm basement wall

future floor slabs and beams provide permanent lateral support

cast insitu diaphragm basement wall

excavation in progress

formation level

Waterproofing Basements ~ basements can be waterproofed by one of three basic methods namely :-

1. Use of **dense** monolithic concrete walls and floor.

2. **Tanking techniques (see pages 203 & 204)**

3. **Drained cavity system (see page 205)**

Dense Monolithic Concrete - the main objective is to form a watertight basement using dense high quality reinforced or prestressed concrete by a combination of good materials, good workmanship, attention to design detail and on site construction methods. If strict control of all aspects is employed a sound watertight structure can be produced but it should be noted that such structures are not always water vapourproof. If the latter is desirable some waterproof coating, lining or tanking should be used. The watertightness of dense concrete mixes depends primarily upon two factors :-

1. Water/cement ratio.

2. Degree of compaction.

The hydration of cement during the hardening process produces heat therefore to prevent early stage cracking the temperature changes within the hardening concrete should be kept to a minimum. The greater the cement content the more is the evolution of heat therefore the mix should contain no more cement than is necessary to fulfil design requirements. Concrete with a free water/cement ratio of 0·5 is watertight and although the permeability is three time more at a ratio of 0·6 it is for practical purposes still watertight but above this ratio the concrete becomes progressively less watertight. For lower water/cement ratios the workability of the mix would have to be increased, usually by adding more cement, to enable the concrete to be fully compacted.

Admixtures - if the ingredients of good design, materials and workmanship are present watertight concrete can be produced without the use of admixtures. If admixtures are used they should be carefully chosen and used to obtain a specific objective:-

1. Water-reducing admixtures - used to improve workability

2. Retarding admixtures - slow down rate of hardening

3. Accelerating admixtures - increase rate of hardening-useful for low temperatures - calcium chloride not suitable for reinforced concrete.

4. Water-repelling admixtures - effective only with low water head, will not improve poor quality or porous mixes.

5. Air-entraining admixtures - increases workability-lowers water content.

# Waterproofing Basements

Joints ~ in general these are formed in basement constructions to provide for movement accommodation (expansion joints) or to create a convenient stopping point in the construction process (construction joints). Joints are lines of weakness which will leak unless carefully designed and constructed therefore they should be simple in concept and easy to construct.

Basement Slabs - these are usually designed to span in two directions and as a consequence have relatively heavy top and bottom reinforcement. To enable them to fulfil their basic functions they usually have a depth in excess of 250 mm. The joints, preferably of the construction type, should be kept to a minimum and if waterbars are specified they must be placed to ensure that complete compaction of the concrete is achieved.

Typical Basement Slab Joint Details ~

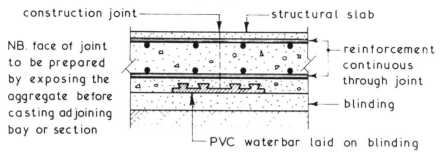

construction joint —————————— structural slab

NB. face of joint to be prepared by exposing the aggregate before casting adjoining bay or section

reinforcement continuous through joint

blinding

PVC waterbar laid on blinding

Basement Walls - joints can be horizontal and/or vertical according to design requirements. A suitable waterbar should be incorporated in the joint to prevent the ingress of water. The top surface of a kicker used in conjunction with single lift pouring if adequately prepared by exposing the aggregate should not require a waterbar but if one is specified it should be either placed on the rear face or consist of a centrally placed mild steel strip inserted into the kicker whilst the concrete is still in a plastic state.

Typical Basement Wall Joint Details ~

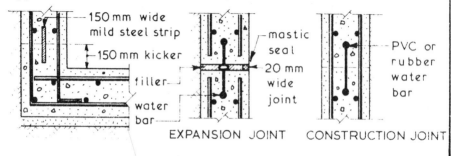

150 mm wide mild steel strip

150 mm kicker

filler

water bar

mastic seal

20 mm wide joint

PVC or rubber water bar

EXPANSION JOINT          CONSTRUCTION JOINT

Mastic Asphalt Tanking ~ the objective of tanking is to provide a continuous waterproof membrane which is applied to the base slab and walls with complete continuity between the two applications. The tanking can be applied externally or internally according to the circumstances prevailing on site. Alternatives to mastic asphalt are polythene sheeting; bituminous compounds; epoxy resin compounds and bitumen laminates.

External Mastic Asphalt Tanking – this is the preferred method since it not only prevents the ingress of water it also protects the main structure of the basement from aggressive sulphates which may be present in the surrounding soil or ground water.

Typical External Tanking Details ~

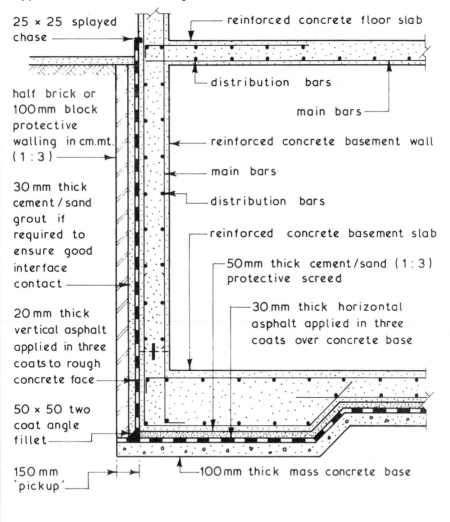

25 × 25 splayed chase

reinforced concrete floor slab

distribution bars

main bars

half brick or 100 mm block protective walling in cm.mt. (1 : 3)

reinforced concrete basement wall

main bars

distribution bars

30 mm thick cement / sand grout if required to ensure good interface contact

reinforced concrete basement slab

50 mm thick cement / sand (1 : 3) protective screed

30 mm thick horizontal asphalt applied in three coats over concrete base

20 mm thick vertical asphalt applied in three coats to rough concrete face

50 × 50 two coat angle fillet

150 mm 'pickup'

100 mm thick mass concrete base

# Waterproofing Basements

Internal Mastic Asphalt Tanking ~ this method should only be adopted if external tanking is not possible since it will not give protection to the main structure and unless adequately loaded may be forced away from the walls and/or floor by hydrostatic pressure. To be effective the horizontal and vertical coats of mastic asphalt must be continuous.

Typical Internal Tanking Details ~

25 × 25 splayed chase

reinforced concrete floor slab

reinforced concrete basement wall

distribution bars

main bars

20 mm thick vertical asphalt applied in three coats to rough concrete face

three coat asphalt collar applied over bitumen primer to extend at least 75 mm on both sides of tanking

sleeve, duct or pipe through wall

50 × 50 two coat angle fillet

30 mm thick cement/sand grout if required to ensure good interface contact

brick or block loading wall

reinforced concrete loading slab

main bars

distribution bars

water bar to kicker

50 mm thick protective screed

reinforced concrete basement slab

75 mm thick blinding

30 mm thick horizontal asphalt applied in three coats over concrete base

NB. Brick basement walls can be built of keyed bricks or the joints can be raked out to a depth of 20 mm to provide key for asphalt tanking

Drained Cavity System ~ this method of waterproofing basements can be used for both new and refurbishment work. The basic concept is very simple in that it accepts that a small amount of water seepage is possible through a monolithic concrete wall and the best method of dealing with such moisture is to collect it and drain it away. This is achieved by building an inner non-load bearing wall to form a cavity which is joined to a floor composed of special triangular tiles laid to falls which enables the moisture to drain away to a sump from which it is either discharged direct or pumped into the surface water drainage system. The inner wall should be relatively vapour tight or alternatively the cavity should be ventilated.

Typical Details ~

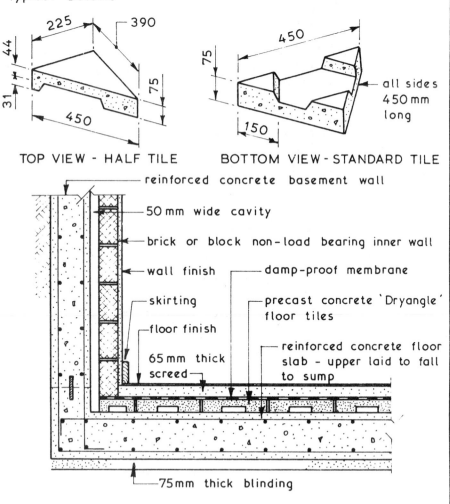

TOP VIEW - HALF TILE      BOTTOM VIEW - STANDARD TILE

- reinforced concrete basement wall
- 50 mm wide cavity
- brick or block non-load bearing inner wall
- wall finish
- damp-proof membrane
- skirting
- precast concrete 'Dryangle' floor tiles
- floor finish
- reinforced concrete floor slab - upper laid to fall to sump
- 65 mm thick screed
- 75mm thick blinding

Excavation ~ to hollow out – in building terms to remove earth to form a cavity in the ground.

Types of Excavation ~

Oversite – the removal of top soil (Building Regulations requirement.)

depth varies from site to site but is usually in a 150 to 300mm range. Top soil contains plant life animal life and decaying matter which makes the soil compressible and therefore unsuitable for supporting buildings.

s u b s o i l

Reduce Level – carried out below oversite level to form a level surface on which to build and can consist of both cutting and filling operations. The level to which the ground is reduced is called the formation level.

cut

original ground level

fill

cut

proposed formation level

NB. Water in Excavations – this should be removed since it can:~
 1. Undermine sides of excavation.
 2. Make it impossible to adequately compact bottom of excavation to receive foundations.
 3. Cause puddling which can reduce the bearing capacity of the subsoil.

Trench Excavations ~ narrow excavations primarily for strip foundations and buried services – excavation can be carried out by hand or machine.

Typical Examples ~

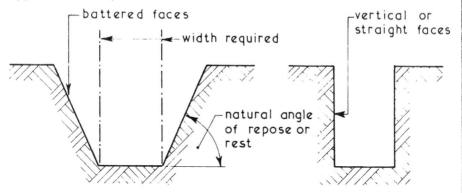

Disadvantage ~ extra cost of over excavating and extra backfilling.

Disadvantage ~ sides of excavation may require some degree of temporary support.

Advantage ~ no temporary support required to sides of excavation.

Advantage ~ minimum amount of soil removed and therefore minimum amount of backfilling.

Pier Holes ~ isolated pits primarily used for foundation pads for columns and piers or for the construction of soakaways.

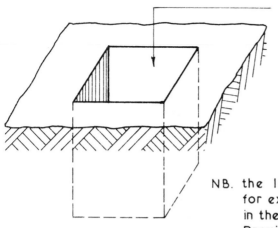

sides of excavation can be battered or straight as described above – deep pier holes may have to be over excavated in plan to provide good access to and good egress from the working area for both men and materials.

NB. the legal safety requirements for excavations are contained in the Construction (General Provisions) Regulations 1961

Site Clearance and Removal of Top Soil ~

On small sites this could be carried out by manual means using hand held tools such as picks, shovels and wheelbarrows.

On all sites mechanical methods could be used the actual plant employed being dependent on factors such as volume of soil involved, nature of site and time elements.

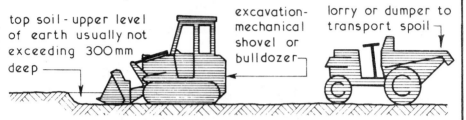

top soil - upper level of earth usually not exceeding 300mm deep

excavation- mechanical shovel or bulldozer

lorry or dumper to transport spoil

Reduced Level Excavations ~

On small sites - hand processes as given above

On all sites mechanical methods could be used dependent on factors given above.

bulldozer for cut and fill operations

mechanical shovel and attendant lorries for cut only operations

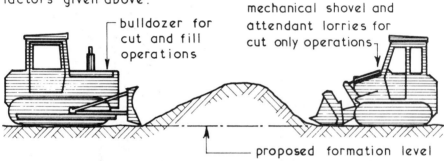

proposed formation level

Trench and Pit Excavations ~

On small sites - hand processes as given above but if depth of excavation exceeds 1·200 some method of removing spoil from the excavation will have to be employed.

On all sites mechanical methods could be used dependent on factors given above.

on large sites a trenching machine could be used

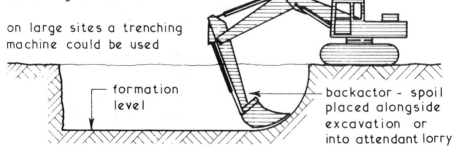

formation level

backactor - spoil placed alongside excavation or into attendant lorry

All subsoils have different abilities in remaining stable during excavation works. Most will assume a natural angle of repose or rest unless given temporary support. The presence of ground water apart from creating difficult working conditions can have an adverse effect on the subsoil's natural angle of repose.

Typical Angles of Repose ~

Excavations cut to a natural angle of repose are called battered.

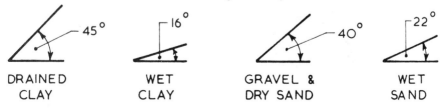

| DRAINED CLAY | WET CLAY | GRAVEL & DRY SAND | WET SAND |
| 45° | 16° | 40° | 22° |

Factors for Temporary Support of Excavations ~

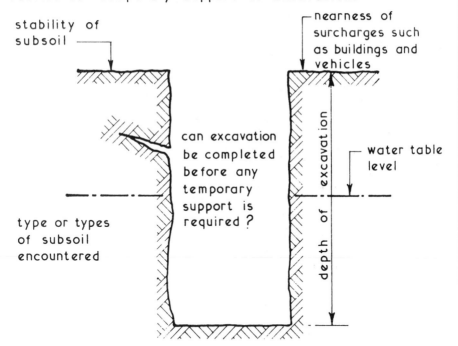

stability of subsoil

nearness of surcharges such as buildings and vehicles

can excavation be completed before any temporary support is required ?

type or types of subsoil encountered

water table level

depth of excavation

Time factors such as period during which excavation will remain open and the time of year when work is carried out.

The minimum legal requirements regarding the support and protection of excavations are set out in the Construction (General Provisions) Regulations 1961.

Temporary Support ~ in the context of excavations this is called timbering irrespective of the actual materials used. If the sides of the excavation are completely covered with timbering it is known as close timbering whereas any form of partial covering is called open timbering.

An adequate supply of timber or other suitable material must be available and used to prevent danger to any person **employed in an excavation over 1·200 deep** from a fall or dislodgement of materials forming the sides of an excavation.

A suitable barrier or fence must be provided to the sides of all **excavations over 2·000 deep or alternatively they must be** securely covered.

Materials must not be placed near to the edge of any excavation, nor must plant be placed or moved near to any excavation so that persons employed in the excavation are endangered.

Typical Example ~

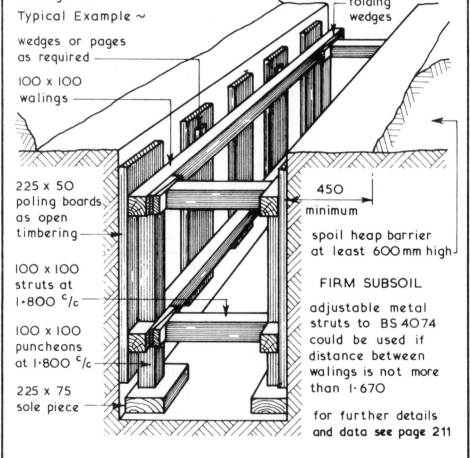

folding wedges

wedges or pages as required

100 x 100 walings

225 x 50 poling boards as open timbering

450 minimum

spoil heap barrier at least 600 mm high

FIRM SUBSOIL

100 x 100 struts at 1·800 $^c/c$

100 x 100 puncheons at 1·800 $^c/c$

225 x 75 sole piece

adjustable metal struts to BS 4074 could be used if distance between walings is not more than 1·670

for further details and data see page 211

Poling Boards ~ a form of temporary support which is placed in position against the sides of excavation after the excavation work has been carried out. Poling boards are placed at centres according to the stability of the subsoils encountered.

Runners ~ a form of temporary support which is driven into position ahead of the excavation work either to the full depth or by a drive and dig technique where the depth of the runner is always lower than that of the excavation.

Trench Sheeting ~ form of runner made from sheet steel with a trough profile – can be obtained with a lapped joint or an interlocking joint.

Water ~ if present or enters an excavation a pit or sump should be excavated below the formation level to act as collection point from which the water can be pumped away.

Typical Example ~

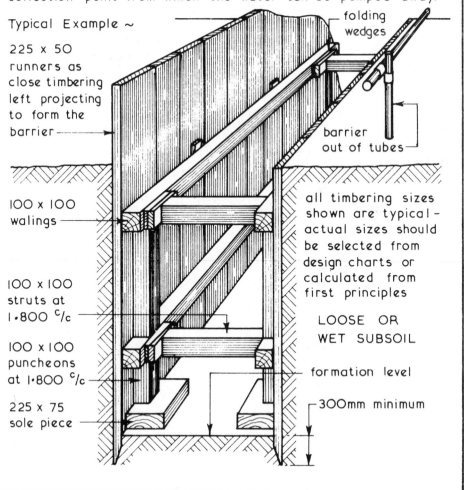

225 x 50 runners as close timbering left projecting to form the barrier

folding wedges

barrier out of tubes

100 x 100 walings

100 x 100 struts at 1·800 $^c/c$

100 x 100 puncheons at 1·800 $^c/c$

225 x 75 sole piece

all timbering sizes shown are typical – actual sizes should be selected from design charts or calculated from first principles

LOOSE OR WET SUBSOIL

formation level

300mm minimum

**Concrete ~** a mixture of cement + fine aggregrate + coarse aggregate + water in controlled proportions and of a suitable quality.

**Cement ~** powder produced from clay and chalk or limestone. In general most concrete is made with ordinary or rapid hardening Portland cement, both types being manufactured to the recommendations of BS 12. Ordinary Portland cement is adequate for most purposes but has a low resistance to attack by acids and sulphates. Rapid hardening Portland cement does not set faster than ordinary Portland cement but it does develop its working strength at a faster rate. For a concrete which must have an acceptable degree of resistance to sulphate attack sulphate resisting Portland cement made to the recommendations of BS 4027 could be specified.

50 kg

BAGS

12 t
to
50 t

SILOS

**Aggregates~** shape, surface texture and grading (distribution of particle sizes) are factors which influence the workability and strength of a concrete mix. Fine aggregates are those materials which pass through a 5mm sieve whereas coarse aggregates are those materials which are retained on a 5mm sieve. Dense aggregrates are those with a density of more than 1200 kg/m$^3$ for coarse aggregates and more than 1250 kg/m$^3$ for fine aggregates and are covered by BS 882 – Coarse and Fine Aggregates from Natural Sources and BS 1047 Air-cooled Blastfurnace Slag Coarse Aggregates for Concrete. Lightweight aggregates include clinker – BS 1165 ; foamed or expanded blastfurnace slag – BS 877 and exfoliated and expanded materials such as vermiculite, perlite, clay and sintered pulverized-fuel ash to BS 3797.

coarse
aggregate

5mm sieve

fine
aggregate

**Water ~** must be clean and free from impurities which are likely to affect the quality or strength of the resultant concrete. Pond, river, canal and sea water should not be used and only water fit for drinking should be specified.

drinking

Cement ~ whichever type of cement is being used it must be properly stored on site to keep it in good condition. The cement must be kept dry since contact with any moisture whether direct or airborne could cause it to set. A rotational use system should be introduced to ensure that the first batch of cement delivered is the first to be used.

Typical Storage Methods ~

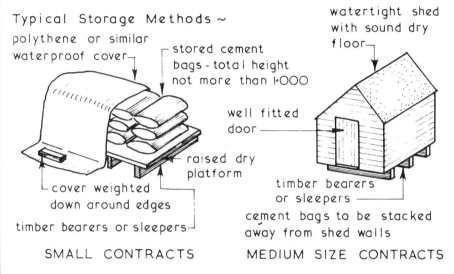

polythene or similar waterproof cover

stored cement bags - total height not more than 1·000

watertight shed with sound dry floor

well fitted door

raised dry platform

cover weighted down around edges

timber bearers or sleepers

SMALL CONTRACTS

timber bearers or sleepers

cement bags to be stacked away from shed walls

MEDIUM SIZE CONTRACTS

LARGE CONTRACTS – for bagged cement watertight shed as above for bulk delivery loose cement a cement storage silo.

Aggregates ~ essentials of storage are to keep different aggregate types and/or sizes separate, store on a clean, hard free draining surface and to keep the stored aggregates clean and free of leaves and rubbish.

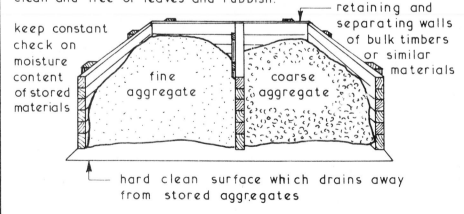

keep constant check on moisture content of stored materials

retaining and separating walls of bulk timbers or similar materials

fine aggregate

coarse aggregate

hard clean surface which drains away from stored aggregates

# Concrete Production—Volume Batching

Concrete Batching ~ a batch is one mixing of concrete and can be carried out by measuring the quantities of materials required by volume or weight. The main aim of both methods is to ensure that all consecutive batches are of the same standard and quality.

Volume Batching ~ concrete mixes are often quoted by ratio such as 1 : 2 : 4 (cement : fine aggregate or sand : coarse aggregate.) A bag of cement weighing 50 kg has a volume of 0·033 m³ therefore for the above mix 2 x 0·033 (0·066 m³) of sand and 4 x 0·033 (0·132 m³) of coarse aggregate is required. To ensure accurate amounts of materials are used for each batch a gauge box should be employed its size being based on convenient handling. Ideally a batch of concrete should be equated to using 1 bag of cement per batch. Assuming a gauge box 300mm deep and 300 mm wide with a volume of half the required sand the gauge box size would be –

volume = length x width x depth = length x 300 x 300

$$\therefore \text{length} = \frac{\text{volume}}{\text{width} \times \text{depth}} = \frac{0 \cdot 033}{0 \cdot 3 \times 0 \cdot 3} = 0 \cdot 366 \text{ m}$$

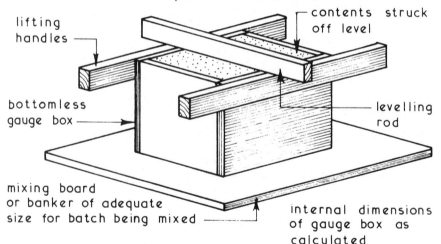

lifting handles

contents struck off level

bottomless gauge box

levelling rod

mixing board or banker of adequate size for batch being mixed

internal dimensions of gauge box as calculated

For the above given mix fill gauge box once with cement, twice with sand and four times with coarse aggregate.

An allowance must be made for the bulking of damp sand which can be as much as 33⅓ %. General rule of thumb unless using dry sand allow for 25% bulking.

Materials should be well mixed dry before adding water.

Weigh Batching ~ this is a more accurate method of measuring materials for concrete than volume batching since it reduces considerably the risk of variation between different batches. The weight of sand is affected very little by its dampness which in turn leads to greater accuracy in proportioning materials. When loading a weighing hopper the materials should be loaded in a specific order –

1. Coarse aggregates – tends to push other materials out and leaves the hopper clean.

2. Cement – this is sandwiched between the other materials since some of the fine cement particles could be blown away if cement is put in last.

3. Sand or Fine Aggregates – put in last to stabilise the fine lightweight particles of cement powder.

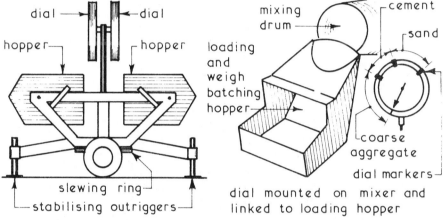

INDEPENDENT WEIGH BATCHER    INTEGRAL WEIGH BATCHER

Typical Densities ~ cement - 1440 kg/m³   sand - 1600 kg/m³
coarse aggregrate - 1440 kg/m³

Water / Cement Ratio ~ water in concrete has two functions -

1. Start the chemical reaction which causes the mixture to set into a solid mass.

2. Give the mix workability so that it can be placed, tamped or vibrated into the required position.

Very little water is required to set concrete (approximately 0·2 w/c ratio) the surplus evaporates leaving minute voids therefore the more water added to the mix to increase its workability the weaker is the resultant concrete. Generally w/c ratios of 0·4 to 0·5 are adequate for most purposes.

Concrete Supply ~ this is usually geared to the demand or the rate at which the mixed concrete can be placed. Fresh concrete should always be used or placed within 30 minutes of mixing to prevent any undue drying out. Under no circumstances should more water be added after the initial mixing.

Small Batches ~ small easily transported mixers with 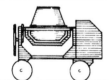 output capacities of up to 100 litres can be used for small and intermittent batches. These mixers are versatile and robust machines which can be used for mixing mortars and plasters as well as concrete.

Medium to Large Batches ~ mixers with output capacities 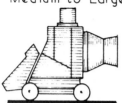 from 100 litres to 10 m³ with either diesel or electric motors. Many models are available with tilting or reversing drum discharge, integral weigh batching and loading hopper and a controlled water supply.

Ready Mixed Concrete ~ used mainly for large concrete batches of up to 6 m³. This method of concrete supply has the advantages of eliminating the need for site space to accommodate storage of materials, mixing plant and the need to employ adequately trained site staff who can constantly produce reliable and consistent concrete  mixes. Ready mixed concrete supply depots also have better facilities and arrangements for producing and supply mixed concrete in winter or inclement weather conditions. In many situations it is possible to place the ready mixed concrete into the required position direct from the delivery lorry via the delivery chute or by feeding it into a concrete pump. The site must be capable of accepting the 20 tonnes laden weight of a typical ready mixed concrete lorry with a turning circle of about 15·000. The supplier will want full details of mix required and the proposed delivery schedule - see BS 5328.

Cofferdams ~ these are temporary enclosures installed in soil or water to prevent the ingress of soil and/or water into the working area with the cofferdam. They are usually constructed from interlocking steel sheet piles which are suitably braced or tied back with ground anchors. Alternatively a cofferdam can be installed using any structural material which will fulfil the required function.

Typical Cofferdam Details ~

steel kicking plate welded to waling

UB end waling bolted to side UB walings through welded on end plates

single skin of interlocking steel sheet piles

corner piece

corner piece

UB side waling

UB side waling

ground level

upper struts and walings

kicking plate

ground level

UB main strut bolted to side waling

UB secondary struts fixed to end walings main struts

end fixing plate welded to strut

lower struts and walings

NB puncheons and wedges to be used as required

formation level

cut off length to suit soil and sheet pile type

for details of sheet pile sections and installation see page 218

Steel Sheet Piling ~ apart from cofferdam work steel sheet can be used as a conventional timbering material in excavations and to form permanent retaining walls. Three common formats of steel sheet piles with interlocking joints are available with a range of section sizes and strengths up to a usual maximum length of 18·000 :-

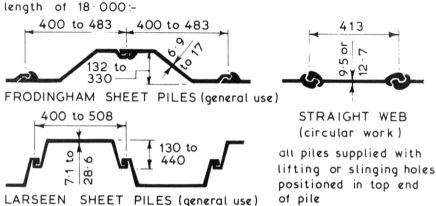

FRODINGHAM SHEET PILES (general use)

LARSEEN SHEET PILES (general use)

STRAIGHT WEB
(circular work)

all piles supplied with lifting or slinging holes positioned in top end of pile

Installing Steel Sheet Piles ~ to ensure that the sheet piles are pitched and installed vertically a driving trestle or guide frame is used. These are usually purpose built to accommodate a panel of 10 to 12 pairs of piles. The piles are lifted into position by a crane and driven by means of a percussion piling hammer or alternatively they can be pushed into the ground by hydraulic rams acting against the weight of the power pack which is positioned over the heads of the pitched piles.

Typical Installation Details ~

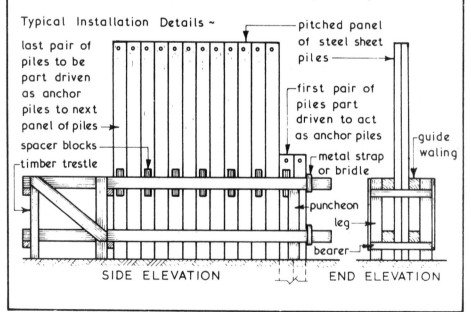

last pair of piles to be part driven as anchor piles to next panel of piles

spacer blocks

timber trestle

pitched panel of steel sheet piles

first pair of piles part driven to act as anchor piles

metal strap or bridle

guide waling

puncheon

leg

bearer

SIDE ELEVATION          END ELEVATION

Caissons ~ these are box-like structures which are similar
in concept to cofferdams but they usually form an integral part
of the finished structure. They can be economically constructed
and installed in water or soil where the depth exceeds 18·000.
There are 4 basic types of caisson namely :-

1. Box Caissons
2. Open Caissons
3. Monolithic Caissons

} usually of precast concrete and used
in water being towed or floated into
position and sunk - land caissons are of
the open type and constructed insitu.

4. Pneumatic Caissons - used in water - see page 220

Typical Caissons Details ~

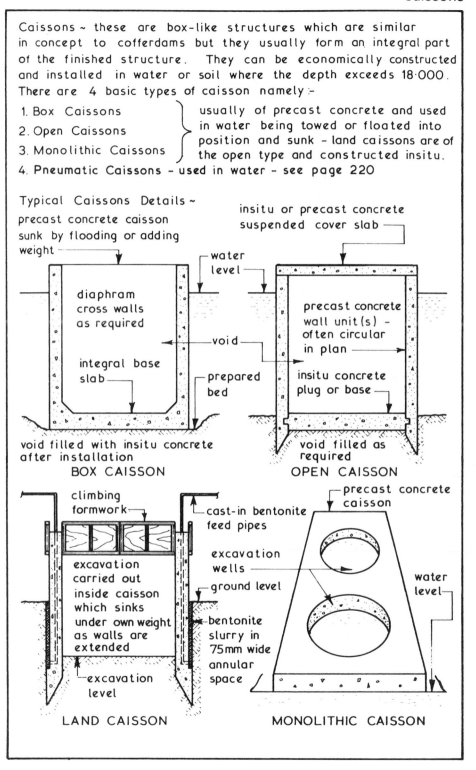

precast concrete caisson
sunk by flooding or adding
weight

insitu or precast concrete
suspended cover slab

water
level

diaphram
cross walls
as required

precast concrete
wall unit(s) -
often circular
in plan

void

integral base
slab

prepared
bed

insitu concrete
plug or base

void filled with insitu concrete
after installation
BOX CAISSON

void filled as
required
OPEN CAISSON

climbing
formwork

cast-in bentonite
feed pipes

excavation
wells

excavation
carried out
inside caisson
which sinks
under own weight
as walls are
extended

ground level

bentonite
slurry in
75mm wide
annular
space

excavation
level

precast concrete
caisson

water
level

LAND CAISSON

MONOLITHIC CAISSON

Pneumatic Caissons ~ these are sometimes called compressed air caissons and are similar in concept to open caissons. They can be used in difficult subsoil conditions below water level and have a pressurised lower working chamber to provide a safe dry working area. Pneumatic caissons can be made of concrete whereby they sink under their own weight or they can be constructed from steel with hollow walls which can be filled with water to act as ballast. These caissons are usually designed to form part of the finished structure.

Typical Pneumatic Caisson Details ~

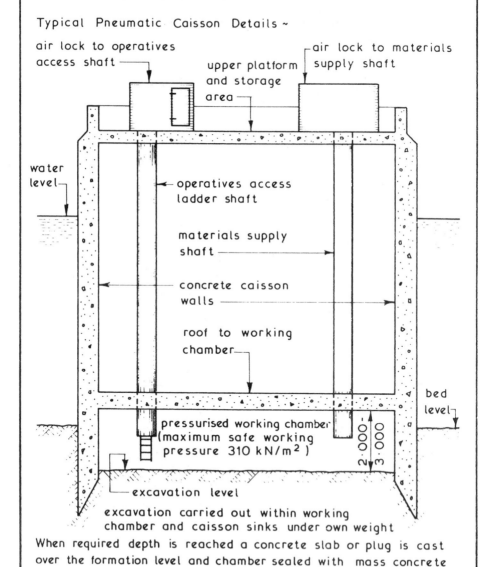

air lock to operatives access shaft

upper platform and storage area

air lock to materials supply shaft

water level

operatives access ladder shaft

materials supply shaft

concrete caisson walls

roof to working chamber

bed level

pressurised working chamber (maximum safe working pressure 310 kN/m²)

2.000 - 3.000

excavation level

excavation carried out within working chamber and caisson sinks under own weight

When required depth is reached a concrete slab or plug is cast over the formation level and chamber sealed with mass concrete

Underpinning ~ the main objective of most underpinning work is to transfer the load carried by a foundation from its existing bearing level to a new level at a lower depth. Underpinning techniques can also be used to replace an existing weak foundation. An underpinning operation may be necessary for one or more of the following reasons :-

1. Uneven Settlement - this could be caused by uneven loading of the building, unequal resistance of the soil, action of tree roots or cohesive soil settlement.

2. Increase in Loading - this could be due to the addition of an extra storey or an increase in imposed loadings such as that which may occur with a change of use.

3. Lowering of Adjacent Ground - usually required when constructing a basement adjacent to existing foundations.

General Precautions ~ before any form of underpinning work is commenced the following precautions should be taken :-

1. Notify adjoining owners of proposed works giving full details and temporary shoring or tying.

2. Carry out a detailed survey of the site, the building to be underpinned and of any other adjoining or adjacent building or structures. A careful record of any defects found should be made and where possible agreed with the adjoining owner(s) before being lodged in a safe place.

3. Indicators or 'tell tales' should be fixed over existing cracks so that any subsequent movements can be noted and monitored.

4. If settlement is the reason for the underpinning works a thorough investigation should be carried out to establish the cause and any necessary remedial work put in hand before any underpinning works are started.

5. Before any underpinning work is started the loads on the building to be underpinned should be reduced as much as possible by removing the imposed loads from the floors and installing any props and/or shoring which is required.

6. Any services which are in the vicinity of the proposed underpinning works should be identified, traced, carefully exposed, supported and protected as necessary.

Underpinning to Walls ~ to prevent fracture, damage or settlement of the wall(s) being underpinned the work should always be carried out in short lengths called legs or bays. The length of these bays will depend upon the following factors :-

1. Total length of wall to be underpinned.

2. Wall loading.

3. General state of repair and stability of wall and foundation to be underpinned.

4. Nature of subsoil beneath existing foundation.

5. Estimated spanning ability of existing foundation.

Generally suitable bay lengths are :-

1·000 to 1·500 for mass concrete strip foundations supporting walls of traditional construction.

1·500 to 3·000 for reinforced concrete strip foundations supporting walls of moderate loading.

In all cases the total sum of the unsupported lengths of wall should not exceed 25% of the total wall length.

The sequence of bays should be arranged so that working in adjoining bays is avoided until one leg of underpinning has been completed, pinned and cured sufficiently to support the wall above.

Typical Underpinning Schedule ~

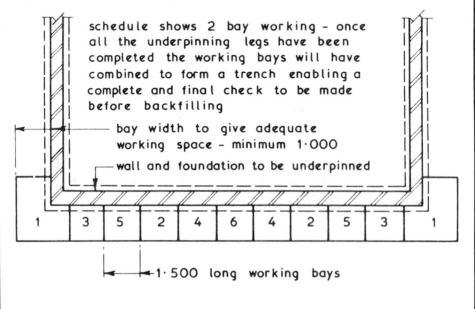

schedule shows 2 bay working - once all the underpinning legs have been completed the working bays will have combined to form a trench enabling a complete and final check to be made before backfilling

bay width to give adequate working space - minimum 1·000

wall and foundation to be underpinned

| 1 | 3 | 5 | 2 | 4 | 6 | 4 | 2 | 5 | 3 | 1 |

1·500 long working bays

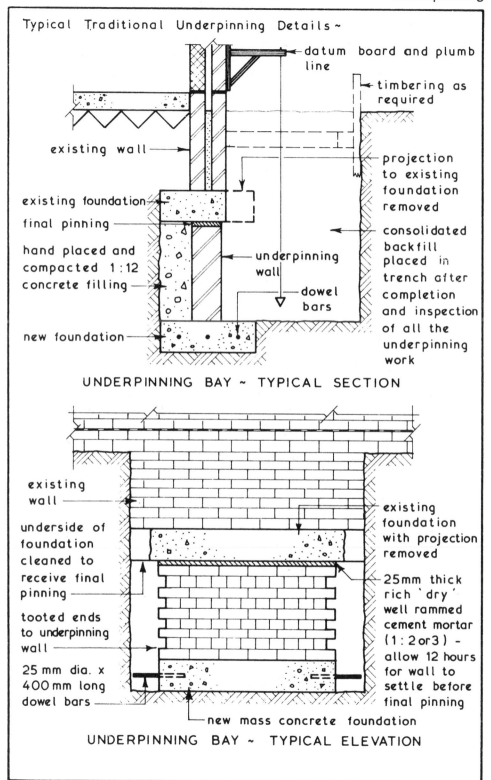

Typical Traditional Underpinning Details ~

datum board and plumb line

timbering as required

existing wall

projection to existing foundation removed

existing foundation

final pinning

consolidated backfill placed in trench after completion and inspection of all the underpinning work

hand placed and compacted 1:12 concrete filling

underpinning wall

dowel bars

new foundation

UNDERPINNING BAY ~ TYPICAL SECTION

existing wall

existing foundation with projection removed

underside of foundation cleaned to receive final pinning

25mm thick rich 'dry' well rammed cement mortar (1:2 or 3) - allow 12 hours for wall to settle before final pinning

tooted ends to underpinning wall

25 mm dia. x 400 mm long dowel bars

new mass concrete foundation

UNDERPINNING BAY ~ TYPICAL ELEVATION

Jack Pile Underpinning ~ this method can be used when the depth of a suitable bearing capacity subsoil is too deep to make traditional underpinning uneconomic. Jack pile underpinning is quiet, vibration free and flexible since the pile depth can be adjusted to suit subsoil conditions encountered. The existing foundations must be in a good condition since they will have to span over the heads of the pile caps which are cast onto the jack pile heads after the hydraulic jacks have been removed.

Typical Details ~

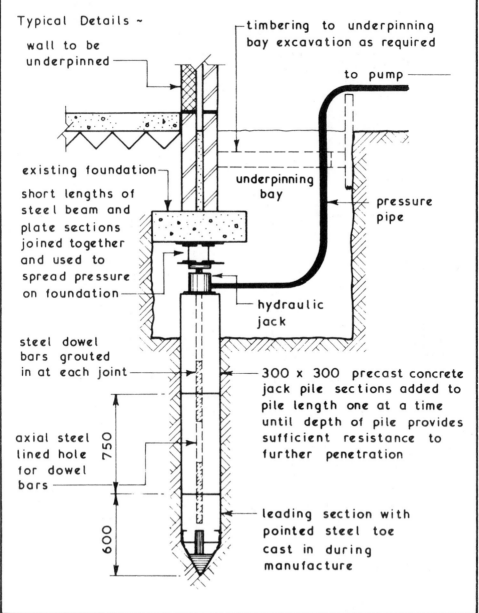

timbering to underpinning bay excavation as required

wall to be underpinned

to pump

existing foundation

short lengths of steel beam and plate sections joined together and used to spread pressure on foundation

underpinning bay

pressure pipe

hydraulic jack

steel dowel bars grouted in at each joint

300 x 300 precast concrete jack pile sections added to pile length one at a time until depth of pile provides sufficient resistance to further penetration

axial steel lined hole for dowel bars

750

600

leading section with pointed steel toe cast in during manufacture

Needle and Pile Underpinning ~ this method of underpinning can be used where the condition of the existing foundation is unsuitable for traditional or jack pile underpinning techniques.  The brickwork above the existing foundation must be in a sound condition since this method relies  on the `arching effect´ of the brick bonding to transmit the wall loads  onto the needles and ultimately to the piles. The piles used with this method are usually small diameter bored piles - see page 169

Typical  Details ~

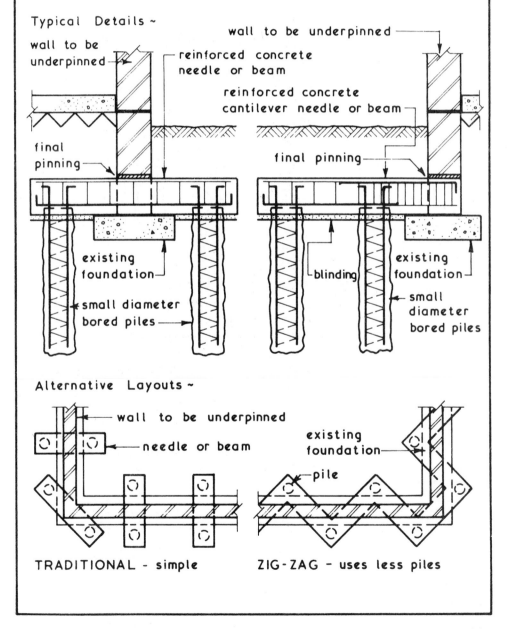

Alternative  Layouts ~

TRADITIONAL - simple            ZIG-ZAG - uses less piles

# Underpinning

'Pynford' Stool Method of Underpinning ~ this method can be used where the existing foundations are in a poor condition and it enables the wall to be underpinned in a continuous run without the need for needles or shoring. The reinforced concrete beam formed by this method may well be adequate to spread the load of the existing wall or it may be used in conjunction with other forms of underpinning such as traditional and jack pile.

Typical Details ~

Stage 1 - holes
formed in wall
to recieve steel
or precast concrete
stools

Stage 2 - stools
inserted and
pinned to soffit
of brickwork
over opening

Stage 3 - brickwork
between pinned stools
removed to leave
wall supported on
pinned stools

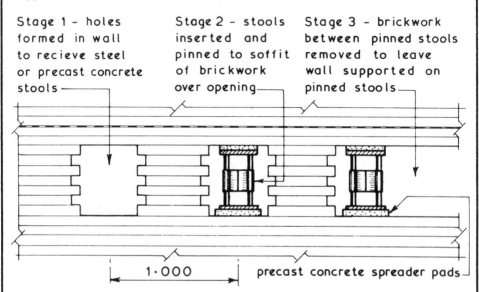

1·000    precast concrete spreader pads

Stage 4 - reinforcement
fabricated and placed
around pinned stools

Stage 5 - formwork
erected and beam
cast

Stage 6 - formwork
removed, beam allowed
to cure before being
pinned to underside
of wall

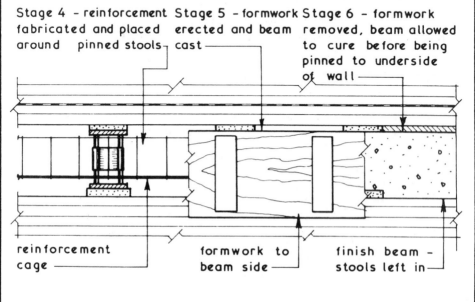

reinforcement
cage

formwork to
beam side

finish beam -
stools left in

Underpinning Columns ~ columns can be underpinned in the same manner as walls using traditional or jack pile methods after the columns have been relieved of their loadings. The beam loads can usually be transferred from the columns by means of dead shores and the actual load of the column can be transferred by means of a pair of beams acting against a collar attached to the base of the column shaft.

Typical Details ~

pair of precast concrete short beams bolted together to form collar

25mm deep chase

pair of precast concrete or steel beams bolted together

hydraulic jack

reinforced concrete column

bored or jack pile

concrete bearing pad as alternative to pile

foundation to be underpinned

kentledge of precast concrete blocks or similar to act as counterweight

steel column

pair of steel channels with web plates welded to steel column

hydraulic jack

pile cap

pair of steel or precast concrete beams bolted together

pair of bored or jack piles

foundation to be underpinned

227

Classification of Water ~ water can be classified by its relative position to or within the ground thus —

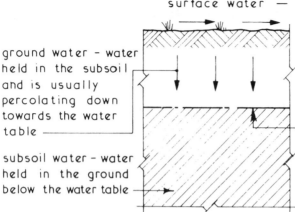

surface water — run off from an earth surface $\simeq 10\%$ of water falling onto surface but on hard paved areas run off is usually 75 to 90 %

ground water - water held in the subsoil and is usually percolating down towards the water table

water table – upper level of water held in the soil which varies with wet and dry periods

subsoil water - water held in the ground below the water table

Problems of Water in the Subsoil ~

1. A high water table could cause flooding during wet periods.

2. Subsoil water can cause problems during excavation works by its natural tendency to flow into the voids created by the excavation activities.

3. It can cause an unacceptable humidity level around finished buildings and structures.

Control of Ground Water ~ this can take one of two forms which are usually referred to as temporary and permanent exclusion —

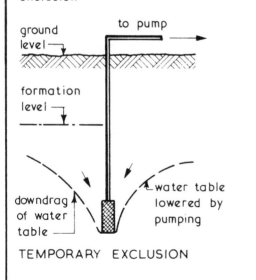

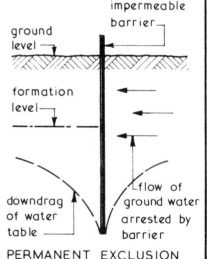

ground level

to pump

formation level

downdrag of water table

water table lowered by pumping

TEMPORARY EXCLUSION

impermeable barrier

ground level

formation level

downdrag of water table

flow of ground water arrested by barrier

PERMANENT EXCLUSION

Permanent Exclusion ~ this can be defined as the insertion of an impermeable barrier to stop the flow of water within the ground.

Temporary Exclusion ~ this can be defined as the lowering of the water table and within the economic depth range of 1·500 can be achieved by subsoil drainage methods, for deeper treatment a pump or pumps are usually involved.

Simple Sump Pumping ~ suitable for trench work and/or where small volumes of water are involved —

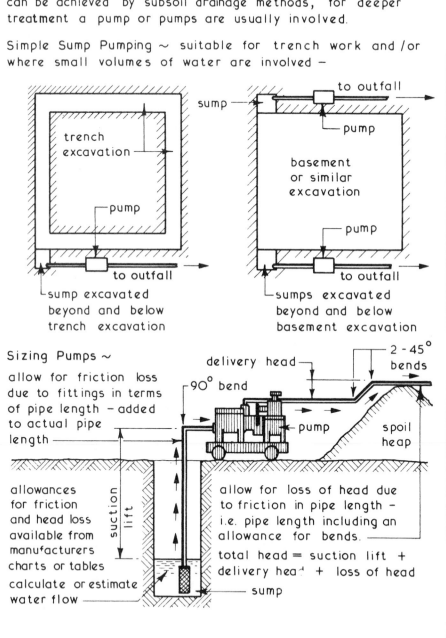

Sizing Pumps ~

allow for friction loss due to fittings in terms of pipe length – added to actual pipe length

allowances for friction and head loss available from manufacturers charts or tables calculate or estimate water flow

allow for loss of head due to friction in pipe length – i.e. pipe length including an allowance for bends.

total head = suction lift + delivery head + loss of head

Jetted Sumps ~ this method achieves the same objectives as **the simple sump methods of dewatering (page 229) but it will** prevent the soil movement which may occur with the simple or open sump methods. In this method a bore-hole is formed in the subsoil by jetting a metal tube into the ground by means of water pressure to the maximum suction lift of the pump being used. A disposable wellpoint connected to a disposable suction pipe is lowered into the tube. The annulus so formed is filled with sand to act as a filtering media, the metal tube being withdrawn after or during the placing of the sand media.

Typical Example ~

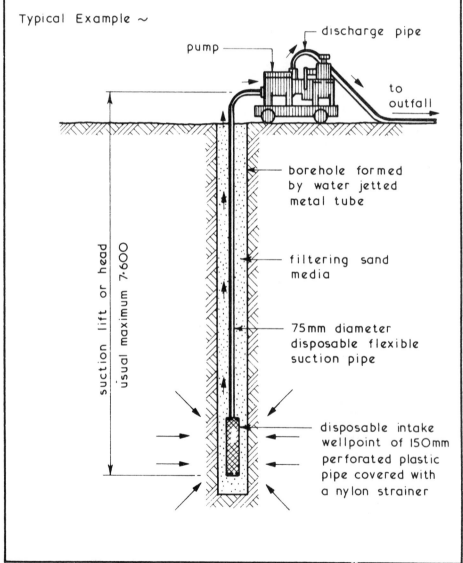

discharge pipe

pump

to outfall

borehole formed by water jetted metal tube

filtering sand media

75mm diameter disposable flexible suction pipe

disposable intake wellpoint of 150mm perforated plastic pipe covered with a nylon strainer

suction lift or head

usual maximum 7·600

Wellpoint Systems ~ method of lowering the water table to a position below the formation level to give a dry working area. The basic principle is to jet into the subsoil a series of wellpoints which are connected to a common header pipe which is connected to a vacuum pump. Wellpoint systems are suitable for most subsoils and can encircle an excavation or be laid progressively alongside as in the case of a trench excavation. If the proposed formation level is below the suction lift capacity of the pump **a multi-stage system can be employed - see page 232**

Typical Details ~

stop valve

150 mm dia. lightweight header pipe

flexible connection pipe

38 mm dia. jetting and riser pipe

retaining collar

75 mm diameter slotted strainer cover

perforated inner tube

riser pipe and wellpoint under suction from pump

up to 1·000

1·300

water flows through strainer cover and perforated tube into riser pipe

jetting shoe

jetting pipe connected to high pressure jetting pump and the water jet emitted from the jetting shoe moves soil particles away enabling wellpoint to sink

rubber ball valve

JETTING

DEWATERING

231

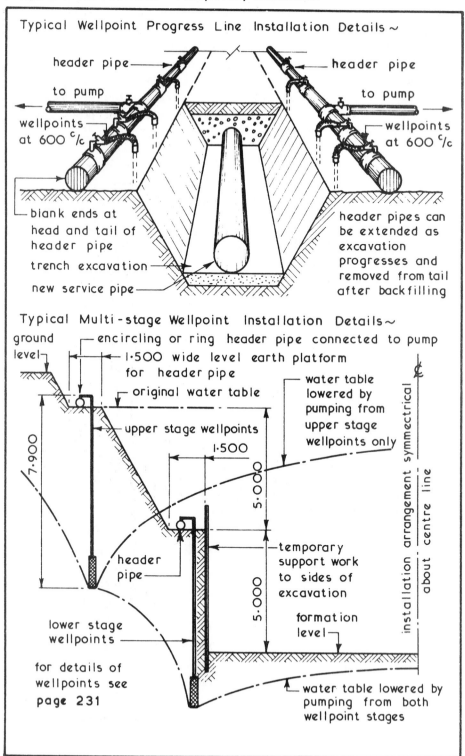

Typical Wellpoint Progress Line Installation Details ~

header pipe

header pipe

to pump

to pump

wellpoints at 600 ᶜ/c

wellpoints at 600 ᶜ/c

blank ends at head and tail of header pipe

trench excavation

new service pipe

header pipes can be extended as excavation progresses and removed from tail after backfilling

Typical Multi-stage Wellpoint Installation Details ~

ground level

encircling or ring header pipe connected to pump

1·500 wide level earth platform for header pipe

original water table

upper stage wellpoints

1·500

water table lowered by pumping from upper stage wellpoints only

7·900

5·000

installation arrangement symmectrical about centre line

header pipe

temporary support work to sides of excavation

5·000

formation level

lower stage wellpoints

for details of wellpoints see **page 231**

water table lowered by pumping from both wellpoint stages

Thin Grouted Membranes ~ these are permanent curtain or cut-off non structural walls or barriers inserted in the ground to enclose the proposed excavation area. They are suitable for silts and sands and can be installed rapidly but they must be adequately supported by earth on both sides. The only limitation is the depth to which the formers can be driven and extracted.

Typical Details ~

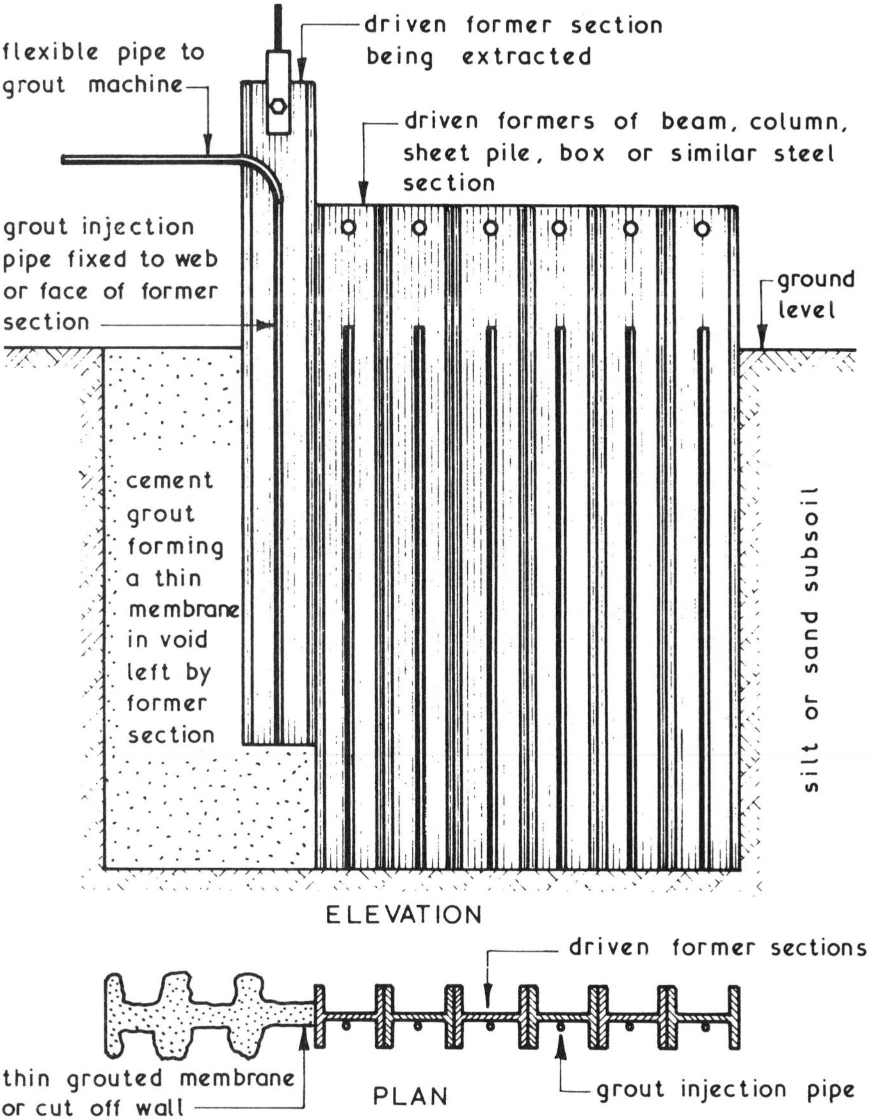

flexible pipe to grout machine

driven former section being extracted

driven formers of beam, column, sheet pile, box or similar steel section

grout injection pipe fixed to web or face of former section

ground level

cement grout forming a thin membrane in void left by former section

silt or sand subsoil

ELEVATION

driven former sections

thin grouted membrane or cut off wall

PLAN

grout injection pipe

233

Contiguous Piling ~ this forms a permanent structural wall of interlocking bored piles. Alternate piles are bored and cast by traditional methods after which the interlocking piles are bored using a special auger or cutter. This system is suitable for most types of subsoil and has the main advantages of being economical on small and confined sites; capable of being formed close to existing foundations and can be installed with the minimum of vibration and noise. Ensuring a complete interlock of all piles over the entire length may be difficult to achieve in practice therefore the exposed face of the piles is usually covered with a mesh or similar fabric and face with rendering or sprayed concrete. Alternatively a reinforced concrete wall could be cast in front of the contiguous piling. This method of ground water control is suitable for structures such as basements, road underpasses and underground car parks.

Typical Details ~

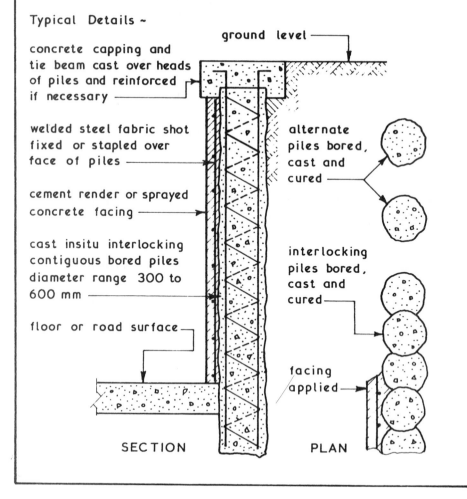

ground level

concrete capping and tie beam cast over heads of piles and reinforced if necessary

welded steel fabric shot fixed or stapled over face of piles

cement render or sprayed concrete facing

cast insitu interlocking contiguous bored piles diameter range 300 to 600 mm

floor or road surface

alternate piles bored, cast and cured

interlocking piles bored, cast and cured

facing applied

SECTION          PLAN

Diaphragm Walls ~ these are structural concrete walls which can be cast insitu (usually by the bentonite slurry method) or constructed using precast concrete components (see page 236) They are suitable for most subsoils and their installation generates only a small amount of vibration and noise making them suitable for works close to existing buildings. The high cost of these walls makes them uneconomic unless they can be incorporated into the finished structure. Diaphragm walls are suitable for basements, underground car parks and similar structures.

Typical Cast Insitu Concrete Diaphragm Wall Details ~

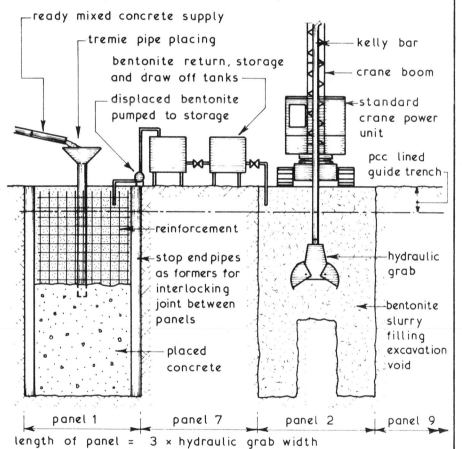

- ready mixed concrete supply
- tremie pipe placing
- bentonite return, storage and draw off tanks
- displaced bentonite pumped to storage
- kelly bar
- crane boom
- standard crane power unit
- pcc lined guide trench
- reinforcement
- stop end pipes as formers for interlocking joint between panels
- placed concrete
- hydraulic grab
- bentonite slurry filling excavation void

| panel 1 | panel 7 | panel 2 | panel 9 |

length of panel = 3 × hydraulic grab width

NB. Bentonite is a controlled mixture of fullers earth and water which produces a mud or slurry which has thixotropic properties and exerts a pressure in excess of earth + hydrostatic pressure present on sides of excavation.

# Ground Water Control—Permanent Exclusion

Precast Concrete Diaphragm Walls ~ these walls have the same applications as their insitu counterparts and have the advantages of factory produced components but lack the design flexibility of cast insitu walls. The panel or post and panel units are installed in a trench filled with a special mixture of bentonite and cement with a retarder to control the setting time. This mixture ensures that the joints between the wall components are effectively sealed. To provide stability the panels or posts are tied to the retained earth with ground anchors.

Typical Precast Concrete Diaphragm Wall Details ~

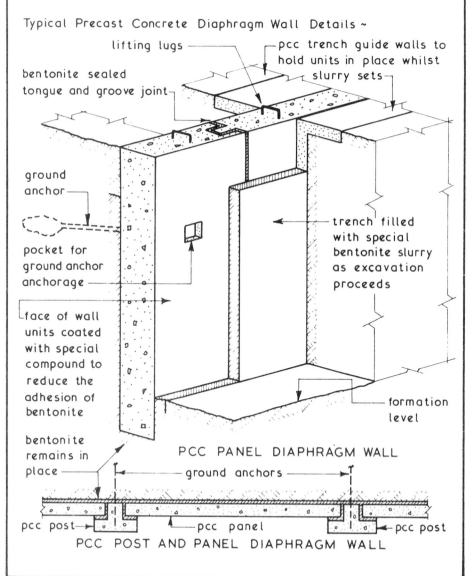

lifting lugs

pcc trench guide walls to hold units in place whilst slurry sets

bentonite sealed tongue and groove joint

ground anchor

pocket for ground anchor anchorage

trench filled with special bentonite slurry as excavation proceeds

face of wall units coated with special compound to reduce the adhesion of bentonite

formation level

bentonite remains in place

PCC PANEL DIAPHRAGM WALL

ground anchors

pcc post

pcc panel

pcc post

PCC POST AND PANEL DIAPHRAGM WALL

Grouting Methods ~ these techniques are used to form a curtain or cut off wall in high permeability soils where pumping methods could be uneconomic. The curtain walls formed by grouting methods are non-structural therefore adequate earth support will be required and in some cases this will be a distance of at least 4·000 from the face of the proposed excavation. Grout mixtures are injected into the soil by pumping the grout at high pressure through special injection pipes inserted in the ground. The pattern and spacing of the injection pipes will depend on the grout type and soil conditions.

Grout Types ~

1. Cement Grouts – mixture of neat cement and water; cement sand up to 1 : 4 or PFA (pulverised fuel ash): cement to a 1 : 1 ratio. Suitable for coarse grained soils and fissured and jointed rock strata.

2. Chemical Grouts – one shot (premixed) or two shot (first chemical is injected followed immediately by second chemical resulting in an immediate reaction) methods can be employed to form a permanent gel in the soil to reduce its permeability and at the same time increase the soil's strength. Suitable for medium to coarse sands and gravels.

3. Resin Grouts – these are similar in application to chemical grouts but have a low viscosity and can therefore penetrate into silty fine sands.

Typical Cement Grouting Details ~

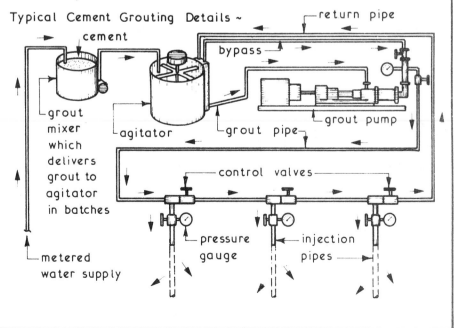

237

Ground Freezing Techniques ~ this method is suitable for all types of saturated soils and rock and for soils with a moisture content in excess of 8% of the voids. The basic principle is to insert into the ground a series of freezing tubes to form an ice wall thus creating an impermeable barrier. The treatment takes time to develop and the initial costs are high therefore it is only suitable for large contracts of reasonable duration. The freezing tubes can be installed vertically for conventional excavations and horizontally for tunnelling works. The usual circulating brines employed are magnesium chloride and calcium chloride with a temperature of -15° to -25°C which would take 10 to 17 days to form an ice wall 1·000 thick. Liquid nitrogen could be used as the freezing medium to reduce the initial freezing period if the extra cost can be justified.

Typical Ground Freezing Details ~

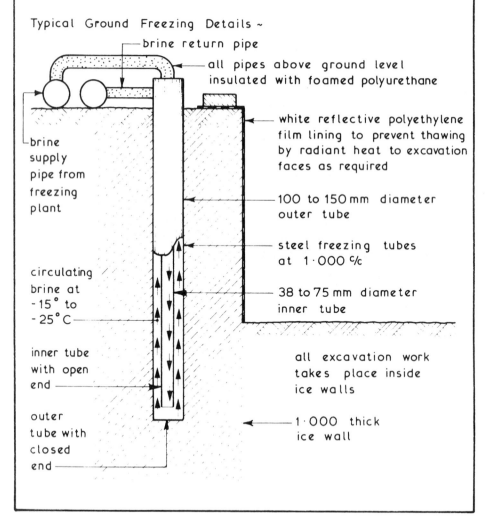

brine return pipe

all pipes above ground level insulated with foamed polyurethane

brine supply pipe from freezing plant

white reflective polyethylene film lining to prevent thawing by radiant heat to excavation faces as required

100 to 150 mm diameter outer tube

steel freezing tubes at 1·000 %

circulating brine at -15° to -25°C

38 to 75 mm diameter inner tube

inner tube with open end

all excavation work takes place inside ice walls

outer tube with closed end

1·000 thick ice wall

Soil Investigation ~ before a decision is made as to the type of foundation which should be used on any particular site a soil investigation should be carried out to establish existing ground conditions and soil properties. The methods which can be employed together with other sources of information such as local knowledge, ordnance survey and geological maps, mining records and aerial photography should be familiar to students at this level. If such an investigation reveals a naturally poor subsoil or extensive filling the designer has several options :-

1. Not to Build - unless a new and suitable site can be found building is only possible if the poor ground is localised and the proposed foundations can be designed around these areas with the remainder of the structure bridging over these positions.

2. Remove and Replace - the poor ground can be excavated, removed and replaced by compacted fills. Using this method there is a risk of differential settlement and generally for depths over 4·000 it is uneconomic.

3. Surcharging - this involves preloading the poor ground with a surcharge of aggregate or similar material to speed up settlement and thereby improve the soil's bearing capacity. Generally this method is uneconomic due to the time delay before actual building operations can commence which can vary from a few weeks to two or more years.

4. Vibration - this is a method of strengthening ground by vibrating a granular soil into compacted stone columns either by using the natural coarse granular soil or by **replacement - see page 240**

5. Dynamic Compaction - this is a method of soil improvement which consists of dropping a heavy weight through a considerable vertical distance to compact the soil and thus improve its bearing capacity and is especially suitable for **granular soils - see page 241**

6. Jet Grouting - this method of consolidating ground can be used in all types of subsoil and consists of lowering a monitor probe into a 150 mm diameter prebored guide hole. The probe has two jets the upper of which blasts water, concentrated by compressed air to force any loose material up the guide to ground level. The lower jet fills the void **with a cement slurry which sets into a solid mass - see page 242**

Ground Vibration ~ the objective of this method is to strengthen the existing soil by rearranging and compacting coarse granular particles to form stone columns with the ground. This is carried out by means of a large poker vibrator which has an effective compacting radius of 1·500 to 2·700. On large sites the vibrator is inserted on a regular triangulated grid pattern with centres ranging from 1·500 to 3·000. In coarse grained soils extra coarse aggregate is tipped into the insertion positions to make up levels as required whereas in clay and other fine particle soils the vibrator is surged up and down enabling the water jetting action to remove the surrounding soft material thus forming a borehole which is backfilled with a coarse granular material compacted insitu by the vibrator. The backfill material is usually of 20 to 70 mm size of uniform grading within the chosen range. Ground vibration is not a piling system but a means of strengthening ground to increase the bearing capacity within a range of 200 to 500 kN/m$^2$.

Typical Details ~

lifting pulley

manifold section

follower section (s)

top jets to assist in removal

vibration isolator

vibrator section containing hydraulic motor with rotating eccentric

compacted stone column

Typical Vibrator Data :-
length - 5·000
weight - 2 tonnes
vibration - 30 to 60 hz.

standard track mounted crane

water and hydraulic hoses

side fins to prevent vibrator twisting

nose cone housing lower jetting nozzle

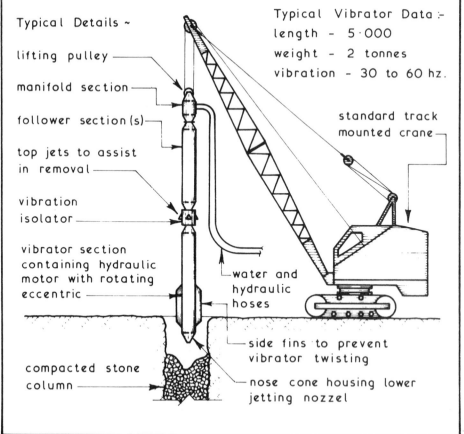

Dynamic Compaction ~ this method of ground improvement consists of dropping a heavy weight from a considerable height and is particularly effective in granular soils. Where water is present in the subsoil trenches should be excavated to allow the water to escape and not collect in the craters formed by the dropped weight. The drop pattern, size of weight and height of drop are selected to suit each individual site but generally 3 or 4 drops are made in each position forming a crater up to 2·500 deep and 5·000 in diameter. Vibration through the subsoil can be a problem with dynamic compaction operations therefore the proximity and condition of nearby buildings must be considered together with the depth, position and condition of existing services on site.

Typical Details ~

NB. Final ground level after compaction treatment and final levelling could be up to 1·500 lower than original ground level

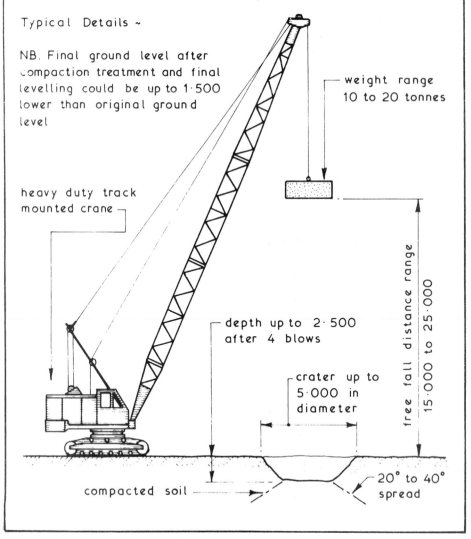

heavy duty track mounted crane

weight range 10 to 20 tonnes

free fall distance range 15·000 to 25·000

depth up to 2·500 after 4 blows

crater up to 5·000 in diameter

compacted soil

20° to 40° spread

Jet Grouting ~ this is a means of consolidating ground by lowering into preformed bore holes a monitor probe. The probe is rotated and the sides of the bore hole are subjected to a jet of pressurized water and air from a single outlet which enlarges and compacts the bore hole sides. At the same time a cement grout is being introduced under pressure to fill the void being created. The water used by the probe and any combined earth is forced up to the surface in the form of a sludge. If the monitor probe is not rotated grouted panels can be formed. The spacing, depth and layout of the bore holes is subject to specialist design.

Typical Details ~

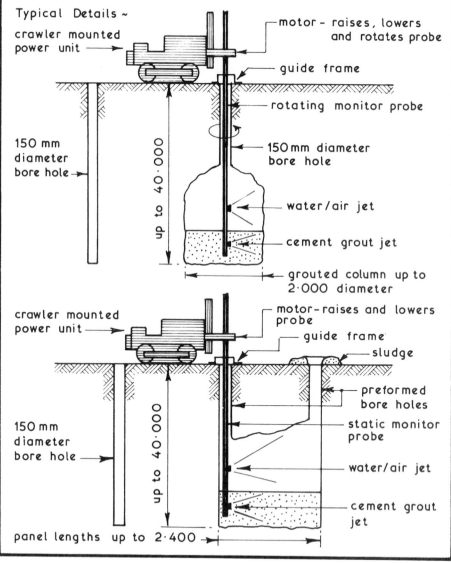

# 5 SUPERSTRUCTURE

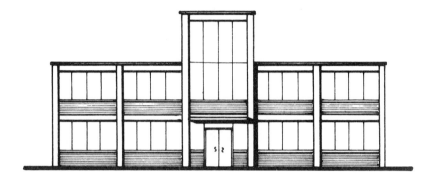

BRICK AND BLOCK WALLS

ARCHES AND OPENINGS

WINDOWS, GLASS AND GLAZING

DOMESTIC AND INDUSTRIAL DOORS

REINFORCED CONCRETE FRAMED STRUCTURES

FORMWORK

PRECAST CONCRETE FRAMES

STRUCTURAL STEELWORK

TIMBER PITCHED AND FLAT ROOFS

LONG SPAN ROOFS

SHELL ROOF CONSTRUCTION

PANEL WALLS AND CURTAIN WALLING

CONCRETE CLADDINGS

PRESTRESSED CONCRETE

SOUND INSULATION

THERMAL INSULATION

STAGE 1

Consideration to be given to the following :~
1. Building type and usage.
2. Building owner's requirements and preferences.
3. Local planning restrictions.
4. Legal restrictions and requirements.
5. Site restrictions.
6. Capital resources.
7. Future policy in terms of maintenance and adaptation.

STAGE 2

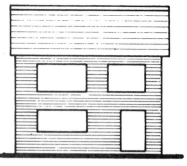

Decide on positions, sizes and shapes of openings.

STAGE 3

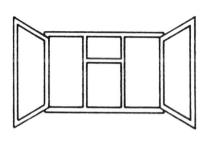

Decide on style, character and materials for openings

STAGE 4

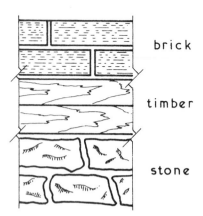

brick

timber

stone

Decide on basic materials for fabric of roof and walls

STAGE 5

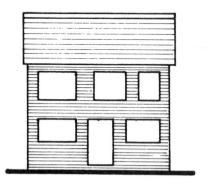

Review all decisions and make changes if required

Bricks ~ these are walling units within a length of 337·5mm, a width of 225mm and a height of 112·5mm. The usual size of bricks in common use is length 215mm, width 112·5mm and height 65mm and like blocks they must be laid in a definite pattern or bond if they are to form a structural wall. Bricks are usually made from clay (BS 3921) or from sand and lime (BS 187) and are available in a wide variety of strengths, types, textures, colours and special shaped bricks to BS 4729.

Typical Details ~

external bricks usually selected for face appearance

damp-proof course

150 min.

65

10

thickness in half brick modules

solid brick wall

ground floor

bricks below ground level must be of a suitable quality

mass concrete foundation

SECTION

corner brick or quoin
quarter brick or queen closer
row of bricks = course

bricks laid to a bond so that no vertical joints in consecutive courses are above one another – bond shown is English Bond.

cut bricks or bats

cut bricks at mid-panel to make up length in zipper bond

cut bricks or bats

cut bricks at mid-panel to make up length in quarter bond

ELEVATIONS

Bonding ~ an arrangement of bricks in a wall, column or pier laid to a set pattern to maintain an adequate lap.

Purposes of Brick Bonding ~

1. Obtain maximum strength whilst distributing the loads to be carried throughout the wall, column or pier.
2. Ensure lateral stability and resistance to side thrusts.
3. Create an acceptable appearance.

Lap Forms ~

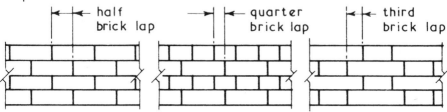

half brick lap     quarter brick lap     third brick lap

HALF BONDING
used in half brick
thick walls built in
stretcher bond

QUARTER BONDING
used in most bonds
built with standard
bricks

THIRD BONDING
used in bonds built
with metric bricks

Simple Bonding Rules ~

1. Bond is set out along length of wall working from each end to ensure that no vertical joints are above one another in consecutive courses.

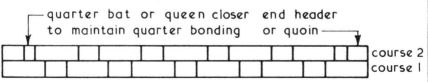

quarter bat or queen closer to maintain quarter bonding     end header or quoin

course 2
course 1

NB all odd numbered courses set out as course 1 and all even numbered courses set out as course 2

2. Walls which are not in exact bond length can be set out thus –

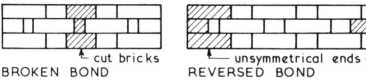

cut bricks
BROKEN BOND

unsymmetrical ends
REVERSED BOND

3. Transverse or cross joints continue unbroken across the width of wall unless stopped by a face stretcher.

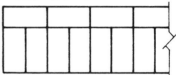

English Bond ~ formed by laying alternate courses of stretchers and headers it is one of the strongest bonds but it will require more facing bricks than other bonds ( 89 facing bricks per m² )

Typical Example ~

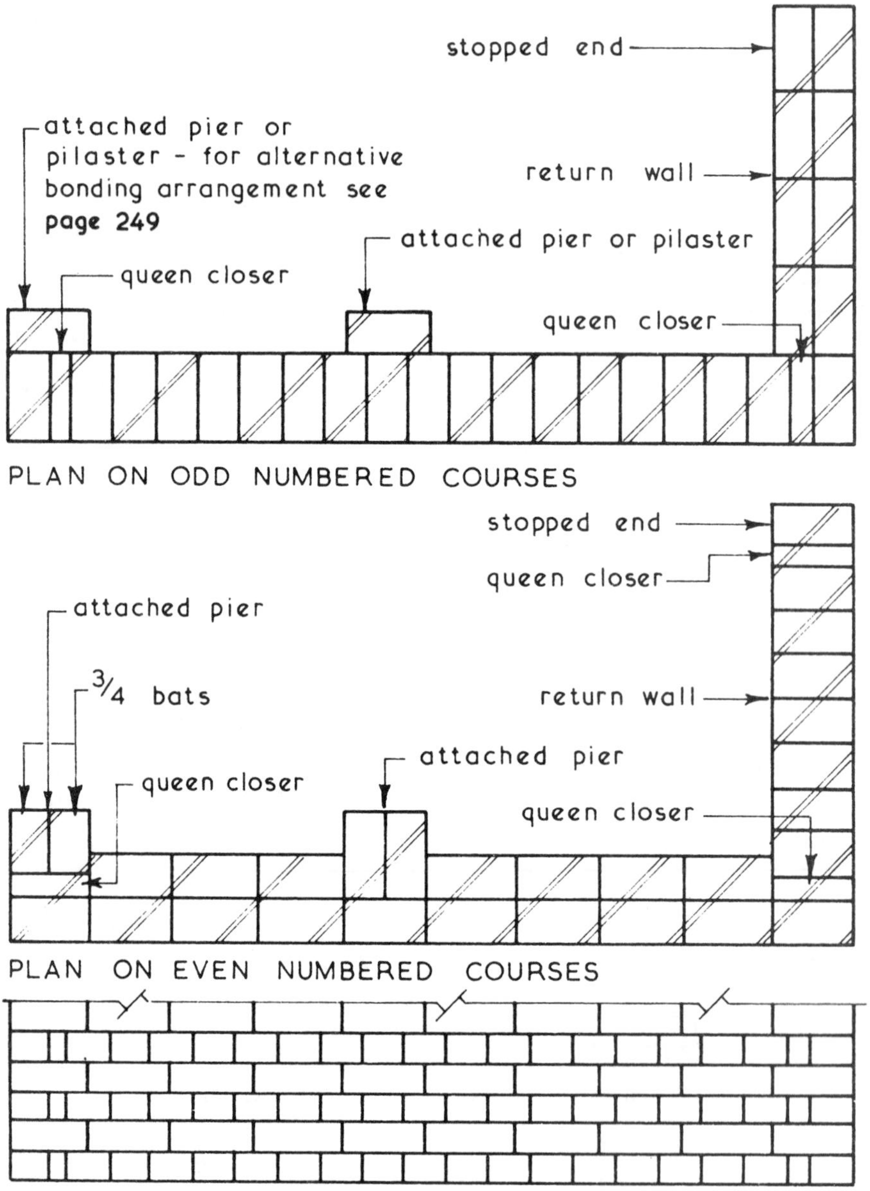

stopped end

attached pier or pilaster - for alternative bonding arrangement see **page 249**

return wall

attached pier or pilaster

queen closer

queen closer

PLAN ON ODD NUMBERED COURSES

stopped end

queen closer

attached pier

3/4 bats

return wall

attached pier

queen closer

queen closer

PLAN ON EVEN NUMBERED COURSES

ELEVATION

Flemish Bond ~ formed by laying headers and stretchers alternately in each course. Not as strong as English bond but is considered to be aesthetically superior uses less facing bricks. (79 facing brick per m²)

Typical Example ~

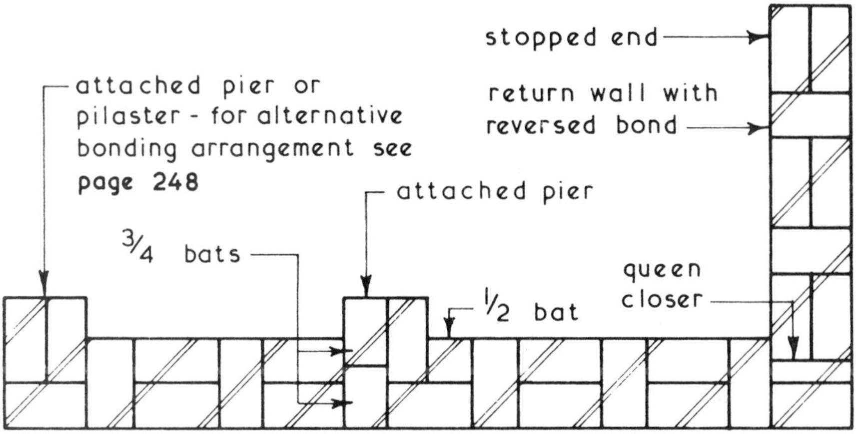

attached pier or pilaster - for alternative bonding arrangement see **page 248**

attached pier

stopped end

return wall with reversed bond

³⁄₄ bats

½ bat

queen closer

PLAN ON ODD NUMBERED COURSES

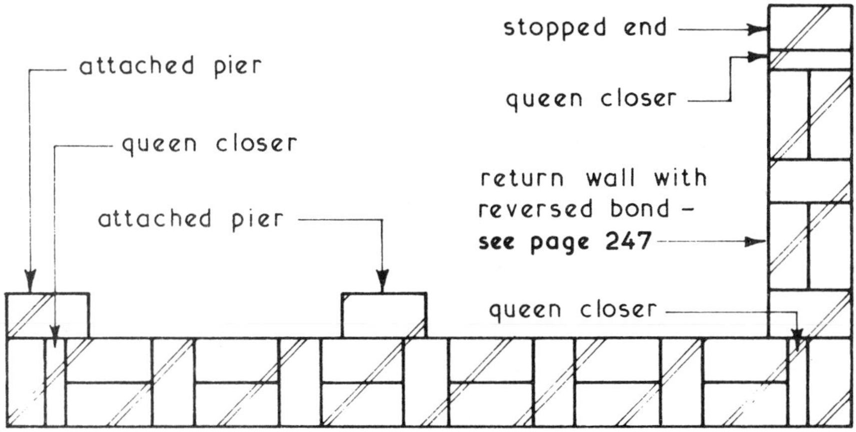

attached pier

queen closer

attached pier

stopped end

queen closer

return wall with reversed bond – **see page 247**

queen closer

PLAN ON EVEN NUMBERED COURSES

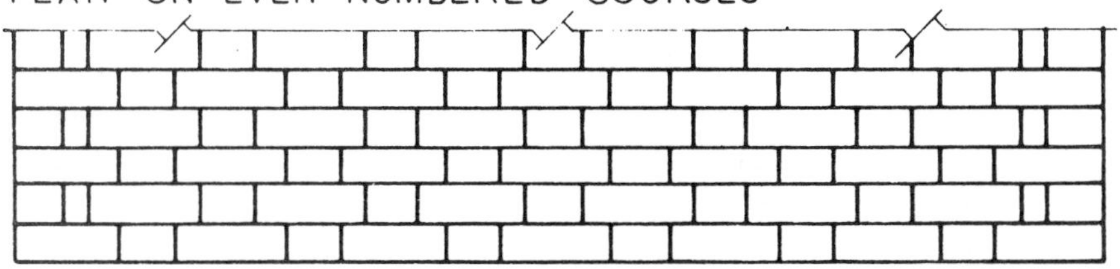

ELEVATION

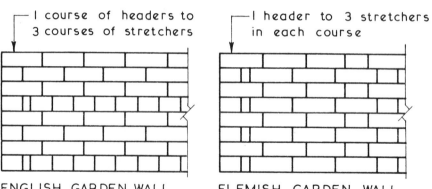

**ENGLISH GARDEN WALL BOND** - gives quick lateral spread of load - uses less facings than English bond.

**FLEMISH GARDEN WALL BOND** - enables a fair face to be kept on both sides of a one brick thick wall.

**ENGLISH CROSS BOND** - header placed next to end stretcher in every other stretcher course which thus staggers stretchers enabling patterns or diapers to be picked out in different texture or coloured bricks.

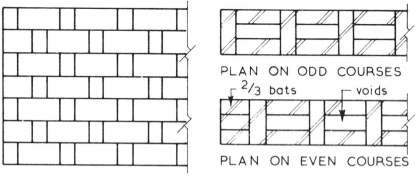

PLAN ON ODD COURSES

PLAN ON EVEN COURSES

**RAT TRAP BOND** - uses brick on edge courses - hollow pockets or voids reduce total weight of wall and by the bricks on edge there is an overall saving of materials.

Attached Piers ~ the main function of an attached pier is to give lateral support to the wall of which it forms part from the base to the top of the wall. It also has the subsidiary function of dividing a wall into distinct lengths whereby each length can be considered as a wall. Generally walls must be tied at end to an attached pier, buttressing or return wall.

Typical Examples~

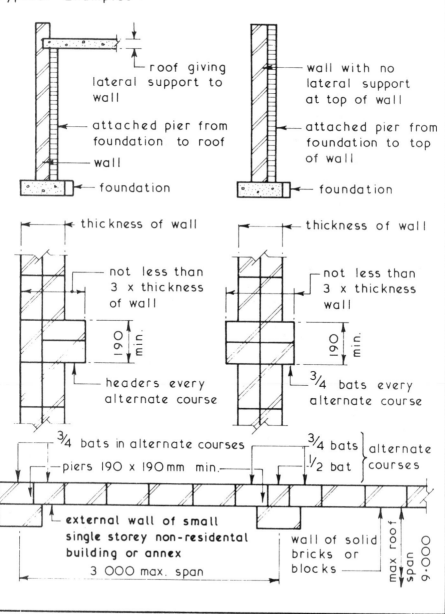

roof giving lateral support to wall

attached pier from foundation to roof

wall

foundation

wall with no lateral support at top of wall

attached pier from foundation to top of wall

foundation

thickness of wall

not less than 3 x thickness of wall

190 min.

headers every alternate course

thickness of wall

not less than 3 x thickness wall

190 min.

³/₄ bats every alternate course

³/₄ bats in alternate courses

piers 190 x 190mm min.

external wall of small single storey non-residental building or annex
3 000 max. span

³/₄ bats ⎫ alternate
½ bat ⎬ courses

wall of solid bricks or blocks

max roof span 9·000

## Solid Block Walls

Blocks ~ these are walling units exceeding in length, width or height the dimensions specified for bricks in BS 3921. Precast concrete blocks should comply with the recommendations set out in BS 6073. Blocks suitable for external solid walls are classified as loadbearing and are required to have a minimum average crushing strength of $2.8 \, N/mm^2$.

Typical Details ~

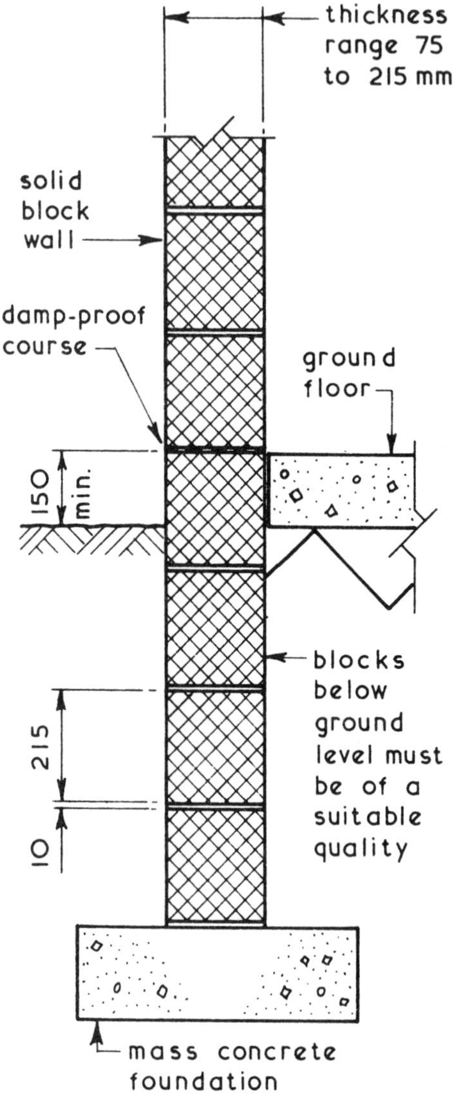

thickness range 75 to 215 mm

solid block wall ──►

damp-proof course ──

ground floor ──

150 min.

215

10

blocks below ground level must be of a suitable quality

mass concrete foundation
SECTION

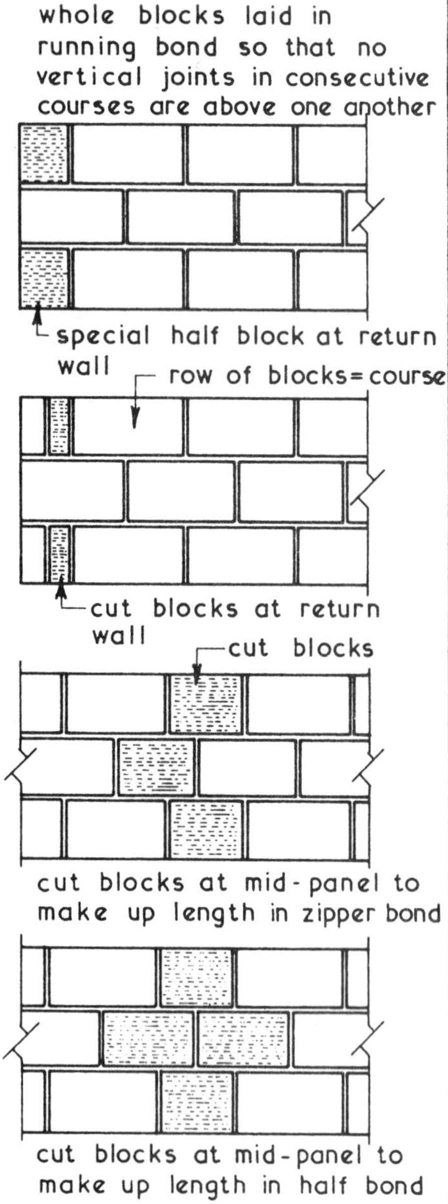

whole blocks laid in running bond so that no vertical joints in consecutive courses are above one another

── special half block at return wall

── row of blocks = course

── cut blocks at return wall

── cut blocks

cut blocks at mid-panel to make up length in zipper bond

cut blocks at mid-panel to make up length in half bond
ELEVATIONS

Cavity Walls ~ these consist of an outer brick or block leaf or skin separated from an inner brick or block leaf or skin by an air space called a cavity. These walls have better thermal insulation and weather resistance properties than a comparable solid brick or block wall and therefore are in general use for the enclosing walls of domestic buildings. The two leaves of a cavity wall are tied together with wall ties at not less than the spacings given in Table C of Approved Document A.

The width of the cavity should be between 50 and 75mm unless vertical twist type ties are used at not more than the centres given in Table C when the cavity width can be between 50 and 100 mm. Cavities should not be ventilated and should be closed with bricks or blocks above eaves level.

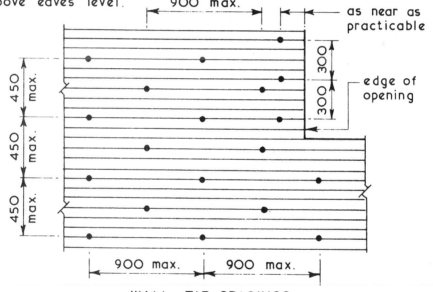

WALL TIE SPACINGS

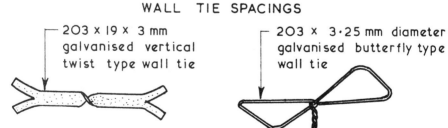

203 x 19 x 3 mm galvanised vertical twist type wall tie

203 x 3·25 mm diameter galvanised butterfly type wall tie

TYPICAL METAL WALL TIES

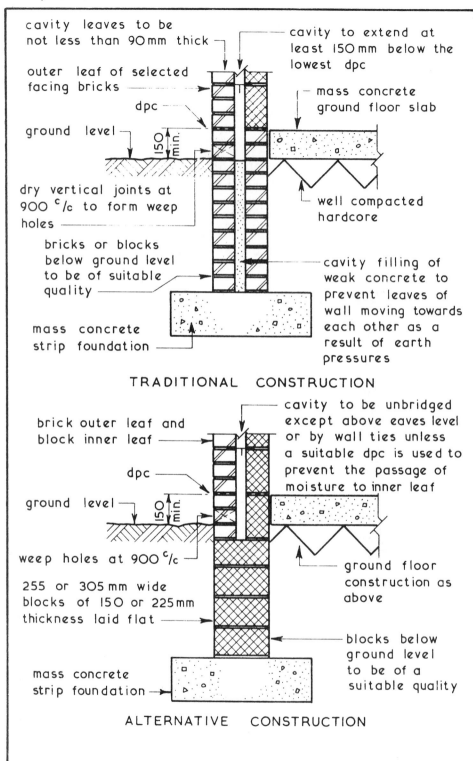

cavity leaves to be
not less than 90 mm thick

cavity to extend at
least 150 mm below the
lowest dpc

outer leaf of selected
facing bricks

mass concrete
ground floor slab

dpc

ground level

150 min.

dry vertical joints at
900 c/c to form weep
holes

well compacted
hardcore

bricks or blocks
below ground level
to be of suitable
quality

cavity filling of
weak concrete to
prevent leaves of
wall moving towards
each other as a
result of earth
pressures

mass concrete
strip foundation

TRADITIONAL CONSTRUCTION

brick outer leaf and
block inner leaf

cavity to be unbridged
except above eaves level
or by wall ties unless
a suitable dpc is used to
prevent the passage of
moisture to inner leaf

dpc

ground level

150 min.

weep holes at 900 c/c

ground floor
construction as
above

255 or 305 mm wide
blocks of 150 or 225 mm
thickness laid flat

blocks below
ground level
to be of a
suitable quality

mass concrete
strip foundation

ALTERNATIVE CONSTRUCTION

Parapet ~ a low wall projecting above the level of a roof, bridge or balcony forming a guard or barrier at the edge. Parapets are exposed to the elements on three faces namely front, rear and top and will therefore need careful design and construction if they are to be durable and reliable.

Typical Details ~

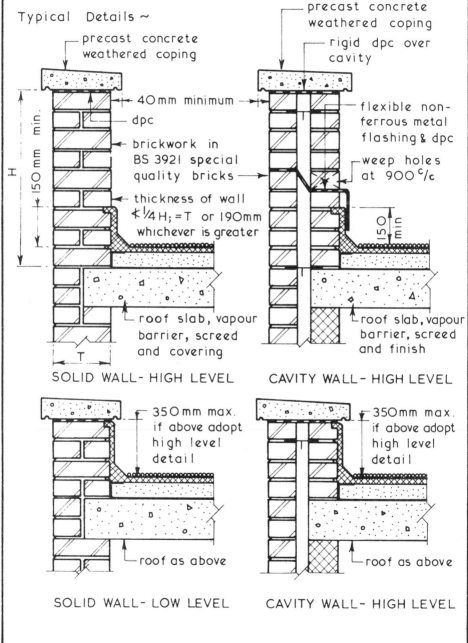

SOLID WALL- HIGH LEVEL

CAVITY WALL- HIGH LEVEL

SOLID WALL- LOW LEVEL

CAVITY WALL- HIGH LEVEL

Function ~ the primary function of any damp-proof course (dpc) or damp-proof membrane (dpm) is to provide an impermeable barrier to the passage of moisture. The three basic ways in which damp-proof courses are used is to ~

1. Resist moisture penetration from below (rising damp.)
2. Resist moisture penetration from above.
3. Resist moisture penetration from horizontal entry.

Typical Examples ~

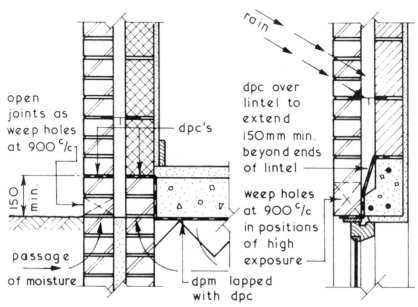

open joints as weep holes at 900 ᶜ/c

dpc's

150 min.

passage of moisture

dpm lapped with dpc

rain

dpc over lintel to extend 150mm min. beyond ends of lintel

weep holes at 900 ᶜ/c in positions of high exposure

**PENETRATION FROM BELOW**
(Ground Floor/External Wall)

**PENETRATION FROM ABOVE**
(Window/Door Head)

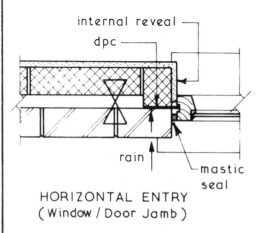

internal reveal

dpc

rain

mastic seal

**HORIZONTAL ENTRY**
(Window/Door Jamb)

SUITABLE DPC MATERIALS

Engineering bricks - BS 3921
**Slates - BS 680**
Lead - BS 1178
Copper - BS 2870
Bitumen based products
Propriety emulsions
Polythene
Pitch polymers
**Mastic asphalt - BS 988 & 6577**
BS 743 - Materials for dpc's

Calculated Brickwork ~ for small and residential buildings up to three storeys high the sizing of load bearing brick walls can be taken from data given in Part C of Approved Document A. The alternative methods for these and other load bearing brick walls is given in CP 111 – Structural Recommendations for Loadbearing Walls.

The main factors governing the loadbearing capacity of brick walls and columns are :-

1. Thickness of wall.

2. Strength of bricks used.

3. Type of mortar used.

4. Slenderness ratio of wall or column.

5. Eccentricity of applied load.

Thickness of Wall ~ this must always be sufficient throughout its entire body to carry the design loads and induced stresses. Other design requirements such as thermal and sound insulation properties must also be taken into account when determining the actual wall thickness to be used.

Effective Thickness ~ this is the assumed thickness of the wall or column used for the purpose of calculating its slenderness ratio. (see page 259)

Typical Examples ~

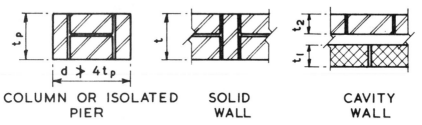

COLUMN OR ISOLATED PIER

SOLID WALL

CAVITY WALL

effective thickness = $t_p$    eff. th. = $t$    eff. th. = $2/3(t_1 + t_2)$

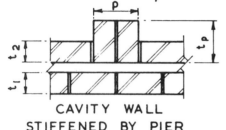

CAVITY WALL STIFFENED BY PIER

effective thickness
$= 2/3(t_1 + $ effective thickness $t_2)$

effective thickness of $t_2$ can be ascertained by multiplying $t_2$ by a stiffening coefficient obtained from Table 1 CP 111

Strength of Bricks ~ due to the wide variation of the raw materials and methods of manufacture bricks can vary greatly in their compressive strength. The compressive strength of a particular type of brick or batch of bricks is taken as the arithmetic mean of a sample of ten bricks tested in accordance with the appropriate British Standard. A typical range for clay bricks would be from 170 to 20 MN/m$^2$ the majority of which would be in the 90 to 20 MN/m$^2$ band. Generally calcium silicate bricks have a lower compressive strength than clay bricks with a typical strength range of 65 to 10 MN/m$^2$.

Strength of Mortars ~ mortars consist of an aggregate (sand) and a binder which is usually cement ; cement plus additives to improve workability or cement and lime. The factors controlling the strength of any particular mix are the ratio of binder to aggregate plus the water : cement ratio. The strength of any particular mix can be ascertained by taking the arithmetic mean of a series of test cubes.

Wall Design Strength ~ the basic stress of any brickwork depends on the crushing strength of the bricks and the type of mortar used to form the wall unit. This relationship can be plotted on a graph using the data given in Table 3a of CP 111 as shown below :-

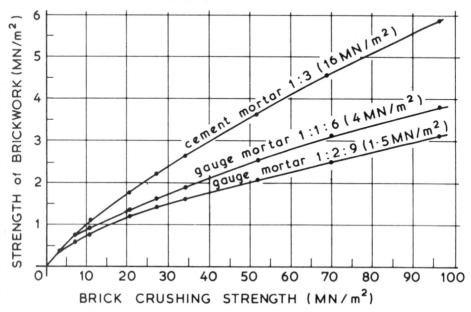

Slenderness Ratio ~ this is the relationship of the effective height to the effective thickness thus :-

$$\text{slenderness ratio} = \frac{\text{effective height}}{\text{effective thickness}} = \frac{h}{t}$$

Effective Height ~ this is the dimension taken to calculate the slenderness ratio as opposed to the actual height.

Typical Examples - actual height = H    effective height = h

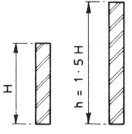 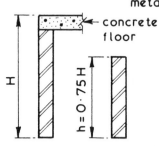 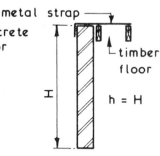

| NO LATERAL SUPPORT AT TOP | CONCRETE FLOOR BEARING ON WALL | FLOOR EFFECTIVELY TIED TO WALL |

Effective Thickness ~ this is the dimension taken to calculate the slenderness ratio as opposed to the actual thickness.

Typical Examples - actual thickness = T    effective thickness = t

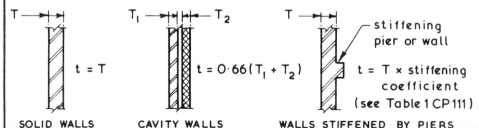

SOLID WALLS    CAVITY WALLS    WALLS STIFFENED BY PIERS

$t = T$     $t = 0.66(T_1 + T_2)$     $t = T \times$ stiffening coefficient (see Table 1 CP 111)

Stress Reduction ~ the permissible stress for a wall is based on the basic stress multiplied by a reduction factor related to the slenderness factor and the eccentricity of the load :-

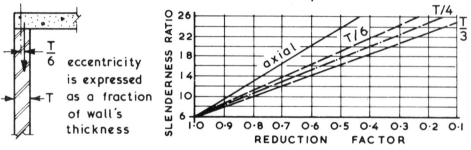

eccentricity is expressed as a fraction of wall's thickness

Supports Over Openings ~ the primary function of any support over an opening is to carry the loads above the opening and transmit them safely to the abutments, jambs or piers on both sides. A support over an opening is usually required since the opening infilling such as a door or window frame will not have sufficient strength to carry the load through its own members.

Types of Support ~

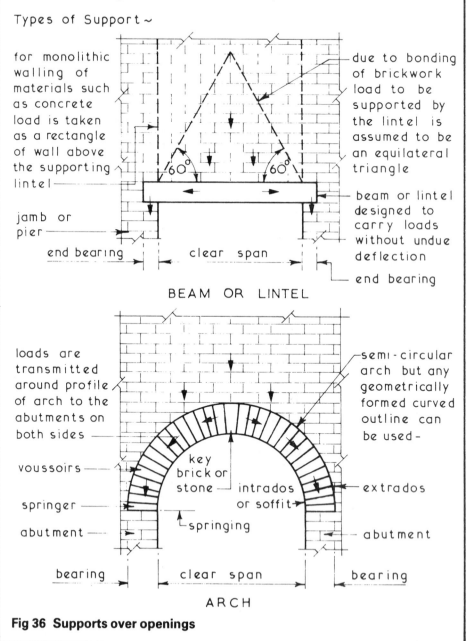

for monolithic walling of materials such as concrete load is taken as a rectangle of wall above the supporting lintel

due to bonding of brickwork load to be supported by the lintel is assumed to be an equilateral triangle

60° 60°

beam or lintel designed to carry loads without undue deflection

jamb or pier

end bearing

clear span

end bearing

**BEAM OR LINTEL**

loads are transmitted around profile of arch to the abutments on both sides

semi-circular arch but any geometrically formed curved outline can be used –

voussoirs

key brick or stone — intrados or soffit→

extrados

springer

abutment

springing

abutment

bearing

clear span

bearing

**ARCH**

**Fig 36 Supports over openings**

Arch Construction ~ by the arrangement of the bricks or stones in an arch over an opening it will be self supporting once the jointing material has set and gained adequate strength. The arch must therefore be constructed over a temporary support until the arch becomes self supporting. The traditional method is to use a framed timber support called a centre. Permanent arch centres are also available for small spans and simple formats.

Typical Arch Formats ~

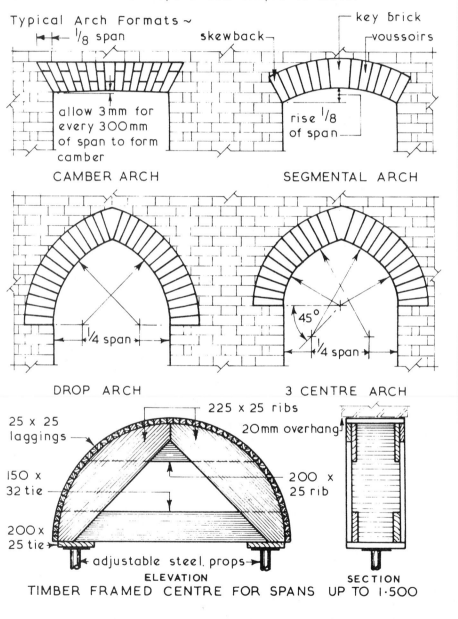

CAMBER ARCH

SEGMENTAL ARCH

DROP ARCH

3 CENTRE ARCH

TIMBER FRAMED CENTRE FOR SPANS UP TO 1·500

Openings ~ these consist of a head, jambs and sill and the different methods and treatments which can be used in their formation is very wide but they are all based on the same concepts.

Typical Head Details ~

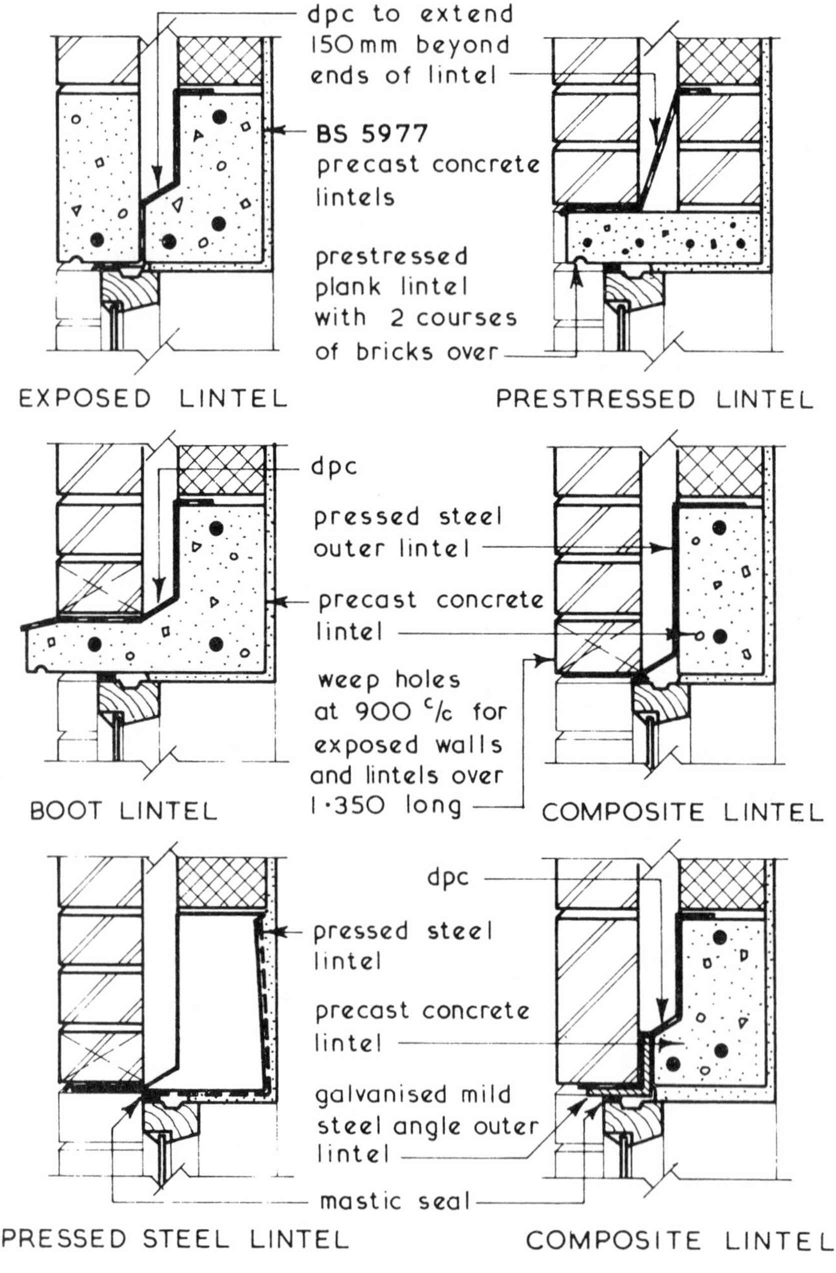

dpc to extend 150 mm beyond ends of lintel

BS 5977 precast concrete lintels

prestressed plank lintel with 2 courses of bricks over

EXPOSED LINTEL

PRESTRESSED LINTEL

dpc

pressed steel outer lintel

precast concrete lintel

weep holes at 900 ᶜ/c for exposed walls and lintels over 1·350 long

BOOT LINTEL

COMPOSITE LINTEL

dpc

pressed steel lintel

precast concrete lintel

galvanised mild steel angle outer lintel

mastic seal

PRESSED STEEL LINTEL

COMPOSITE LINTEL

Jambs ~ these may be bonded as in solid walls or unbonded as in cavity walls. The latter must have some means of preventing the ingress of moisture from the outer leaf to the inner leaf and hence the interior of the building.

Typical Jamb Details ~

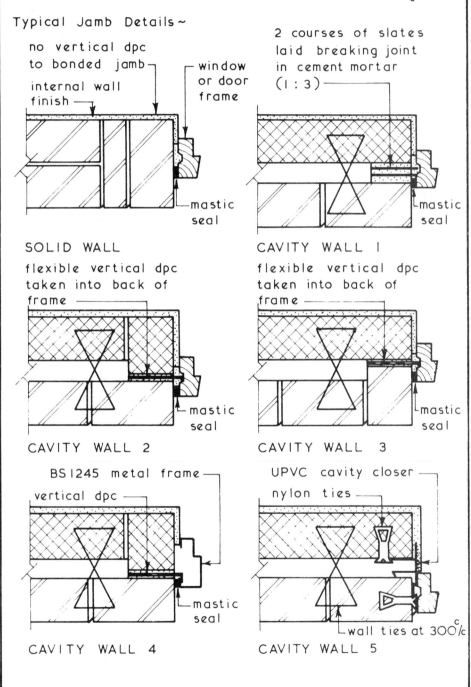

no vertical dpc to bonded jamb

internal wall finish

window or door frame

mastic seal

**SOLID WALL**

2 courses of slates laid breaking joint in cement mortar (1 : 3)

mastic seal

**CAVITY WALL 1**

flexible vertical dpc taken into back of frame

mastic seal

**CAVITY WALL 2**

flexible vertical dpc taken into back of frame

mastic seal

**CAVITY WALL 3**

BS1245 metal frame

vertical dpc

mastic seal

**CAVITY WALL 4**

UPVC cavity closer

nylon ties

wall ties at 300%c

**CAVITY WALL 5**

Sills ~ the primary function of any sill is to collect the rainwater which has run down the face of the window or door and shed it clear of the wall below.

Typical Sill Details ~

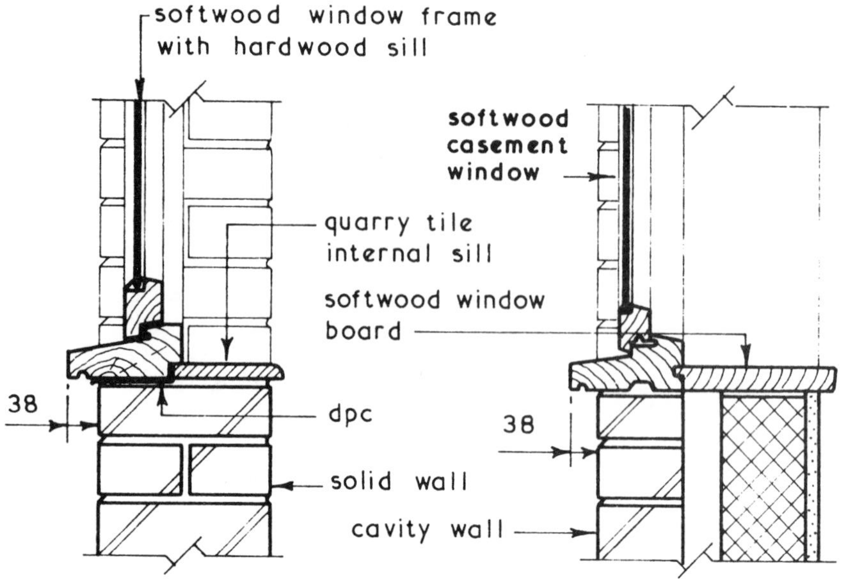

softwood window frame with hardwood sill

softwood casement window

quarry tile internal sill

softwood window board

dpc

solid wall

cavity wall

38

TIMBER SILL 1

38

TIMBER SILL 2

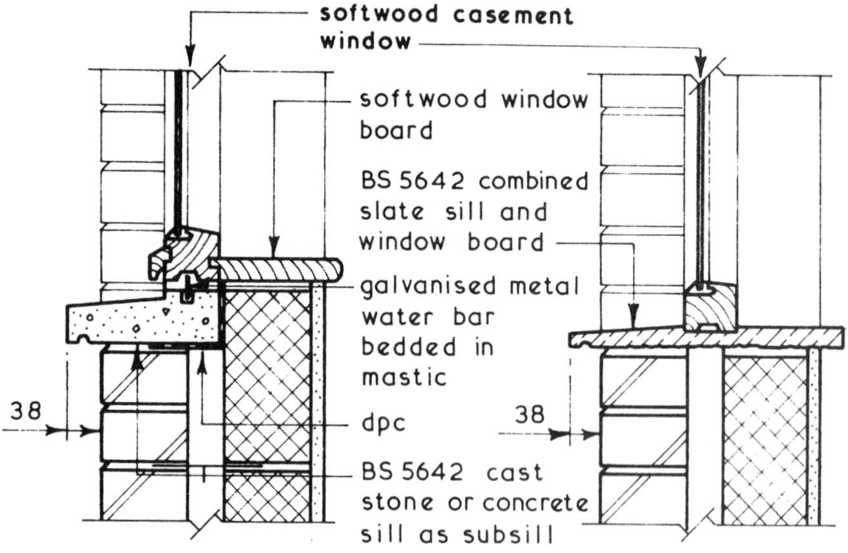

softwood casement window

softwood window board

BS 5642 combined slate sill and window board

galvanised metal water bar bedded in mastic

dpc

BS 5642 cast stone or concrete sill as subsill

38

CAST STONE SUBSILL

38

SLATE SILL

264

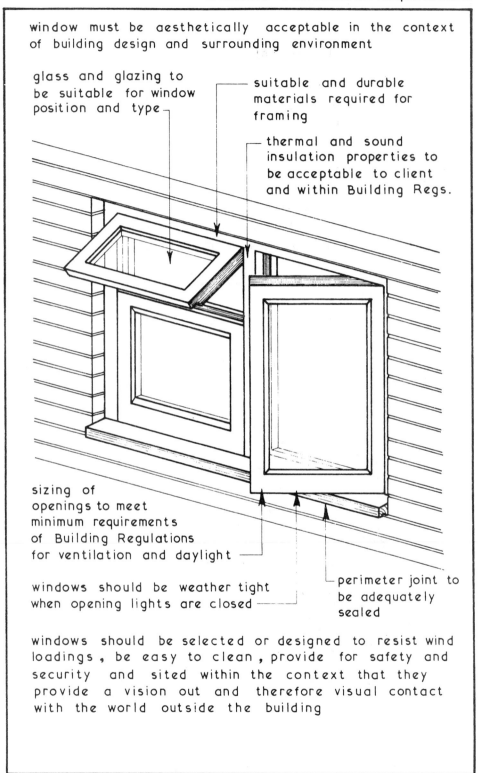

window must be aesthetically acceptable in the context of building design and surrounding environment

glass and glazing to be suitable for window position and type

suitable and durable materials required for framing

thermal and sound insulation properties to be acceptable to client and within Building Regs.

sizing of openings to meet minimum requirements of Building Regulations for ventilation and daylight

windows should be weather tight when opening lights are closed

perimeter joint to be adequately sealed

windows should be selected or designed to resist wind loadings, be easy to clean, provide for safety and security and sited within the context that they provide a vision out and therefore visual contact with the world outside the building

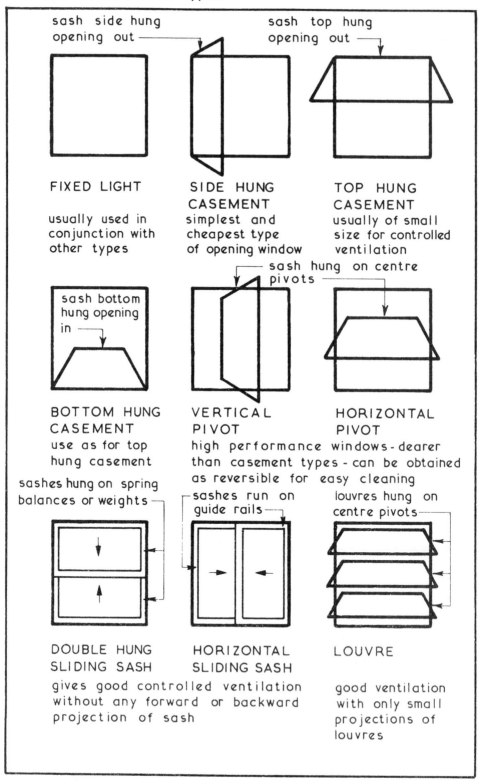

sash side hung
opening out

sash top hung
opening out

**FIXED LIGHT**

usually used in
conjunction with
other types

**SIDE HUNG
CASEMENT**

simplest and
cheapest type
of opening window

**TOP HUNG
CASEMENT**

usually of small
size for controlled
ventilation

sash bottom
hung opening
in

sash hung on centre
pivots

**BOTTOM HUNG
CASEMENT**

use as for top
hung casement

**VERTICAL
PIVOT**

**HORIZONTAL
PIVOT**

high performance windows - dearer
than casement types - can be obtained
as reversible for easy cleaning

sashes hung on spring
balances or weights

sashes run on
guide rails

louvres hung on
centre pivots

**DOUBLE HUNG
SLIDING SASH**

gives good controlled ventilation
without any forward or backward
projection of sash

**HORIZONTAL
SLIDING SASH**

**LOUVRE**

good ventilation
with only small
projections of
louvres

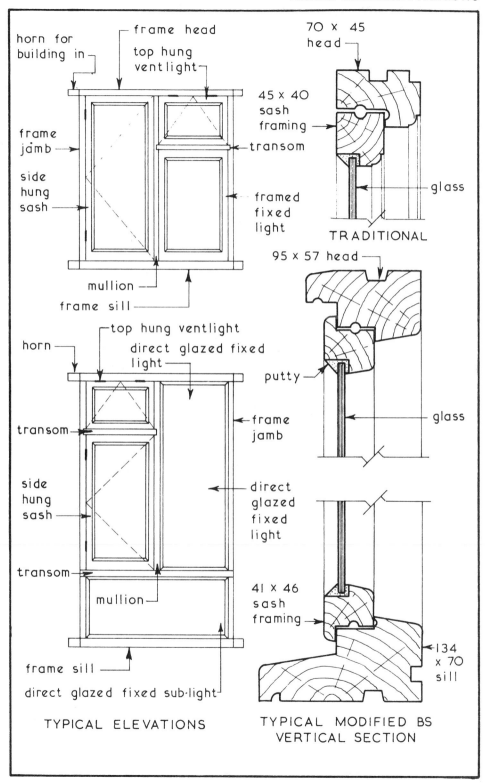

**TRADITIONAL**

**TYPICAL ELEVATIONS**

**TYPICAL MODIFIED BS VERTICAL SECTION**

Metal Windows ~ these can be obtained in steel (BS 6510) or in aluminium alloy (BS 4873). Steel windows are cheaper in initial cost than aluminium alloy but have higher maintenance costs over their anticipated life, both can be obtained fitted into timber subframes. Generally they give a larger glass area for any given opening size than similar timber windows but they can give rise to condensation on the metal components.

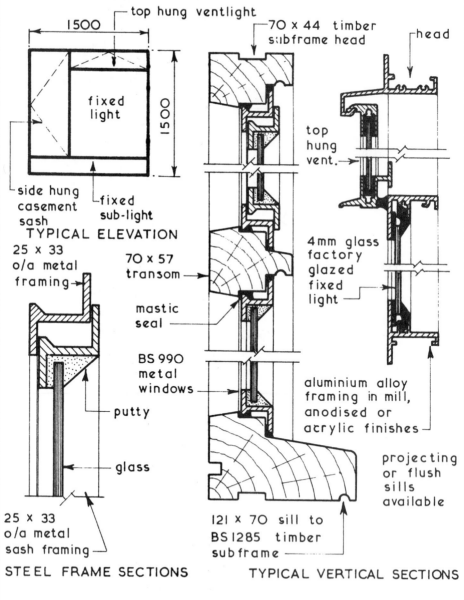

1500

top hung ventlight

fixed light

1500

side hung casement sash

fixed sub-light

**TYPICAL ELEVATION**

25 x 33 o/a metal framing

70 x 57 transom

mastic seal

BS 990 metal windows

putty

glass

25 x 33 o/a metal sash framing

**STEEL FRAME SECTIONS**

70 x 44 timber subframe head

head

top hung vent.

4mm glass factory glazed fixed light

aluminium alloy framing in mill, anodised or acrylic finishes

projecting or flush sills available

121 x 70 sill to BS 1285 timber subframe

**TYPICAL VERTICAL SECTIONS**

Timber Windows ~ wide range of ironmongery available which can be factory fitted or supplied and fixed on site.

Metal Windows ~ ironmongery usually supplied with and factory fitted to the windows.

Typical Examples ~

malleable iron, curly tail pattern

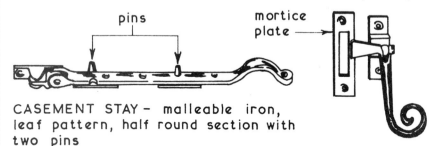

pins

mortice plate

CASEMENT STAY – malleable iron, leaf pattern, half round section with two pins
Sizes: 200; 250 and 300 mm

CASEMENT FASTENER

hot pressed aluminium, plain end pattern

CASEMENT STAY – cast aluminium, plain end pattern with one pin
Sizes: 250 and 300 mm

wedge plate

CASEMENT FASTENER

box staple

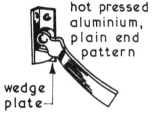

hot pressed brass

CASEMENT STAY – steel and brass, sliding screw down pattern
Sizes: 250 and 300 mm

VENTLIGHT CATCH

used with bottom hung ventlights

CASEMENT STAY – steel, stayput pattern
Arm Sizes: 100; 140 and 175 mm

malleable iron or brass
Sizes: 150  175 and  200mm

QUADRANT STAY

# Sliding Sash Windows

Sliding Sash Windows ~ these are an alternative format to the conventional side hung casement windows and can be constructed as a vertical or double hung sash window or as a horizontal sliding window in timber, metal, plastic or in any combination of these materials. The performance and design functions of providing daylight, ventilation, vision out, etc., are the same as those given for traditional windows in **Windows - Performance Requirements on page 265**

Typical Double Hung Weight Balanced Window Details ~

21mm thick pulley head

plywood parting slip suspended from pulley head

50mm long angle blocks

upper sash

92 x 21 head outside lining

70 x 14 head inside lining

48 x 41 sash framing

19 x 14 removable staff bead

lower sash

41 x 22 glazing bar

glass

**ELEVATION**

21 x 8 parting bead

sash cord

plywood back lining

49 x 24 splayed and rebated meeting rails

70 x 14 jamb inside lining

staff bead

horn

sash cord

21 x 8 parting bead

41 x 22 glazing bar

glass

48 x 41 sash framing

21mm thick pulley stile

57 x 41 bottom rail

33 x 19 draught stop

92 x 21 jamb outside lining

135 x 60 flush sill

sash weights - access through pocket piece in pulley stile

**JAMB DETAIL**

**VERTICAL SECTION**

Double Hung Sash Windows ~ these vertical sliding sash windows come in two formats when constructed in timber. The weight balanced format is shown on page 270 the alternative spring balanced type is illustrated below. Both formats are usually designed and constructed to the recommendations set out in BS 644 Part 2.

Typical Double Hung Spring Balanced Window Details ~

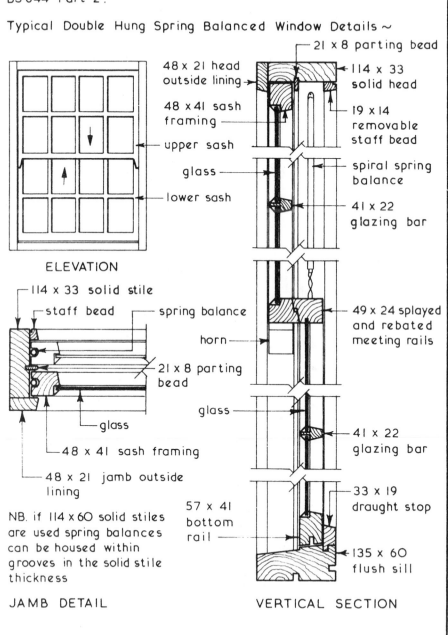

ELEVATION

JAMB DETAIL

VERTICAL SECTION

21 x 8 parting bead

48 x 21 head outside lining

114 x 33 solid head

48 x 41 sash framing

19 x 14 removable staff bead

upper sash

glass

spiral spring balance

lower sash

41 x 22 glazing bar

114 x 33 solid stile

staff bead

spring balance

49 x 24 splayed and rebated meeting rails

horn

21 x 8 parting bead

glass

glass

41 x 22 glazing bar

48 x 41 sash framing

48 x 21 jamb outside lining

33 x 19 draught stop

NB. if 114 x 60 solid stiles are used spring balances can be housed within grooves in the solid stile thickness

57 x 41 bottom rail

135 x 60 flush sill

# Sliding Sash Windows

Horizontally Sliding Sash Windows ~ these are an alternative format to the vertically sliding or double hung sash windows shown on pages 270 & 271 and can be constructed in timber, metal plastic or combinations of these materials with single or double glazing. A wide range of arrangements are available with two or more sliding sashes which can have a ventlight incorporated in the outer sliding sash.

Typical Horizontally Sliding Sash Window Details ~

timber subframe

window frame bedded in mastic

nylon slipper shoes and polypropylene pile seals

head of timber subframe

anodised aluminium framing

integral pull handle

outer sash — inner sash

## ELEVATION

factory glazed anodised aluminium horizon sliding sash

outer sash

integral pull handle

window frame bedded in mastic

inner sliding sash

weather sealed meeting stiles

outer sliding sash

timber subframe

inner sash

bushed nylon rollers

sill of timber subframe

## VERTICAL SECTION

weather seal

## HORIZONTAL SECTION

Pivot Windows ~ like other windows these are available in timber, metal, plastic or in combinations of these materials.

They can be constructed with centre jamb pivots enabling the sash to pivot or rotate in the horizontal plane or alternatively the pivots can be fixed in the head and sill of the frame so that the sash rotates in the vertical plane.

Typical Example ~

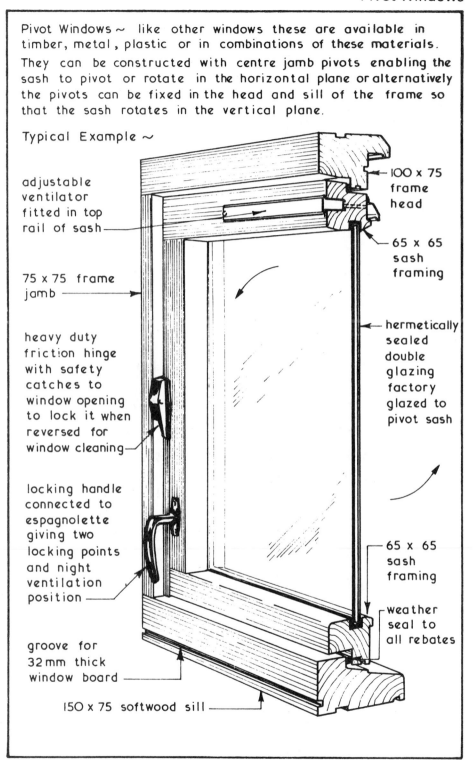

adjustable ventilator fitted in top rail of sash

100 x 75 frame head

65 x 65 sash framing

75 x 75 frame jamb

heavy duty friction hinge with safety catches to window opening to lock it when reversed for window cleaning

hermetically sealed double glazing factory glazed to pivot sash

locking handle connected to espagnolette giving two locking points and night ventilation position

65 x 65 sash framing

weather seal to all rebates

groove for 32mm thick window board

150 x 75 softwood sill

Bay Windows ~ these can be defined as any window with side lights which projects in front of the external wall and is supported by a sill height wall. Bay windows not supported by a sill height wall are called oriel windows. They can be of any window type, constructed from any of the usual window materials and are available in three plan formats namely square, splay and circular or segmental. Timber corner posts can be boxed, solid or jointed the latter being the common method.

Typical Examples ~

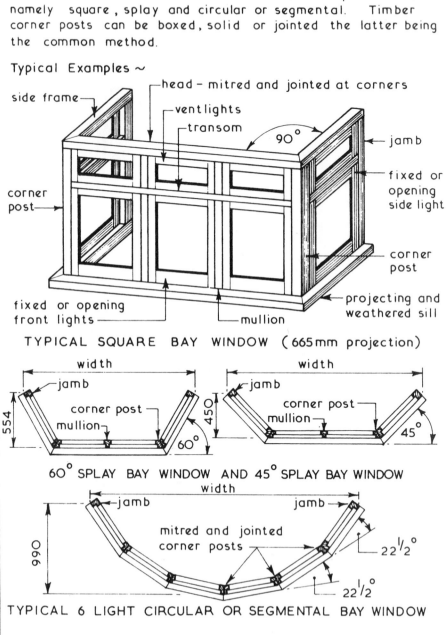

TYPICAL SQUARE BAY WINDOW (665mm projection)

60° SPLAY BAY WINDOW AND 45° SPLAY BAY WINDOW

TYPICAL 6 LIGHT CIRCULAR OR SEGMENTAL BAY WINDOW

Schedules~ the main function of a schedule is to collect together all the necessary information for a particular group of components such as windows, doors and drainage inspection chambers. There is no standard format for schedules but they should be easy to read, accurate and contain all the necessary information for their purpose. Schedules are usually presented in a tabulated format which can be related to and read in conjunction with the working drawings.

Typical Example ~

| WINDOW SCHEDULE — Sheet 1 of 1 | | | | Drawn By: RC | | Date: 14/4/81 | Rev. | |
|---|---|---|---|---|---|---|---|---|
| Contract Title & Number: Lane End Farm - H341/80 | | | | | | Drg. Nos. C(31) 450-7 | | |
| Location or Number | Type | Material | Overall Size | Glass | Ironmongery | Sill | |
| | | | | | | External | Internal |
| 1 & 2 | 9FCV4 - Subframe - | steel softwood | 910 x 1214 970 x 1275 | 146 x 1140 632 x 553 670 x 594 3mm float | supplied with casements | 2 cos. plain tiles | 150 x 150 x 15 quarry tiles |
| 3, 4, 5 & 6 | 240V - | softwood | 1206 x 1206 | 480 x 280 580 x 700 480 x 1030 3mm float | ditto | ditto | 25 mm thick softwood |
| 7 | Purpose made - Drg. No. C(31)-457 | softwood | 1770 x 1600 | 460 x 200 1080 x 300 460 x 1040 1080 x 1140 3mm float | 1 - 200 mm 1 - 300 mm al. stays 1 - al. alloy fastener | sill of frame | ditto |

Glass ~ this material is produced by fusing together soda, lime and silica with other minor ingredients such as magnesia and alumina. A number of glass types are available for domestic work and these include :-

Clear Float ~ used where clear undistorted vision is required. Available thicknesses range from 3mm to 25mm.

Clear Sheet ~ suitable for all clear glass areas but because the two faces of the glass are never perfectly flat or parallel some distortion of vision usually occurs. This type of glass is gradually being superseded by the clear float glass. Available thicknesses range from 3mm to 6mm.

Translucent Glass ~ these are patterned glasses most having one patterned surface and one relatively flat surface. The amount of obscurity and diffusion obtained depend on the type and nature of pattern. Available thicknesses range from 4mm to 6mm for patterned glasses and from 5mm to 10mm for rough cast glasses.

Wired Glass ~ obtainable as a clear polished wired glass or as a rough cast wired glass with a nominal thickness of 7mm. Generally used where a degree of fire resistance is required. Georgian wired glass has a 12mm square mesh whereas the hexagonally wired glass has a 20mm mesh.

Choice of Glass ~ the main factors to be considered are :-
1. Resistance to wind loadings.    2. Clear vision required.
3. Privacy.   4. Security.   5. Fire resistance.   6. Aesthetics.

Glazing Terminology ~

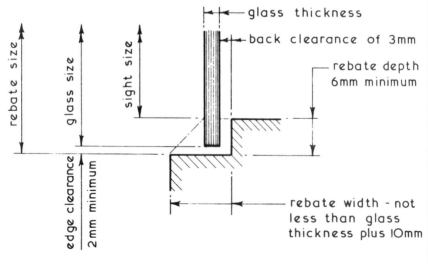

Glazing ~ the act of fixing glass into a frame or surround.
In domestic work this is usually achieved by locating the
glass in a rebate and securing it with putty or beading
and should be carried out in accordance with the
**recommendations** contained in **BS 6262**

Timber Surrounds ~ linseed oil putty to BS 544 – rebate
to be clean, dry and primed before glazing is carried out.
Putty should be protected with paint within two weeks
of application.

Metal Surrounds ~ metal casement putty if metal surround
is to be painted – if surround is not to be painted a
non-setting compound should be used.

Typical Glazing Details ~

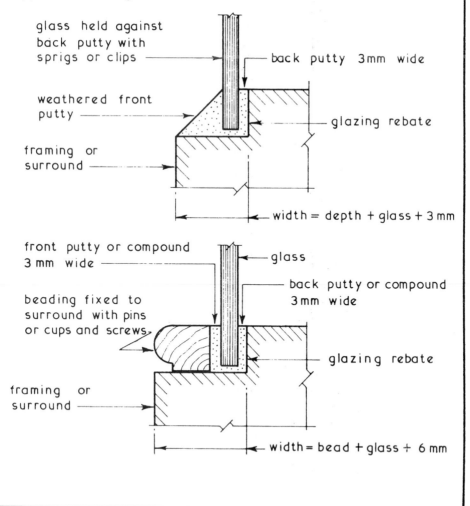

glass held against
back putty with
sprigs or clips ———— back putty 3mm wide

weathered front
putty ———— glazing rebate

framing or
surround ————

width = depth + glass + 3 mm

front putty or compound
3 mm wide ———— glass

back putty or compound
3 mm wide

beading fixed to
surround with pins
or cups and screws

glazing rebate

framing or
surround ————

width = bead + glass + 6 mm

Double Glazing ~ as its name implies this is where two layers of glass are used instead of the traditional single layer. Double glazing can be used to reduce the rate of heat loss through windows and glazed doors or it can be employed to reduce the sound transmission through windows. In the context of thermal insulation this is achieved by having a small air space within the range of 6 to 20 mm between the two layers of glass. The hermetically sealed double glazing unit with an air space of 20mm containing dehydrated air to prevent internal misting gives good results. If metal frames are used these should have a thermal break incorporated in their design. All opening sashes in a double glazing system should be fitted with adequate weather seals to reduce the rate of heat loss through the opening clearance gap.

In the context of sound insulation three factors affect the performance of double glazing. Firstly good installation to ensure airtightness, secondly the weight of glass used and thirdly the size of air space between the layers of glass. The heavier the glass used the better the sound insulation and the air space needs to be within the range of 50 to 300 mm. Absorbent lining to the reveals within the air space will also improve the sound insulation properties of the system.

Typical Examples ~

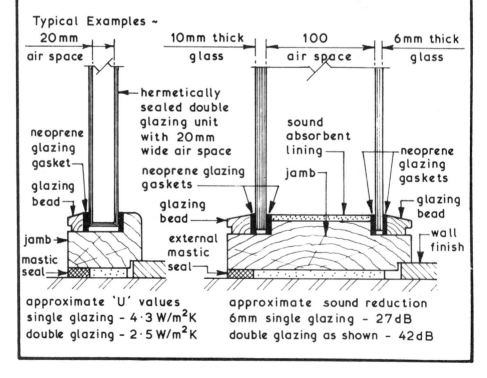

| 20mm air space | 10mm thick glass | 100 air space | 6mm thick glass |

hermetically sealed double glazing unit with 20mm wide air space

neoprene glazing gasket

glazing bead

jamb

mastic seal

neoprene glazing gaskets

glazing bead

external mastic seal

sound absorbent lining

jamb

neoprene glazing gaskets

glazing bead

wall finish

approximate 'U' values
single glazing - 4·3 W/m²K
double glazing - 2·5 W/m²K

approximate sound reduction
6mm single glazing - 27dB
double glazing as shown - 42dB

Doors ~ can be classed as external or internal. External doors are usually thicker and more robust in design than internal doors since they have more functions to fulfil.

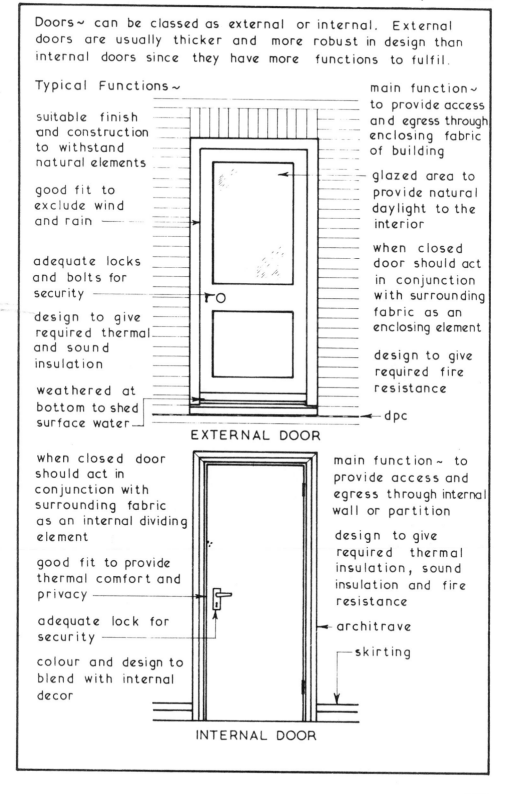

Typical Functions ~

suitable finish and construction to withstand natural elements

good fit to exclude wind and rain

adequate locks and bolts for security

design to give required thermal and sound insulation

weathered at bottom to shed surface water

main function ~ to provide access and egress through enclosing fabric of building

glazed area to provide natural daylight to the interior

when closed door should act in conjunction with surrounding fabric as an enclosing element

design to give required fire resistance

dpc

**EXTERNAL DOOR**

when closed door should act in conjunction with surrounding fabric as an internal dividing element

good fit to provide thermal comfort and privacy

adequate lock for security

colour and design to blend with internal decor

main function ~ to provide access and egress through internal wall or partition

design to give required thermal insulation, sound insulation and fire resistance

architrave

skirting

**INTERNAL DOOR**

External Doors ~ these are available in a wide variety of types and styles in timber, aluminium alloy or steel. The majority of external doors are however made from timber, the metal doors being mainly confined to fully glazed doors such as 'patio doors'.

Typical Examples of External Doors ~

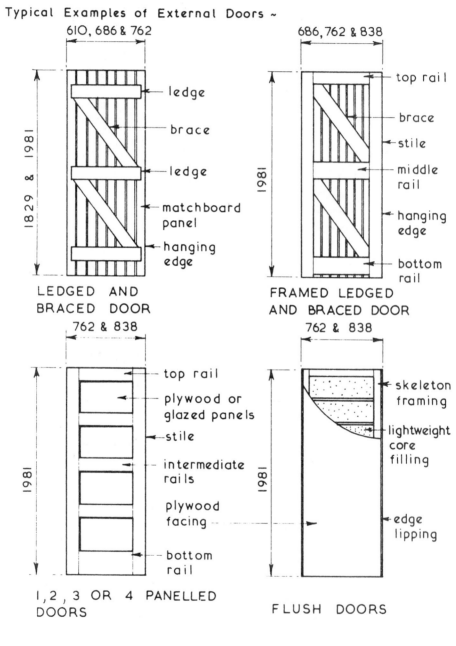

610, 686 & 762

ledge
brace
ledge
matchboard panel
hanging edge

1829 & 1981

LEDGED AND BRACED DOOR

686, 762 & 838

top rail
brace
stile
middle rail
hanging edge
bottom rail

1981

FRAMED LEDGED AND BRACED DOOR

762 & 838

top rail
plywood or glazed panels
stile
intermediate rails
plywood facing
bottom rail

1981

1, 2, 3 OR 4 PANELLED DOORS

762 & 838

skeleton framing
lightweight core filling
edge lipping

1981

FLUSH DOORS

Typical examples of purpose made and non-standard external doors~

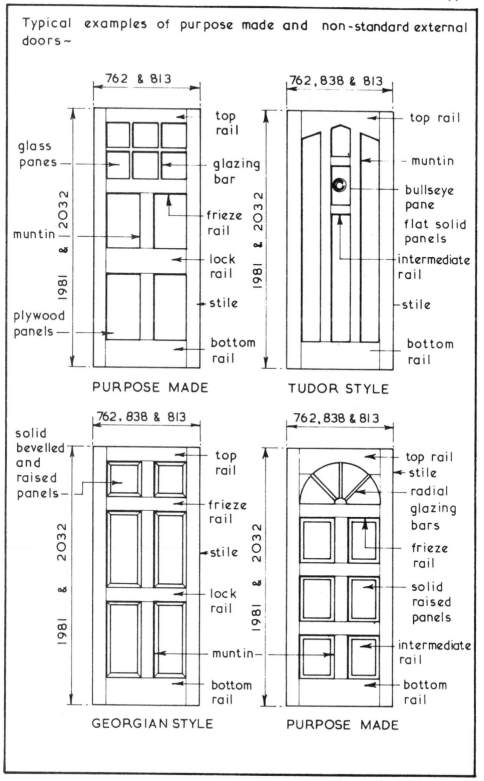

762 & 813

glass panes

1981 & 2032

muntin

plywood panels

top rail

glazing bar

frieze rail

lock rail

stile

bottom rail

**PURPOSE MADE**

762, 838 & 813

1981 & 2032

top rail

muntin

bullseye pane

flat solid panels

intermediate rail

stile

bottom rail

**TUDOR STYLE**

762, 838 & 813

solid bevelled and raised panels

2032 & 1981

top rail

frieze rail

stile

lock rail

muntin

bottom rail

**GEORGIAN STYLE**

762, 838 & 813

1981 & 2032

top rail

stile

radial glazing bars

frieze rail

solid raised panels

intermediate rail

bottom rail

**PURPOSE MADE**

Door Frames ~ these are available for all standard external door frames and can be obtained with a fixed solid or glazed panel above a door height transom. Door frames are available for doors opening inwards or outwards. Most door frames are made to the recommendations set out in BS 1567.

Typical Example ~

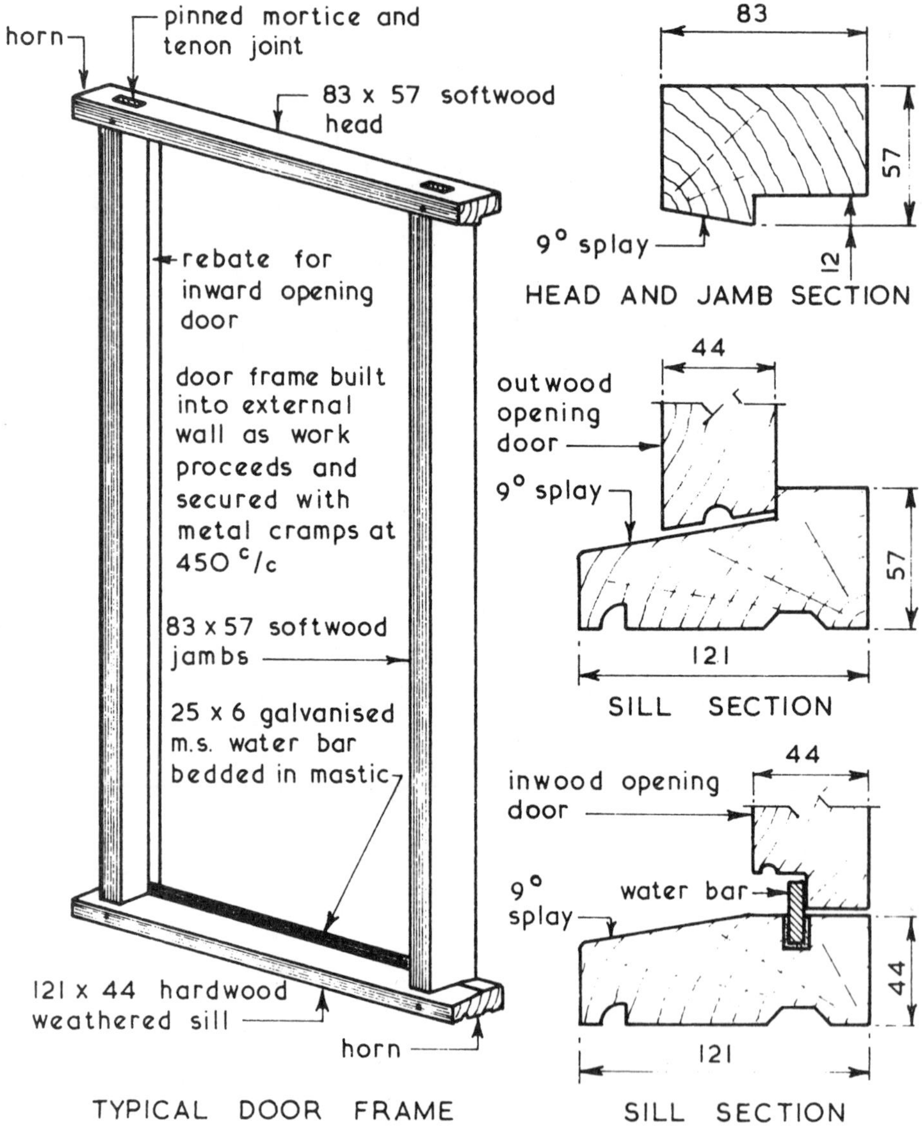

horn

pinned mortice and tenon joint

83 x 57 softwood head

rebate for inward opening door

door frame built into external wall as work proceeds and secured with metal cramps at 450 $^c$/c

83 x 57 softwood jambs

25 x 6 galvanised m.s. water bar bedded in mastic

121 x 44 hardwood weathered sill

horn

TYPICAL DOOR FRAME

83

57

9° splay

12

HEAD AND JAMB SECTION

outwood opening door

9° splay

44

57

121

SILL SECTION

inwood opening door

9° splay

water bar

44

44

121

SILL SECTION

Door Ironmongery ~ available in a wide variety of materials, styles and finishes but will consist of essentially the same components:-

Hinges or Butts - these are used to fix the door to its frame or lining and to enable it to pivot about its hanging edge.

Locks, Latches and Bolts - the means of keeping the door in its closed position and providing the required degree of security. The handles and cover plates used in conjunction with locks and latches are collectively called door furniture.

Letter Plates - fitted in external doors to enable letters etc., to be deposited through the door.

Other items include Finger and Kicking Plates which are used to protect the door fabric where there is high usage, Draught Excluders to seal the clearance gap around the edges of the door and Security Chains to enable the door to be partially opened and thus retain some security.

Typical Examples ~

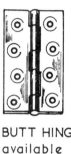

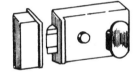

**LETTER PLATE**
can be obtained with
an integral knocker

**RIM NIGHT LATCH**
fitted on internal face

**BUTT HINGE**
available in
steel, brass
nylon and
aluminium

**RIM LOCK**
fitted on internal face

**BARREL BOLT**
available in steel
brass and aluminium

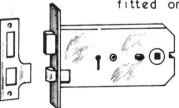

**MORTICE LOCK**
fitted within body of door

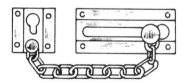

**SECURITY DOOR CHAIN**
extruded brass with welded
brassed steel chain

Industrial Doors ~ these doors are usually classified by their method of operation and construction. There is a very wide range of doors available and the choice should be based on the following considerations :-

1. Movement - vertical or horizontal.

2. Size of opening.

3. Position and purpose of door(s).

4. Frequency of opening and closing door(s).

5. Manual or mechanical operation.

6. Thermal and/or sound insulation requirements.

7. Fire resistance requirements.

Typical Industrial Door Types ~

1. Straight Sliding -

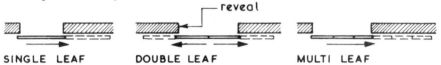

SINGLE LEAF    DOUBLE LEAF    MULTI LEAF

These types can be top hung with a bottom guide roller or hung with bottom rollers and top guides - see page 285

2 Sliding / Folding -

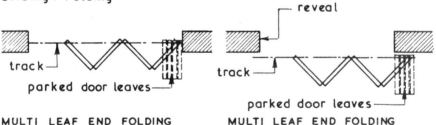

MULTI LEAF END FOLDING    MULTI LEAF END FOLDING
HUNG BETWEEN REVEALS       HUNG BEHIND OPENING

These types can be top hung with a bottom guide roller or hung with bottom rollers and top guides - see page 286

3. Shutters -

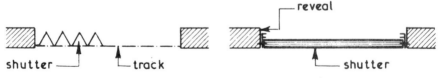

shutter ───── track

HORIZONTAL FOLDING SHUTTER    ROLLER SHUTTER

Shutters can be installed between, behind or in front of the reveals - see page 287

Straight Sliding Doors ~ these doors are easy to operate, economic to maintain and present no problems for the inclusion of a wicket gate. They do however take up wall space to enable the leaves to be parked in the open position. The floor guide channel associated with top hung doors can become blocked with dirt causing a malfunction of the sliding movement whereas the rollers in bottom track doors can seize up unless regularly lubricated and kept clean. Straight sliding doors are available with either manual or mechanical operation.

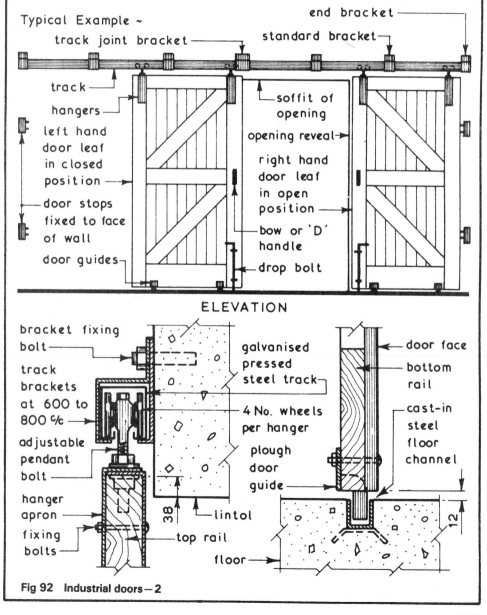

Fig 92    Industrial doors — 2

Sliding / Folding Doors ~ these doors are an alternative format to the straight sliding door types and have the same advantages and disadvantages except that the parking space required for the opened door is less than that for straight sliding doors.   Sliding / folding are usually manually operated and can be arranged in groups of 2 to 8 leaves.

Typical Example ~

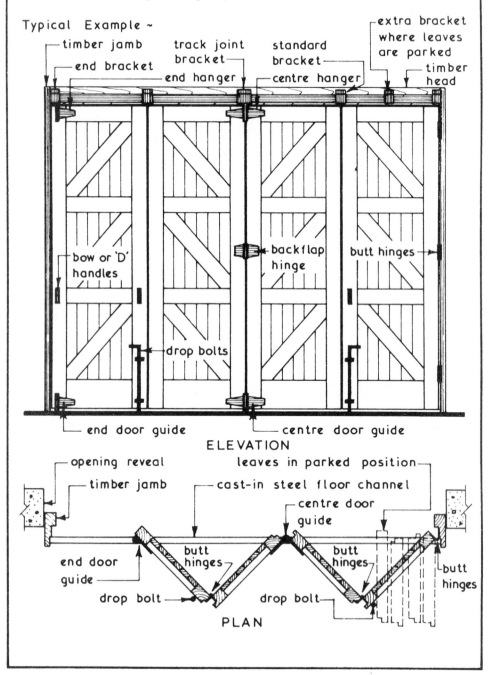

ELEVATION

PLAN

Shutters ~ horizontal folding shutters are similar in operation to sliding/folding doors but are composed of smaller leaves and present the same problems. Roller shutters however do not occupy any wall space but usually have to be fully opened for access. They can be manually operated by means of a pole when the shutters are self coiling, operated by means of an endless chain winding gear or mechanically raised and lowered by an electric motor but in all cases they are slow to open and close. Vision panels cannot be incorporated in the roller shutter but it is possible to include a small wicket gate or door in the design.

Typical Details ~

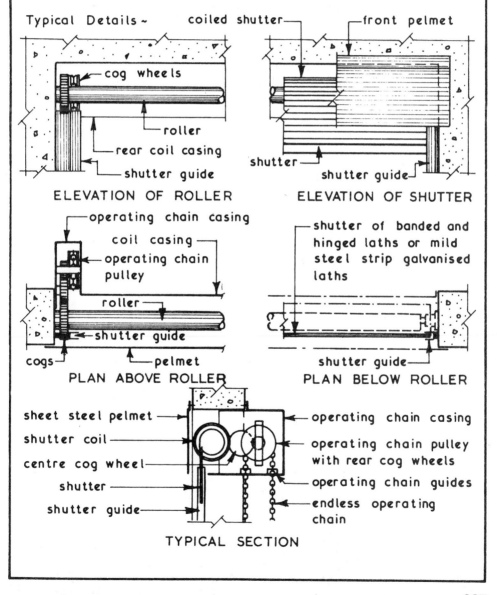

ELEVATION OF ROLLER

ELEVATION OF SHUTTER

PLAN ABOVE ROLLER

PLAN BELOW ROLLER

TYPICAL SECTION

# Crosswall Construction

Crosswall Construction ~ this is a form of construction where load bearing walls are placed at right angles to the lateral axis of the building, the front and rear walls being essentially non-load bearing cladding. Crosswall construction is suitable for buildings up to 5 storeys high where the floors are similar and where internal separating or party walls are required such as in blocks of flats or maisonettes. The intermediate floors span longitudinally between the crosswalls providing the necessary lateral restraint and if both walls and floors are of cast insitu reinforced concrete the series of 'boxes' so formed is sometimes called box frame construction. Great care must be taken in both design and construction to ensure that the junctions between the non-load bearing claddings and the crosswalls are weathertight. If a pitched roof is to be employed with the ridge parallel to the lateral axis an edge beam will be required to provide a seating for the trussed or common rafters and to transmit the roof loads to the crosswalls.

Typical Crosswall Arrangement Details ~

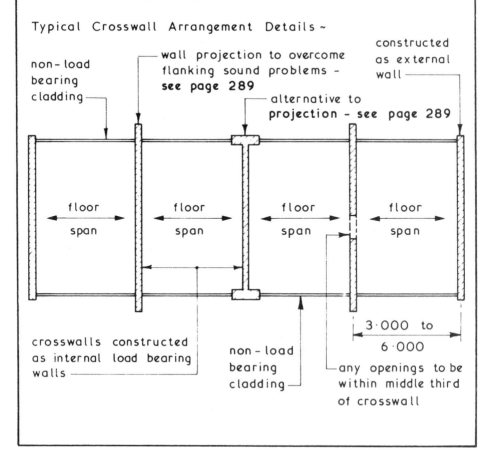

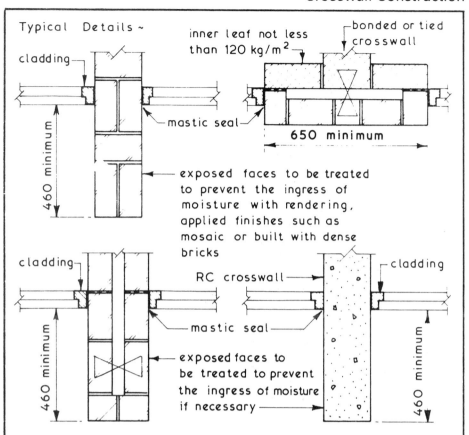

Typical Details ~

cladding

inner leaf not less than 120 kg/m²

bonded or tied crosswall

mastic seal

650 minimum

460 minimum

exposed faces to be treated to prevent the ingress of moisture with rendering, applied finishes such as mosaic or built with dense bricks

cladding

RC crosswall

cladding

mastic seal

exposed faces to be treated to prevent the ingress of moisture if necessary

460 minimum

460 minimum

Advantages of Crosswall Construction :-

1. Load bearing and non-load bearing components can be standardised and in some cases prefabricated giving faster construction times.

2. Fenestration between crosswalls unrestricted.

3. Crosswalls although load bearing need not be weather resistant as is the case with external walls.

Disadvantages of Crosswall Construction :-

1. Limitations of possible plans.

2. Need for adequate lateral ties between crosswalls.

3. Need to weather adequately projecting crosswalls.

Floors :-

An insitu solid reinforced concrete floor will provide the greatest rigidity, all other form must be adequately tied to walls

Simply Supported Slabs ~ these are slabs which rest on a bearing and for design purposes are not considered to be fixed to the support and are therefore , in theory, free to lift. In practice however they are restrained from unacceptable lifting by their own self weight plus any loadings.

Concrete Slabs ~ concrete is a material which is strong in compression and weak in tension and if the member is overloaded its tensile resistance may be exceeded leading to structural failure.

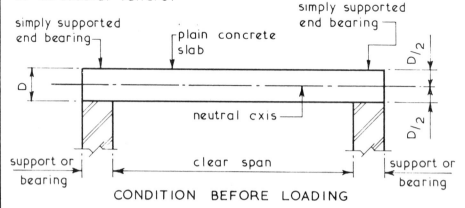

CONDITION BEFORE LOADING

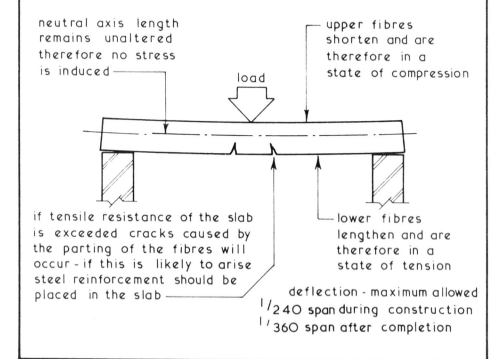

neutral axis length remains unaltered therefore no stress is induced

upper fibres shorten and are therefore in a state of compression

load

if tensile resistance of the slab is exceeded cracks caused by the parting of the fibres will occur - if this is likely to arise steel reinforcement should be placed in the slab

lower fibres lengthen and are therefore in a state of tension

deflection - maximum allowed
$1/240$ span during construction
$1/360$ span after completion

Reinforcement~ generally in the form of steel bars which are used to provide the tensile strength which plain concrete lacks. The number, diameter, spacing . shape and type of bars to be used have to be designed, the process of which is beyond the scope of this text. Reinforcement is placed as near to the outside fibres as practicable, a cover of concrete over the reinforcement is required to protect the steel bars from corrosion and to provide a degree of fire resistance. Slabs which are square in plan are considered to be spanning in two directions and therefore main reinforcing bars are used both ways whereas slabs which are rectangular in plan are considered to span across the shortest distance and main bars are used in this direction only with smaller diameter distribution bars placed at right angles forming a mat or grid.

Typical Details ~

simply supported RC slab

main bars in both directions

15mm minimum cover of concrete over main reinforcement

span equal in both directions

### SQUARE SLAB

simply supported RC slab

distribution bars

main bars

15mm minimum cover of concrete over main reinforcement

shortest span

### RECTANGULAR SLAB

Construction ~ whatever method of construction is used the construction sequence will follow the same pattern -

1. Assemble and erect formwork.
2. Prepare and place reinforcement.
3. Pour and compact or vibrate concrete.
4. Strike and remove formwork in stages as curing proceeds.

Typical Example ~

edge formwork

concrete poured and compacted or vibrated around reinforcement

main reinforcement - cover maintained by plastic or similar spacers - see Detail 'A'

decking of suitable material such as plywood with all joints sealed or taped to prevent grout loss

distribution bars - position maintained by wire binding or clips - see Detail 'A'

surface finish as specified

adjustable steel or timber props at centres to suit spanning ability of joists

joists supporting decking spaced at centres to suit spanning ability of decking

tying wire or clip

plastic spacer

distribution bar

main bars

DETAIL 'A'

telescopic steel floor centres with sheet steel decking giving clear spans between support walls

ALTERNATIVE DECKING SUPPORT

Beams ~ these are horizontal load bearing members which are classified as either main beams which transmit floor and secondary beam loads to the columns or secondary beams which transmit floor loads to the main beams. Concrete being a material which has little tensile strength needs to be reinforced to resist the induced tensile stresses which can be in the form of ordinary tension or diagonal tension (shear). The calculation of the area, diameter, type, position and number of reinforcing bars required is one of the functions of a structural engineer.

Typical RC Beam Details ~

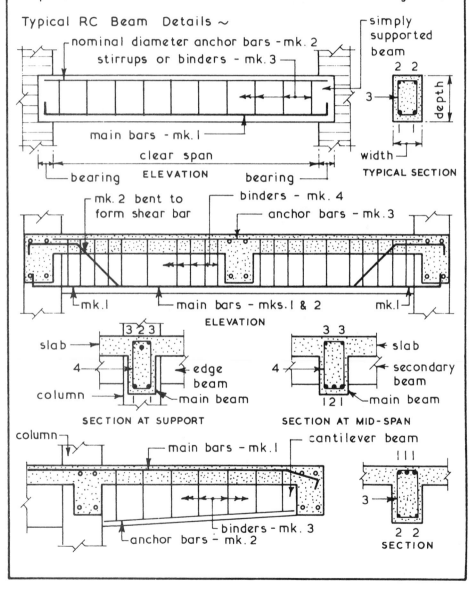

293

Columns ~ these are the vertical load bearing members of the structural frame which transmits the beam loads down to the foundations. They are usually constructed in storey heights and therefore the reinforcement must be lapped to to provide structural continuity.

Typical RC Column Details ~

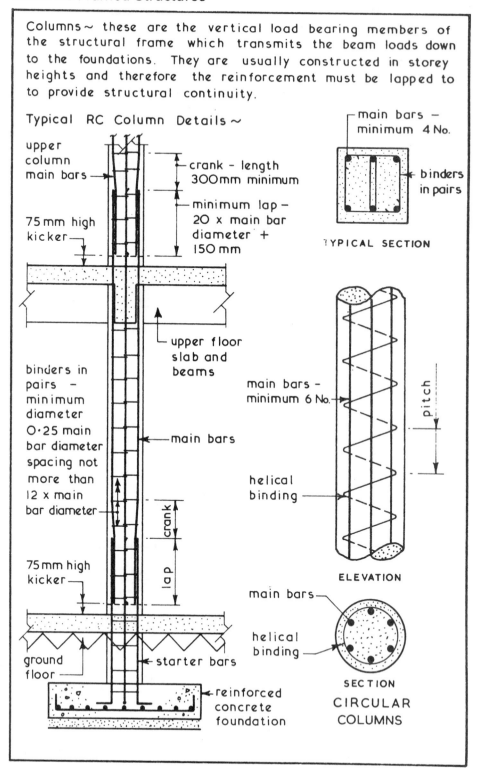

upper column main bars

crank – length 300mm minimum

minimum lap – 20 x main bar diameter + 150 mm

75mm high kicker

upper floor slab and beams

binders in pairs – minimum diameter 0·25 main bar diameter spacing not more than 12 x main bar diameter

main bars

75mm high kicker

crank

lap

ground floor

starter bars

reinforced concrete foundation

main bars – minimum 4 No.

binders in pairs

TYPICAL SECTION

main bars – minimum 6 No.

pitch

helical binding

ELEVATION

main bars

helical binding

SECTION

CIRCULAR COLUMNS

Basic Formwork ~ concrete when first mixed is a fluid and therefore to form any concrete member the wet concrete must be placed in a suitable mould to retain its shape, size and position as it sets. It is possible with some forms of concrete foundations to use the sides of the excavation as the mould but in most cases when casting concrete members a mould will have to be constructed on site. These moulds are usually called formwork. It is important to appreciate that the actual formwork is the reverse shape of the concrete member which is to be cast.

Basic Principles ~

formwork sides can be designed to offer all the necessary resistance to the imposed pressures as a single member or alternatively they can be designed to use a thinner material which is adequately strutted — for economic reasons the latter method is usually employed

grout tight joints

formwork soffits can be designed to offer all the necessary resistance to the imposed loads as a single member or alternatively they can be designed to a thinner material which is adequately propped — for economic reasons the latter method is usually employed

wet concrete - density is greater than that of the resultant set and dry concrete

formwork sides — limits width and shape of wet concrete and has to resist the hydrostatic pressure of the wet concrete which will diminish to zero within a matter of hours depending on setting and curing rate

formwork base or soffit - limits depth and shape of wet concrete and has to resist the initial dead load of the wet concrete and later the dead load of the dry set concrete until it has gained sufficient strength to support its own dead weight which is usually several days after casting depending on curing rate.

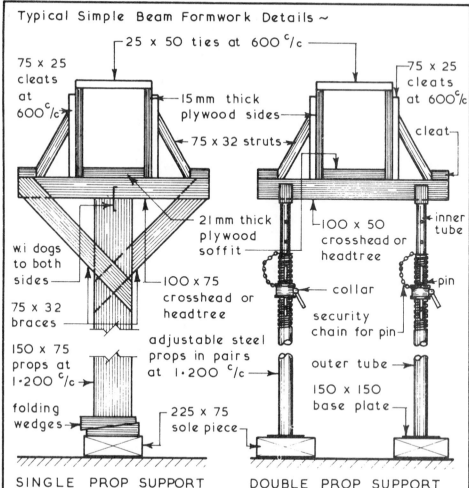

Typical Simple Beam Formwork Details ~

25 x 50 ties at 600 c/c

75 x 25 cleats at 600 c/c

75 x 25 cleats at 600 c/c

15mm thick plywood sides

75 x 32 struts

cleat

21mm thick plywood soffit

inner tube

w.i dogs to both sides

100 x 50 crosshead or headtree

75 x 32 braces

100 x 75 crosshead or headtree

collar

security chain for pin

pin

150 x 75 props at 1·200 c/c

adjustable steel props in pairs at 1·200 c/c

outer tube

folding wedges

225 x 75 sole piece

150 x 150 base plate

**SINGLE PROP SUPPORT**

**DOUBLE PROP SUPPORT**

Erecting Formwork

1. Props positioned and levelled through.
2. Soffit placed, levelled and position checked.
3. Side forms placed, their position checked before being fixed.
4. Strutting position and fixed.
5. Final check before casting.

Suitable Formwork Materials~ timber, steel and special plastics.

Striking or Removing Formwork

1. Side forms as soon as practicable usually within hours of casting this allows drying air movements to take place around the setting concrete.
2. Soffit formwork as soon as practicable usually within days but as a precaution some props are left in position until concrete member is self supporting.

Beam Formwork ~ this is basically a three sided box supported and propped in the correct position and to the desired level. The beam formwork sides have to retain the wet concrete in the required shape and be able to withstand the initial hydrostatic pressure of the wet concrete whereas the formwork soffit apart from retaining the concrete has to support the initial load of the wet concrete and finally the set concrete until it has gained sufficient strength to be self supporting. It is essential that all joints in the formwork are constructed to prevent the escape of grout which could result in honeycombing and/or feather edging in the cast beam. The removal time for the formwork will vary with air temperature, humidity and consequent curing rate.

Typical Details ~

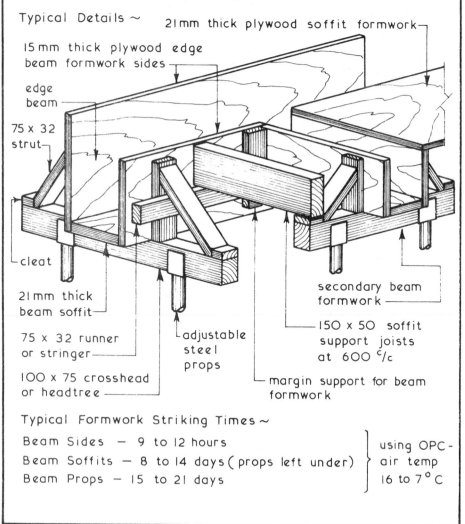

21mm thick plywood soffit formwork

15 mm thick plywood edge beam formwork sides

edge beam

75 x 32 strut

cleat

21 mm thick beam soffit

75 x 32 runner or stringer

100 x 75 crosshead or headtree

adjustable steel props

secondary beam formwork

150 x 50 soffit support joists at 600 $^{c}/_{c}$

margin support for beam formwork

Typical Formwork Striking Times ~

Beam Sides — 9 to 12 hours
Beam Soffits — 8 to 14 days (props left under)
Beam Props — 15 to 21 days

using OPC - air temp 16 to 7 °C

297

Column Formwork ~ this consists of a vertical mould of the desired shape and size which has to retain the wet concrete and resist the initial hydrostatic pressure caused by the wet concrete. To keep the thickness of the formwork material to a minimum horizontal clamps or yokes are used at equal centres for batch filling and at varying centres for complete filling in one pour. The head of the column formwork can be used to support the incoming beam formwork which gives good top lateral restraint but results in complex formwork alternatively the column can be cast to the underside of the beams and at a later stage a collar of formwork can be clamped around the cast column to complete casting and support the incoming beam formwork. Column forms are located at the bottom around a 75 to 100mm high concrete plinth or kicker which has the dual function of location and preventing grout loss from the bottom of the column formwork.

Typical Details ~

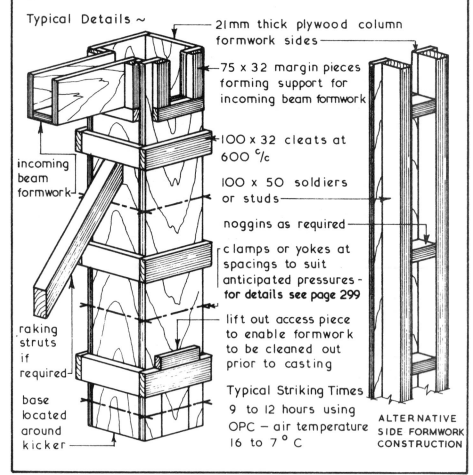

21mm thick plywood column formwork sides

75 x 32 margin pieces forming support for incoming beam formwork

100 x 32 cleats at 600 c/c

100 x 50 soldiers or studs

noggins as required

clamps or yokes at spacings to suit anticipated pressures - for details see page 299

lift out access piece to enable formwork to be cleaned out prior to casting

incoming beam formwork

raking struts if required

base located around kicker

Typical Striking Times 9 to 12 hours using OPC – air temperature 16 to 7° C

ALTERNATIVE SIDE FORMWORK CONSTRUCTION

Column Yokes ~ these are obtainable as a metal yoke or clamp or they can be purpose made from timber.

Typical Examples ~

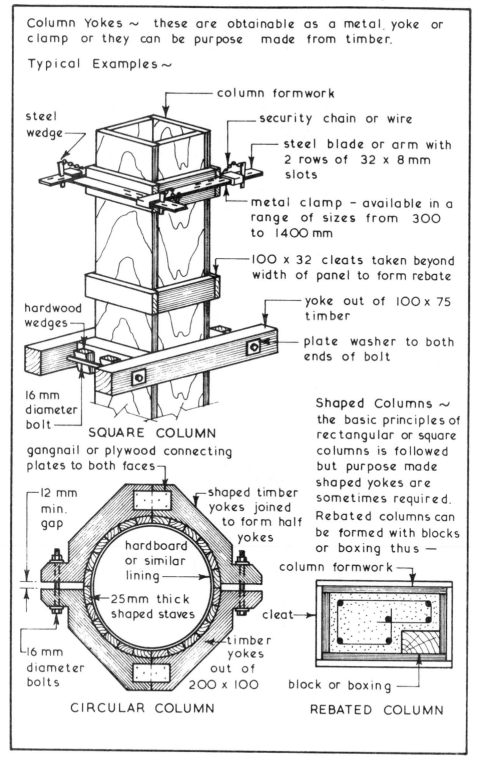

column formwork

steel wedge

security chain or wire

steel blade or arm with 2 rows of 32 x 8mm slots

metal clamp – available in a range of sizes from 300 to 1400 mm

100 x 32 cleats taken beyond width of panel to form rebate

yoke out of 100 x 75 timber

hardwood wedges

plate washer to both ends of bolt

16 mm diameter bolt

**SQUARE COLUMN**

gangnail or plywood connecting plates to both faces

12 mm min. gap

shaped timber yokes joined to form half yokes

hardboard or similar lining

25mm thick shaped staves

16 mm diameter bolts

timber yokes out of 200 x 100

cleat

**CIRCULAR COLUMN**

Shaped Columns ~ the basic principles of rectangular or square columns is followed but purpose made shaped yokes are sometimes required. Rebated columns can be formed with blocks or boxing thus —

column formwork

block or boxing

**REBATED COLUMN**

299

Precast Concrete Frames ~ these frames are suitable for single storey and low rise applications, the former usually in the form of portal frames which are normally studied separately. Precast concrete frames provide the skeleton for the building and can be clad externally and finished internally by all the traditional methods. The frames are usually produced as part of a manufacturer's standard range of designs and are therefore seldom purpose made due mainly to the high cost of the moulds.

Advantages :-

1. Frames are produced under factory controlled conditions resulting in a uniform product of both quality and accuracy.

2. Repetitive casting lowers the cost of individual members.

3. Off site production releases site space for other activities.

4. Frames can be assembled in cold weather and generally by semi-skilled labour.

Disadvantages :-

1. Although a wide choice of frames is available from various manufacturer's these systems lack the design flexibility of cast insitu purpose made frames.

2. Site planning can be limited by manufacturer's delivery and unloading programmes and requirements.

3. Lifting plant of a type and size not normally required by traditional construction methods may be needed.

Typical Site Activities ~

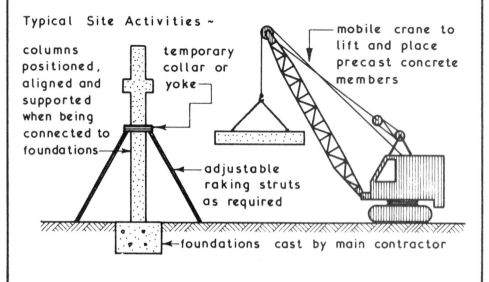

columns positioned, aligned and supported when being connected to foundations

temporary collar or yoke

mobile crane to lift and place precast concrete members

adjustable raking struts as required

foundations cast by main contractor

Foundation Connections ~ the preferred method of connection is to set the column into a pocket cast into a reinforced concrete pad foundation and is suitable for light to medium loadings.  Where heavy column loadings are encountered it may be necessary to use a steel base plate secured to the reinforced concrete pad foundation with holding down bolts.

Typical Details ~

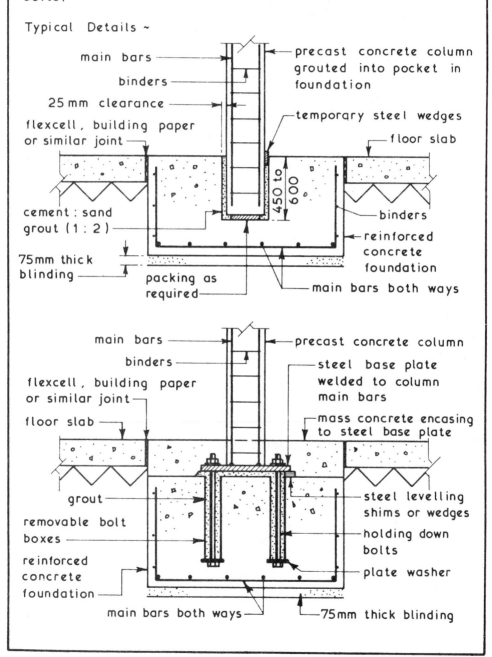

301

# Precast Concrete Frames

Column to Column Connection ~ precast columns are usually cast in one length and can be up to four storeys in height. They are either reinforced with bar reinforcement or they are prestressed according to the loading conditions. If column to column are required they are usually made at floor levels above the beam to column connections and can range from a simple dowel connection to a complex connection involving insitu concrete.

Typical Details ~

dowel bar passes through beam into lower column →

upper column with mortice in lower end to receive dowel bar

grouting holes

column supported on shims until grout has set - joint is then dry packed with cement mortar

joint packed with dry cement mortar

rebated edge beam

lower column with mortice in top end to receive dowel bar which is grouted in

upper column →

4 No. threaded studs cast into upper column

levelling nuts →

exposed and lapped main reinforcing bars →

insitu concrete →

mild steel bearing plate welded to main reinforcing bars of lower column

back nut →

lower column →

exposed binders →

upper column

insitu concrete

dry joint

lower column

Beam to Column Connections ~ as with the column to column connections (see page 302) the main objective is to provide structural continuity at the junction. This is usually achieved by one of two basic methods:-

1. Projecting bearing haunches cast onto the columns with a projecting dowel or stud bolt to provide both location and fixing.

2. Steel to steel fixings which are usually in form of a corbel or bracket projecting from the column providing a bolted connection to a steel plate cast into the end of the beam.

Typical Details ~

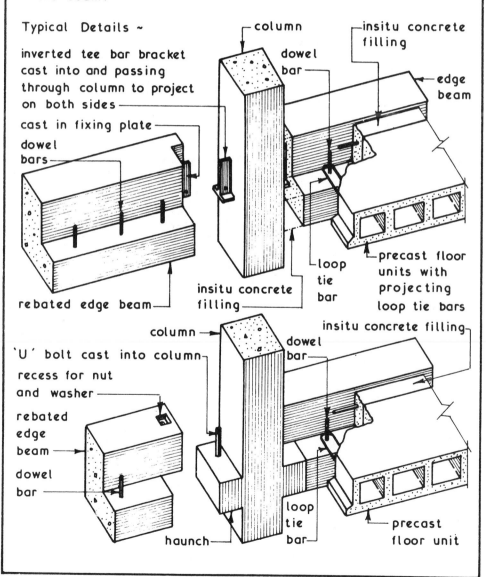

inverted tee bar bracket cast into and passing through column to project on both sides

cast in fixing plate

dowel bars

rebated edge beam

insitu concrete filling

column

dowel bar

insitu concrete filling

edge beam

loop tie bar

precast floor units with projecting loop tie bars

column

'U' bolt cast into column

recess for nut and washer

rebated edge beam

dowel bar

dowel bar

insitu concrete filling

haunch

loop tie bar

precast floor unit

Structural Steelwork ~ the standard sections available for use in structural steelwork are given in BS 4 and BS 4848. These standards give a wide range of sizes and weights to enable the designer to formulate an economic design.

Typical Standard Steelwork Sections ~

### UNIVERSAL BEAMS
203 x 133 x 25 kg /m to
914 x 419 x 388 kg/m

### UNIVERSAL COLUMNS
152 x 152 x 23 kg/m to
356 x 406 x 634 kg/m

### JOISTS
76 x 51 x 6·67 kg/m to
203 x 102 x 25·33 kg/m

### CHANNELS
76 x 38 x 6·7 kg/m to
432 x 102 x 66·54 kg /m

### HOLLOW SECTIONS
50 x 25 x 2·92 kg/m
to 457 x 355 x 156 kg /m

### EQUAL ANGLES
25 x 25 x 1·11 kg/m to
200 x 200 x 71·1 kg/m

### UNEQUAL ANGLES
40 x 25 x 1·91 kg /m to
200 x 150 x 47·1 kg /m

### HOLLOW SECTIONS
27 dia. x 1·89 kg/m to
457 dia. x 138 kg/m

NB. Sizes given are serial or nominal, for actual sizes see relevant BS.

Structural Steelwork Connections ~ these are either shop or site connections according to where the fabrication takes place. Most site connections are bolted whereas shop connections are very often carried out by welding. The design of structural steelwork members and their connections is the province of the structural engineer who selects the type and number of bolts or the size and length of weld to be used according to the connection strength to be achieved

Typical Connection Examples ~

3mm wide expansion gap

universal beams

seating cleats

site connection

universal column

shop connections

**SIMPLE CONNECTION**

universal column

top cleats

site connections

shop connection

erection cleat

universal beam

web cleats if required

**SEMI-RIGID CONNECTION**

site fillet welds

universal column

erection cleat

ground level

site connection

universal beam

shop fillet weld

**RIGID CONNECTION**

150mm thick minimum concrete encasing

shop fillet welds

holding down bolts grouted after final levelling

steel levelling wedges or shims

steel base plate bolted and grouted to RC foundation

removable bolt box of foamed plastic, plywood, PVC tube, etc.,

100 x 100 plate washers

**COLUMN TO FOUNDATION CONNECTION**

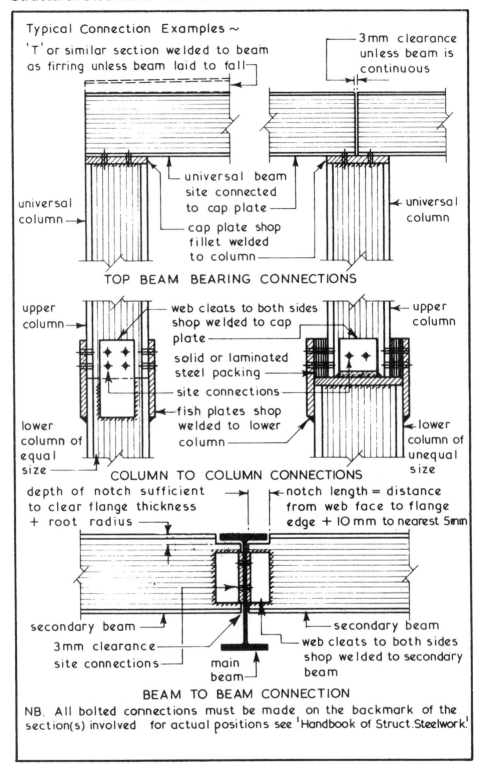

Typical Connection Examples ~

'T' or similar section welded to beam as firring unless beam laid to fall

3mm clearance unless beam is continuous

universal column →

universal beam site connected to cap plate

cap plate shop fillet welded to column

universal column

## TOP BEAM BEARING CONNECTIONS

upper column →

web cleats to both sides shop welded to cap plate

solid or laminated steel packing

site connections

fish plates shop welded to lower column

upper column

lower column of equal size

lower column of unequal size

## COLUMN TO COLUMN CONNECTIONS

depth of notch sufficient to clear flange thickness + root radius

notch length = distance from web face to flange edge + 10 mm to nearest 5mm

secondary beam

3mm clearance

site connections

main beam

secondary beam

web cleats to both sides shop welded to secondary beam

## BEAM TO BEAM CONNECTION

NB. All bolted connections must be made on the backmark of the section(s) involved   for actual positions see 'Handbook of Struct.Steelwork.'

Fire Resistance of Structural Steelwork ~ although steel is a non-combustible material with negligible surface spread of flame properties it does not behave very well under fire conditions. During the initial stages of a fire the steel will actually gain in strength but this reduces to normal at a steel temperature range of 250 to 400 °C and continues to decrease until the steel temperature reaches 550 °C when it has lost most of its strength. Since the temperature rise during a fire is rapid most structural steelwork will need protection to give it a specific degree of fire resistance in **terms of time. Part B of the Building Regulations sets out** the minimum requirements related to building usage and **size, BRE report 'Guidelines for the construction of fire resisting structural elements'** gives acceptable methods.

Typical Examples for a 2 Hour Fire Resistance ~

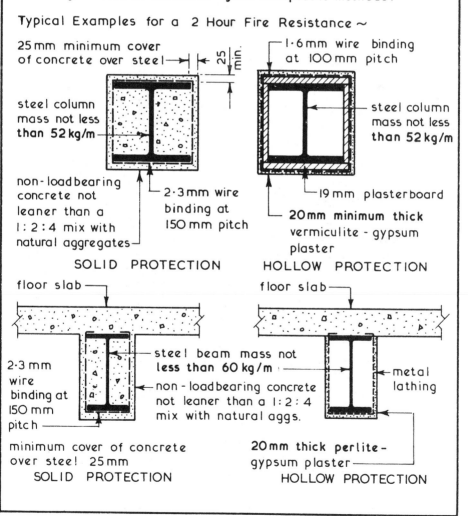

25 mm minimum cover of concrete over steel

steel column mass not less **than 52 kg/m**

non-loadbearing concrete not leaner than a 1:2:4 mix with natural aggregates

2·3 mm wire binding at 150 mm pitch

SOLID PROTECTION

1·6 mm wire binding at 100 mm pitch

steel column mass not less **than 52 kg/m**

19 mm plasterboard

**20 mm minimum thick** vermiculite-gypsum plaster

HOLLOW PROTECTION

floor slab

2·3 mm wire binding at 150 mm pitch

minimum cover of concrete over steel 25 mm

SOLID PROTECTION

floor slab

steel beam mass not **less than 60 kg/m**

non-loadbearing concrete not leaner than a 1:2:4 mix with natural aggs.

metal lathing

**20 mm thick perlite-** gypsum plaster

HOLLOW PROTECTION

# Portal Frames

Portal Frames ~ these can be defined as two dimensional rigid frames which have the basic characteristic of a rigid joint between the column and the beam. The main objective of this form of design is to reduce the bending moment in the beam thus allowing the frame to act as one structural unit. The transfer of stresses from the beam to the column can result in a rotational movement at the foundation which can be overcome by the introduction of a pin or hinge joint. The pin or hinge will allow free rotation to take place at the point of fixity whilst transmitting both load and shear from one member to another. In practice a true 'pivot' is not always required but there must be enough movement to ensure that the rigidity at the point of connection is low enough to overcome the tendency of rotational movement.

Typical Single Storey Portal Frame Formats ~

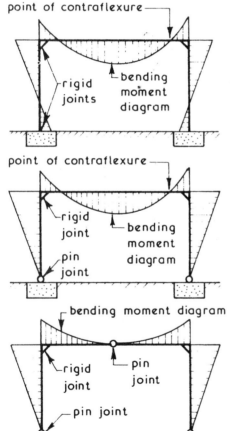

point of contraflexure

rigid joints

bending moment diagram

point of contraflexure

rigid joint

bending moment diagram

pin joint

bending moment diagram

rigid joint

pin joint

pin joint

**FIXED or RIGID PORTAL FRAME :-**

all joints or connections are rigid giving lower bending moments than other formats. Used for small to medium span frames where moments at foundations are not excessive.

**TWO PIN PORTAL FRAME :-**

pin joints or hinges used at foundation connections to eliminate tendency of base to rotate. Used where high base moments and weak ground are encountered.

**THREE PIN PORTAL FRAME :-**

pin joints or hinges used at foundation connections and at centre of beam which reduces bending moment in beam but increases deflection. Used as an alternative to a 2 pin frame.

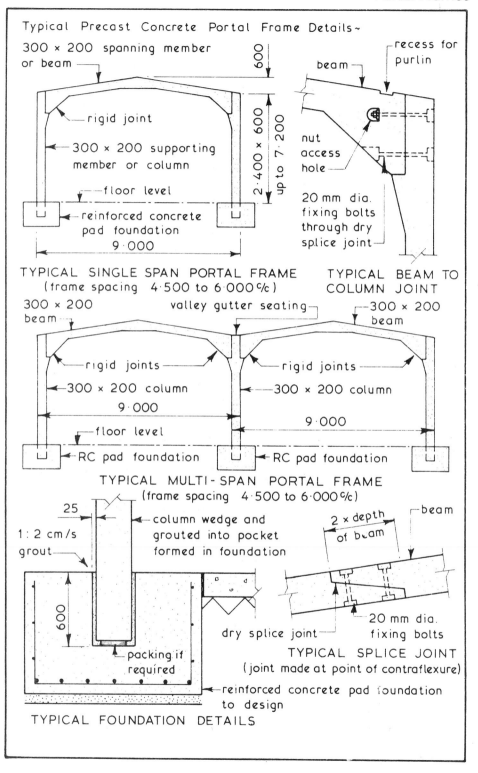

Typical Precast Concrete Portal Frame Details~

300 × 200 spanning member or beam

600

**TYPICAL SINGLE SPAN PORTAL FRAME**
(frame spacing 4·500 to 6·000 %)

rigid joint

300 × 200 supporting member or column

floor level

reinforced concrete pad foundation

9·000

2·400 × 600 up to 7·200

recess for purlin

beam

nut access hole

20 mm dia. fixing bolts through dry splice joint

**TYPICAL BEAM TO COLUMN JOINT**

300 × 200 beam

valley gutter seating

300 × 200 beam

rigid joints

300 × 200 column

9·000

floor level

RC pad foundation

rigid joints

300 × 200 column

9·000

RC pad foundation

**TYPICAL MULTI-SPAN PORTAL FRAME**
(frame spacing 4·500 to 6·000 %)

25

1:2 cm/s grout

600

column wedge and grouted into pocket formed in foundation

packing if required

reinforced concrete pad foundation to design

**TYPICAL FOUNDATION DETAILS**

2 × depth of beam

beam

dry splice joint

20 mm dia. fixing bolts

**TYPICAL SPLICE JOINT**
(joint made at point of contraflexure)

309

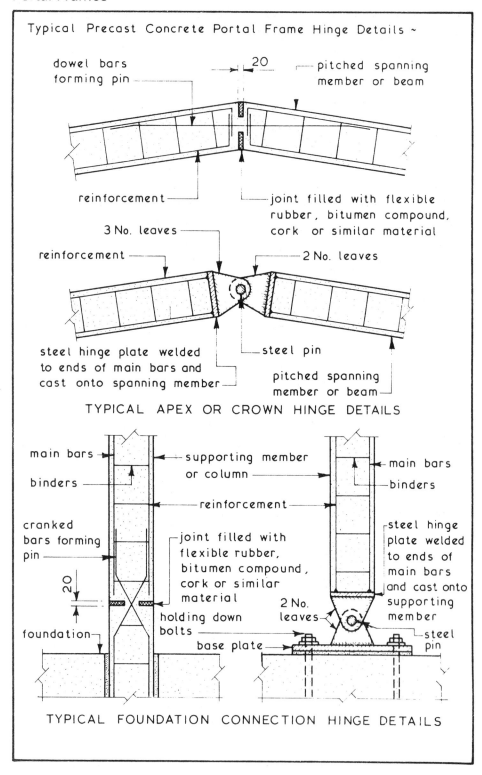

Typical Precast Concrete Portal Frame Hinge Details ~

dowel bars forming pin

20

pitched spanning member or beam

reinforcement

joint filled with flexible rubber, bitumen compound, cork or similar material

3 No. leaves

reinforcement

2 No. leaves

steel hinge plate welded to ends of main bars and cast onto spanning member

steel pin

pitched spanning member or beam

TYPICAL APEX OR CROWN HINGE DETAILS

main bars

binders

supporting member or column

main bars

binders

reinforcement

cranked bars forming pin

joint filled with flexible rubber, bitumen compound, cork or similar material

20

holding down bolts

foundation

base plate

2 No. leaves

steel hinge plate welded to ends of main bars and cast onto supporting member

steel pin

TYPICAL FOUNDATION CONNECTION HINGE DETAILS

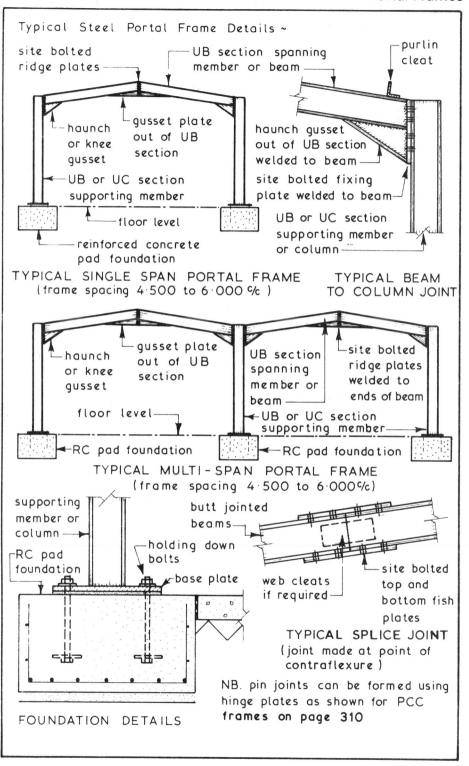

Typical Steel Portal Frame Details ~

site bolted ridge plates

UB section spanning member or beam

purlin cleat

haunch or knee gusset

gusset plate out of UB section

UB or UC section supporting member

floor level

reinforced concrete pad foundation

**TYPICAL SINGLE SPAN PORTAL FRAME**
(frame spacing 4·500 to 6·000 %)

haunch gusset out of UB section welded to beam

site bolted fixing plate welded to beam

UB or UC section supporting member or column

**TYPICAL BEAM TO COLUMN JOINT**

haunch or knee gusset

gusset plate out of UB section

floor level

RC pad foundation

UB section spanning member or beam

site bolted ridge plates welded to ends of beam

UB or UC section supporting member

RC pad foundation

**TYPICAL MULTI-SPAN PORTAL FRAME**
(frame spacing 4·500 to 6·000%)

supporting member or column

RC pad foundation

holding down bolts

base plate

butt jointed beams

web cleats if required

site bolted top and bottom fish plates

**TYPICAL SPLICE JOINT**
(joint made at point of contraflexure)

NB. pin joints can be formed using hinge plates as shown for PCC frames on page 310

FOUNDATION DETAILS

Laminated Timber ~ sometimes called `Gluelam` and is the process of building up beams, ribs, arches, portal frames and other structural units by gluing together layers of timber boards so that the direction of the grain of each board runs parallel with the longitudinal axis of the member being fabricated.

Laminates ~ these are the layers of board and may be jointed in width and length.

Joints ~

Width - joints in consecutive layers should lap twice the board thickness or one quarter of its width whichever is the greater.

Length - scarf and finger joints can be used. Scarf joints should have a minimum slope of 1 in 12 but this can be steeper (say 1 in 6) in the compression edge of a beam :-

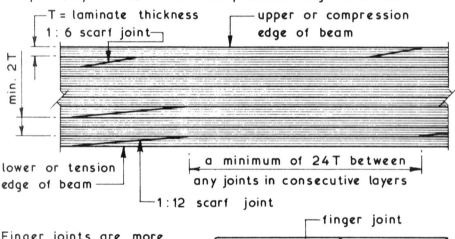

T = laminate thickness
1 : 6 scarf joint
min. 2 T
upper or compression edge of beam
lower or tension edge of beam
a minimum of 24 T between any joints in consecutive layers
1 : 12 scarf joint

Finger joints are more economical in the use of timber than the scarf joints

finger joint
laminated timber beam

Moisture Content ~ timber should have a moisture content equal to that which the member will reach in service and this is known as its equilibrium moisture content; for most buildings this will be between 11 and 15%. Generally at the time of gluing timber should not exceed $15 \pm 3\%$ in moisture content.

Vertical Laminations ~ not often used for structural laminated timber members and is unsatisfactory for curved members.

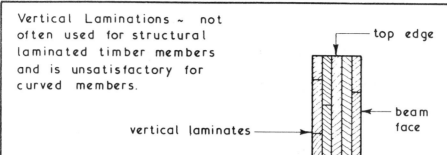

top edge

vertical laminates

beam face

Horizontal Laminations ~ most popular method for all types of laminated timber members. The stress diagrams below show that laminates near the upper edge are subject to a compressive stress whilst those near the lower edge to a tensile stress and those near the neutral axis are subject to shear stress.

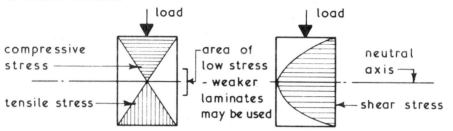

load

compressive stress

tensile stress

area of low stress - weaker laminates may be used

load

neutral axis

shear stress

Flat sawn timber shrinks twice as much as quarter sawn timber therefore flat and quarter sawn timbers should not be mixed in the same member since the different shrinkage rates will cause unacceptable stresses to occur on the glue lines.

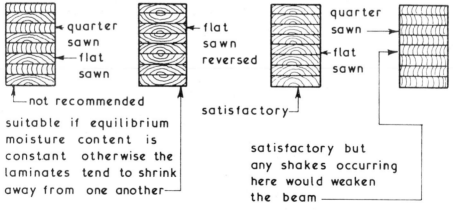

quarter sawn
flat sawn
not recommended
suitable if equilibrium moisture content is constant otherwise the laminates tend to shrink away from one another

flat sawn reversed
satisfactory

quarter sawn
flat sawn
satisfactory but any shakes occurring here would weaken the beam

Planing ~ before gluing laminates should be planed so that the depth of the planer cutter marks are not greater than 0·025mm. This requirement can be met by limiting the pitch of cutter marks to 3mm.

313

Glueing ~ this should be carried out within 48 hours of the planing operation to reduce the risk of the planed surfaces becoming contaminated or case hardened (for suitable adhesives see page 315). Just before gluing up the laminates they should be checked for 'cupping.' The amount of cupping allowed depends upon the thickness and width of the laminates and has a range of 1·5 to 0·75 mm.

Laminate Thickness ~ no laminate should be more than 50mm thick since seasoning up to this thickness can be carried out economically and there is less chance of any individual laminate having excessive cross grain strength.

Straight Members – laminate thickness is determined by the depth of the member, there must be enough layers to allow the end joints (i.e. scarf or finger joints – see page 312) to be properly staggered.

Curved Members – laminate thickness is determined by the radius to which the laminate is to be bent and the species together with the quality of the timber being used. Generally the maximum laminate thickness should be 1/150 of the sharpest curve radius although with some softwoods 1/100 may be used.

Typical Laminated Timber Curved Member ~

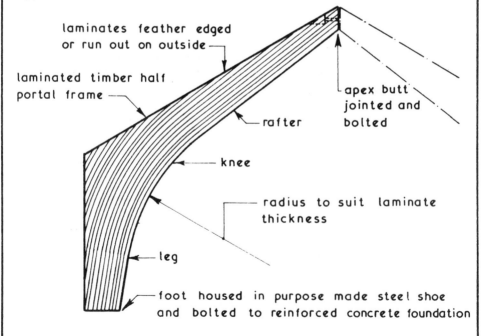

laminates feather edged or run out on outside

laminated timber half portal frame

apex butt jointed and bolted

rafter

knee

radius to suit laminate thickness

leg

foot housed in purpose made steel shoe and bolted to reinforced concrete foundation

Adhesives ~ although timber laminates are carefully machined, the minimum of cupping permitted and efficient cramping methods employed it is not always possible to obtain really tight joints between the laminates. One of the important properties of the adhesive is therefore that it should be gap filling. The maximum permissible gap being 1·25mm.

There are four adhesives suitable for laminated timber work which have the necessary gap filling property and they are namely :-

1. Casein - this is made from sour milk to the requirements of BS 5442. It is a cold setting adhesive in the form of a powder which is mixed with water, it has a tendency to stain timber and is only suitable for members used in dry conditions of service.

2. Urea Formaldehyde - this is a cold setting resin glue formulated to BS 1204 type MR/GF (moisture resistant / gap filling.) Although moisture resistant it is not suitable for prolonged exposure in wet conditions and there is a tendency for the glue to lose its strength in temperatures above 40°C such as when exposed to direct sunlight. The use of this adhesive is usually confined to members used in dry, unexposed conditions of service. This adhesive will set under temperatures down to 10°C.

3. Resorcinol Formaldehyde - this is a cold setting glue formulated to BS 1204 type WBP/GF (weather and boilproof / gap filling.) It is suitable for members used in external situations but is relatively expensive. This adhesive will set under temperatures down to 15°C and does not lose its strength at high temperatures.

4. Phenol Formaldehyde - this is a similar glue to resorcinol formaldehyde but is a warm setting adhesive requiring a temperature of above 86°C in order to set. A mixture called phenol / resorcinol formaldehyde is available and is sometimes used having similar properties to but less expensive than resorcinol formaldehyde but needs a setting temperature of at least 23°C.

Preservative Treatment - this can be employed if required, provided that the pressure impregnated preservative used is selected with regard to the adhesive being employed.

Multi-storey Structures ~ these buildings are usually designed for office, hotel or residential use and contain the means of vertical circulation in the form of stairs and lifts occupying up to 20% of the floor area. These means of circulation can be housed within a core inside the structure and this can be used to provide a degree of restraint to sway due to lateral wind pressures (see page 317).

Typical Basic Multi-storey Structure Types ~

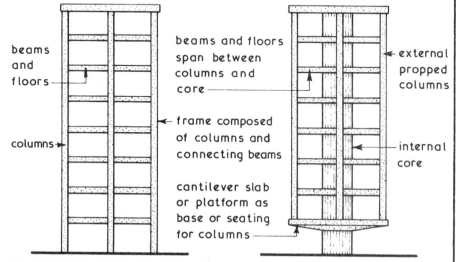

beams and floors

beams and floors span between columns and core

columns

frame composed of columns and connecting beams

external propped columns

internal core

cantilever slab or platform as base or seating for columns

TRADITIONAL FRAMED STRUCTURES    PROPPED STRUCTURES

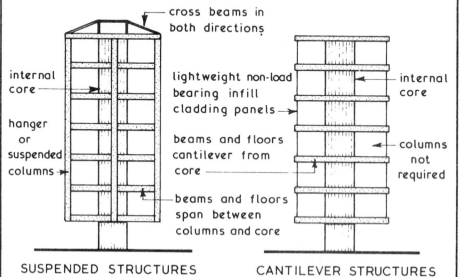

cross beams in both directions

internal core

lightweight non-load bearing infill cladding panels

internal core

hanger or suspended columns

beams and floors cantilever from core

columns not required

beams and floors span between columns and core

SUSPENDED STRUCTURES        CANTILEVER STRUCTURES

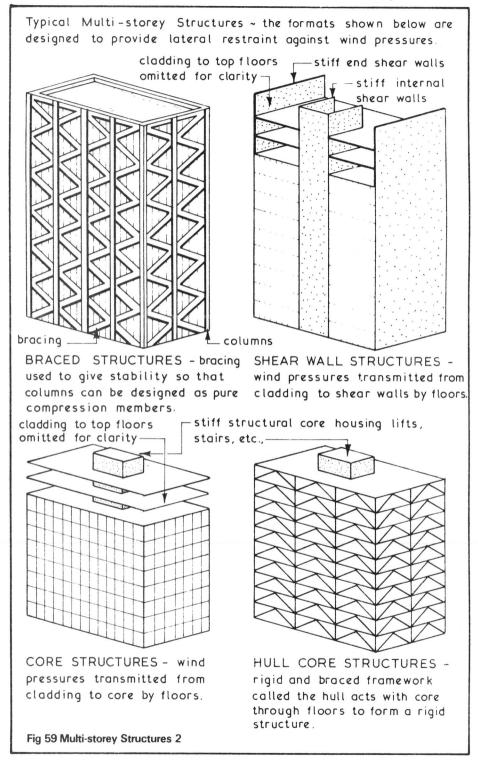

Typical Multi-storey Structures ~ the formats shown below are designed to provide lateral restraint against wind pressures.

cladding to top floors omitted for clarity

stiff end shear walls

stiff internal shear walls

bracing _____ columns

**BRACED STRUCTURES** - bracing used to give stability so that columns can be designed as pure compression members.

**SHEAR WALL STRUCTURES** - wind pressures transmitted from cladding to shear walls by floors.

cladding to top floors omitted for clarity

stiff structural core housing lifts, stairs, etc.,

**CORE STRUCTURES** - wind pressures transmitted from cladding to core by floors.

**HULL CORE STRUCTURES** - rigid and braced framework called the hull acts with core through floors to form a rigid structure.

**Fig 59 Multi-storey Structures 2**

317

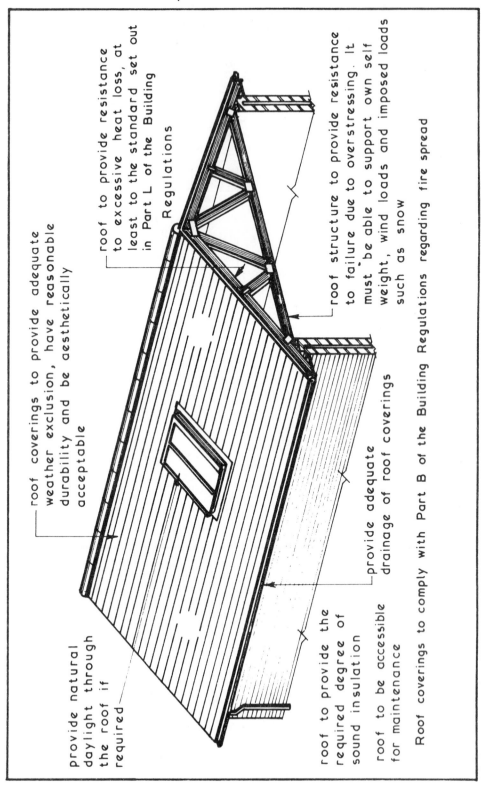

roof to provide resistance to excessive heat loss, at least to the standard set out in Part L of the Building Regulations

roof coverings to provide adequate weather exclusion, have reasonable durability and be aesthetically acceptable

roof structure to provide resistance to failure due to overstressing. It must be able to support own self weight, wind loads and imposed loads such as snow

provide adequate drainage of roof coverings

provide natural daylight through the roof if required

roof to provide the required degree of sound insulation

roof to be accessible for maintenance

Roof coverings to comply with Part B of the Building Regulations regarding fire spread

Roofs~ these can be classified as either:-
        Flat - pitch from 0° to 10°
        Pitched - pitch over 10°
It is worth noting that for design purposes roof pitches over 70° are classified as walls.

Roofs can be designed in many different forms and in combinations of these forms some of which would not be suitable and/or economic for domestic properties.

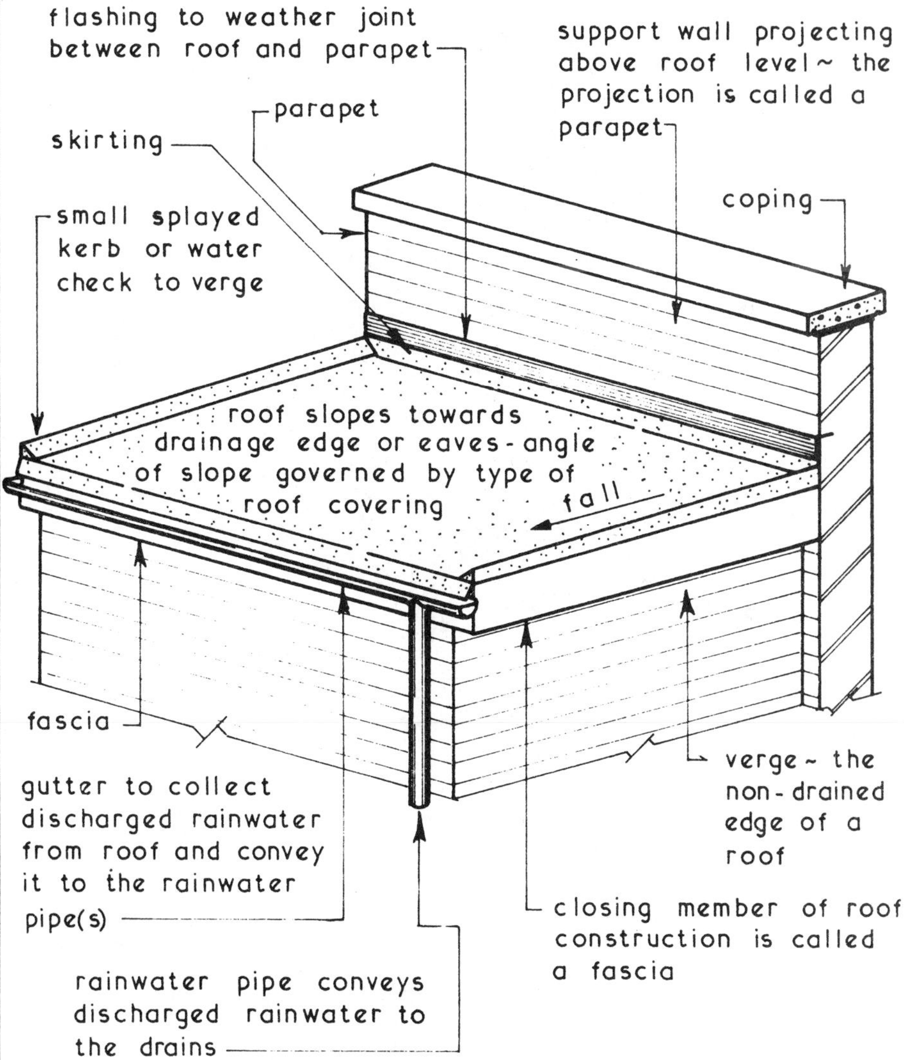

flashing to weather joint between roof and parapet

support wall projecting above roof level~ the projection is called a parapet

parapet

skirting

coping

small splayed kerb or water check to verge

roof slopes towards drainage edge or eaves-angle of slope governed by type of roof covering

fall

fascia

verge~ the non-drained edge of a roof

gutter to collect discharged rainwater from roof and convey it to the rainwater pipe(s)

closing member of roof construction is called a fascia

rainwater pipe conveys discharged rainwater to the drains

FLAT ROOFS

# Basic Roof Forms

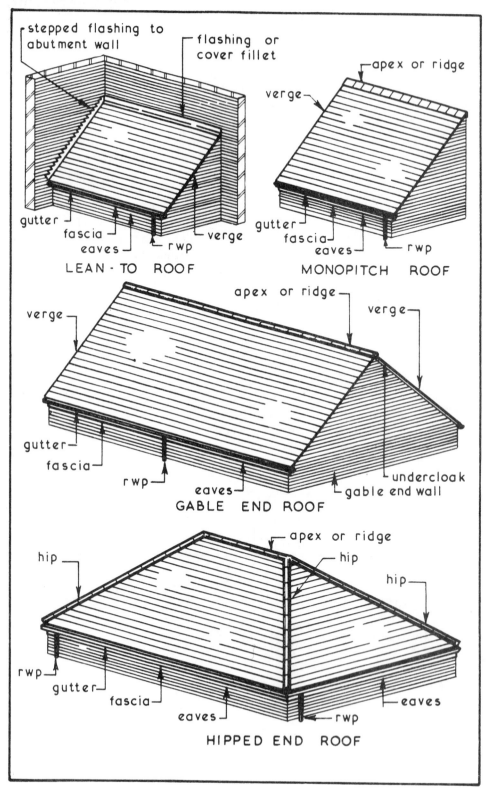

stepped flashing to
abutment wall

flashing or
cover fillet

apex or ridge

verge

gutter
fascia
eaves
rwp

verge

LEAN · TO ROOF

gutter
fascia
eaves
rwp

MONOPITCH ROOF

apex or ridge

verge

verge

gutter
fascia

rwp

eaves

undercloak
gable end wall

GABLE END ROOF

apex or ridge

hip

hip

hip

rwp
gutter

fascia

eaves

eaves

rwp

HIPPED END ROOF

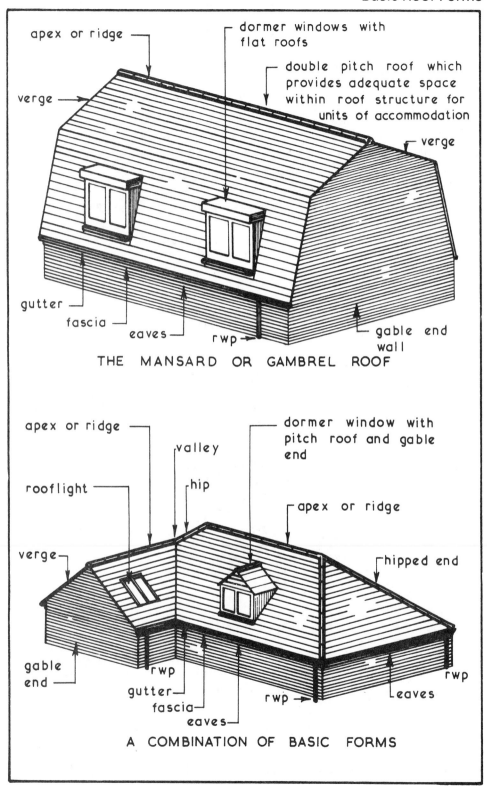

apex or ridge

dormer windows with flat roofs

double pitch roof which provides adequate space within roof structure for units of accommodation

verge

verge

gutter

fascia

eaves

rwp

gable end wall

THE MANSARD OR GAMBREL ROOF

apex or ridge

dormer window with pitch roof and gable end

valley

rooflight

hip

apex or ridge

verge

hipped end

gable end

rwp

gutter

fascia

rwp

eaves

rwp

eaves

A COMBINATION OF BASIC FORMS

321

Pitched Roofs ~ the primary functions of any domestic roof are to -
I. Provide an adequate barrier to the penetration of the elements.
2. Maintain the internal environment by providing an adequate resistance to heat loss.

A roof is in a very exposed situation and must therefore be designed and constructed in such a manner as to -
I. Safely resist all imposed loadings such as snow and wind.
2. Be capable of accommodating thermal and moisture movements.
3. Be durable so as to give a satisfactory performance and reduce maintenance to a minimum.

Component Parts of a Pitched Roof ~

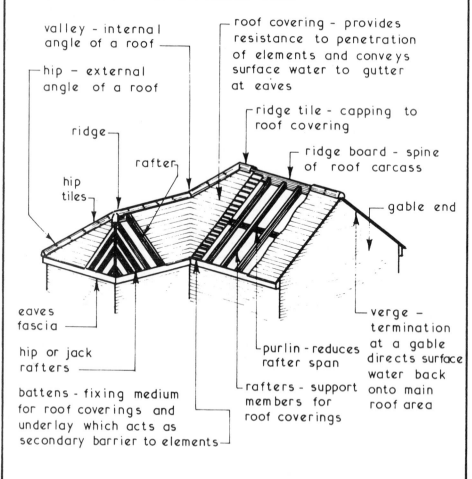

valley - internal angle of a roof

hip - external angle of a roof

roof covering - provides resistance to penetration of elements and conveys surface water to gutter at eaves

ridge

rafter

ridge tile - capping to roof covering

ridge board - spine of roof carcass

hip tiles

hip

gable end

eaves fascia

hip or jack rafters

battens - fixing medium for roof coverings and underlay which acts as secondary barrier to elements

purlin - reduces rafter span

rafters - support members for roof coverings

verge - termination at a gable directs surface water back onto main roof area

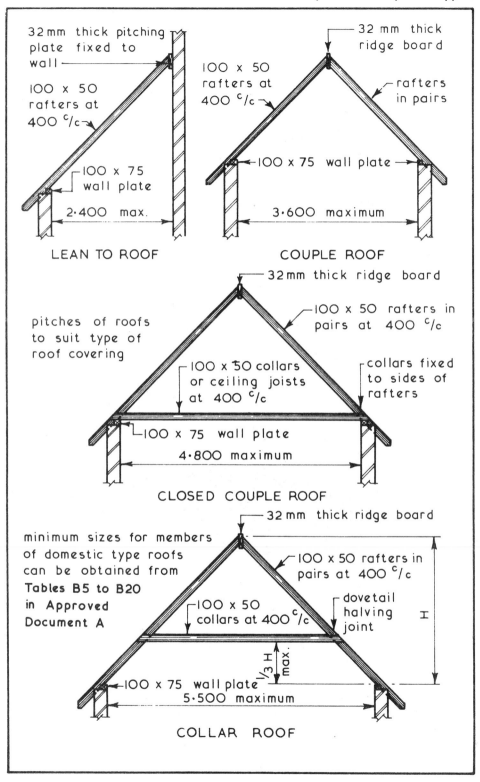

32 mm thick pitching plate fixed to wall

100 x 50 rafters at 400 $^c$/$_c$

100 x 75 wall plate

2·400 max.

LEAN TO ROOF

32 mm thick ridge board

100 x 50 rafters at 400 $^c$/$_c$

rafters in pairs

100 x 75 wall plate

3·600 maximum

COUPLE ROOF

32mm thick ridge board

100 x 50 rafters in pairs at 400 $^c$/$_c$

pitches of roofs to suit type of roof covering

100 x 50 collars or ceiling joists at 400 $^c$/$_c$

collars fixed to sides of rafters

100 x 75 wall plate

4·800 maximum

CLOSED COUPLE ROOF

32 mm thick ridge board

minimum sizes for members of domestic type roofs can be obtained from **Tables B5 to B20 in Approved Document A**

100 x 50 rafters in pairs at 400 $^c$/$_c$

100 x 50 collars at 400 $^c$/$_c$

dovetail halving joint

H

1/3 H max.

100 x 75 wall plate

5·500 maximum

COLLAR ROOF

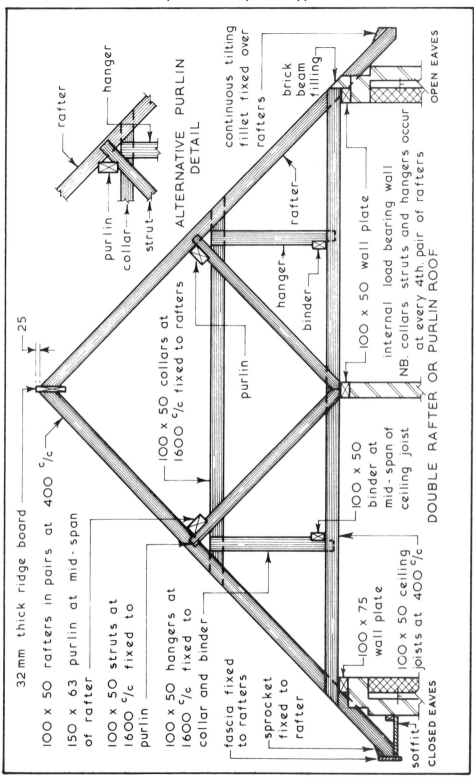

32 mm thick ridge board

100 x 50 rafters in pairs at 400 ℃

150 x 63 purlin at mid-span of rafter

100 x 50 struts at 1600 ℃ fixed to purlin

100 x 50 hangers at 1600 ℃ fixed to collar and binder

fascia fixed to rafters

sprocket fixed to rafter

soffit

CLOSED EAVES

25

rafter

hanger

purlin

collar

strut

ALTERNATIVE PURLIN DETAIL

continuous tilting fillet fixed over rafters

brick beam filling

OPEN EAVES

rafter

hanger

binder

100 x 50 wall plate

internal load bearing wall

NB. collars struts and hangers occur at every 4th. pair of rafters

DOUBLE RAFTER OR PURLIN ROOF

100 x 50 collars at 1600 ℃ fixed to rafters

purlin

100 x 50 binder at mid-span of ceiling joist

100 x 75 wall plate

100 x 50 ceiling joists at 400 ℃

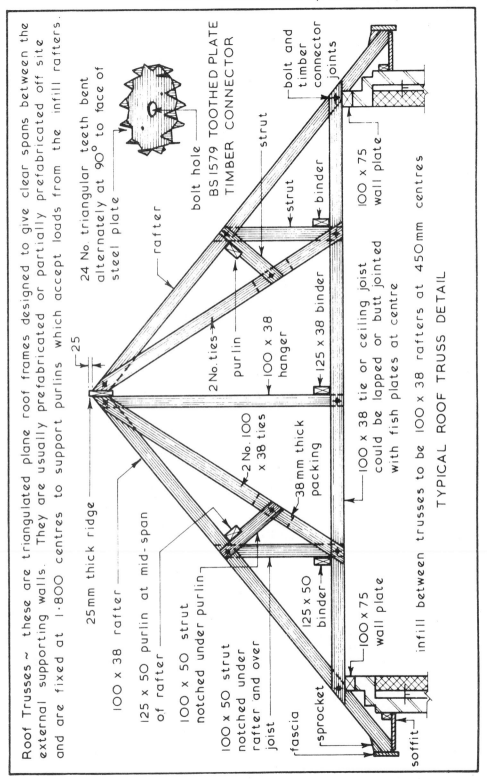

Roof Trusses ~ these are triangulated plane roof frames designed to give clear spans between the external supporting walls. They are usually prefabricated or partially prefabricated off site and are fixed at 1·800 centres to support purlins which accept loads from the infill rafters.

24 No. triangular teeth bent alternately at 90° to face of steel plate

bolt hole

BS1579 TOOTHED PLATE TIMBER CONNECTOR

25mm thick ridge

100 x 38 rafter

125 x 50 purlin at mid-span of rafter

100 x 50 strut notched under purlin

100 x 50 strut notched under rafter and over joist

fascia

sprocket

rafter

2 No. ties

purlin

2 No. 100 x 38 ties

38mm thick packing

125 x 50 binder

100 x 75 wall plate

strut

strut

binder

100 x 38 hanger

125 x 38 binder

100 x 38 tie or ceiling joist could be lapped or butt jointed with fish plates at centre

100 x 75 wall plate

bolt and timber connector joints

infill between trusses to be 100 x 38 rafters at 450mm centres

soffit

TYPICAL ROOF TRUSS DETAIL

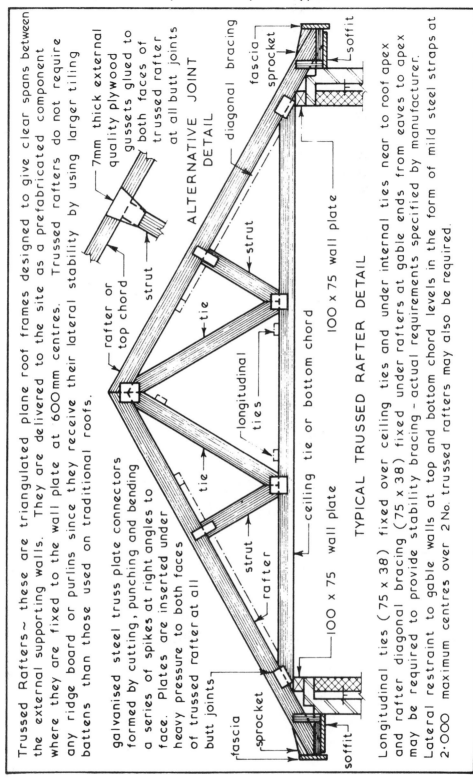

Trussed Rafters~ these are triangulated plane roof frames designed to give clear spans between the external supporting walls. They are delivered to the site as a prefabricated component where they are fixed to the wall plate at 600mm centres. Trussed rafters do not require any ridge board or purlins since they receive their lateral stability by using larger tiling battens than those used on traditional roofs.

galvanised steel truss plate connectors formed by cutting, punching and bending a series of spikes at right angles to face. Plates are inserted under heavy pressure to both faces of trussed rafter at all butt joints

7mm thick external quality plywood gussets glued to both faces of trussed rafter at all butt joints

fascia
sprocket
soffit

100 x 75 wall plate

rafter
strut

tie

strut
rafter

ceiling tie or bottom chord

100 x 75 wall plate

longitudinal ties

tie

rafter or top chord

strut

ALTERNATIVE JOINT DETAIL

diagonal bracing

fascia
sprocket

soffit

100 x 75 wall plate

TYPICAL TRUSSED RAFTER DETAIL

Longitudinal ties (75 x 38) fixed over ceiling ties and under internal ties near to roof apex and rafter diagonal bracing (75 x 38) fixed under rafters at gable ends from eaves to apex may be required to provide stability bracing - actual requirements specified by manufacturer. Lateral restraint to gable walls at top and bottom chord levels in the form of mild steel straps at 2.000 maximum centres over 2 No. trussed rafters may also be required.

Roof Underlays ~ sometimes called sarking or roofing felt provides the barrier to the entry of wind, snow and rain blown between the tiles or slates it also prevents the entry of water from capillary action.

Suitable Materials ~

Bitumen fibre based felts ⎫ supplied in rolls 1m wide
Bitumen asbestos based felts ⎬ x 10 or 20m long to the
Bitumen glass fibre based felts ⎭ recommendations of BS 747

Sheathing and Hair felts – supplied in rolls 810 mm wide x 25 m long to the recommendations of BS 747

Plastic Sheeting underlays - these are lighter, require less storage space, have greater flexibility at low temperatures and high resistance to tearing but have a greater risk to the formation of condensation than the BS 747 felts and should not be used on roof pitches below 20$°$

Typical Details ~

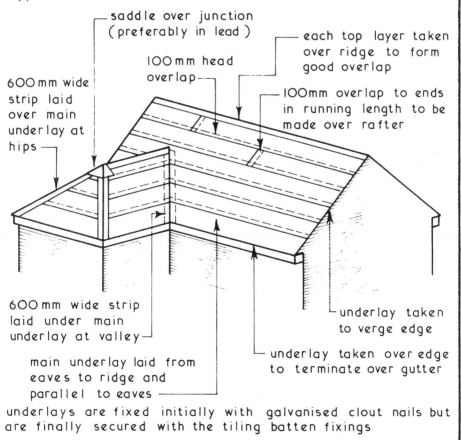

saddle over junction (preferably in lead)

100mm head overlap

600mm wide strip laid over main underlay at hips

each top layer taken over ridge to form good overlap

100mm overlap to ends in running length to be made over rafter

600 mm wide strip laid under main underlay at valley

main underlay laid from eaves to ridge and parallel to eaves

underlay taken to verge edge

underlay taken over edge to terminate over gutter

underlays are fixed initially with galvanised clout nails but are finally secured with the tiling batten fixings

Double Lap Tiles~ these are the traditional tile covering for pitched roofs and are available made from clay and concrete and are usually called plain tiles. Plain tiles have a slight camber in their length to ensure that the tail of the tile will bed and not ride on the tile below. There is always at least two layers of tiles covering any part of the roof. Each tile has at least two nibs on the underside of its head so that it can be hung on support battens nailed over the rafters. Two nail holes provide the means of fixing the tile to the batten, in practice only every 4th. course of tiles is nailed unless the roof exposure is high. Double lap tiles are laid to a bond so that the edge joints between the tiles are in the centre of the tiles immediately below and above the course under consideration.

Typical Plain Tile Details ~

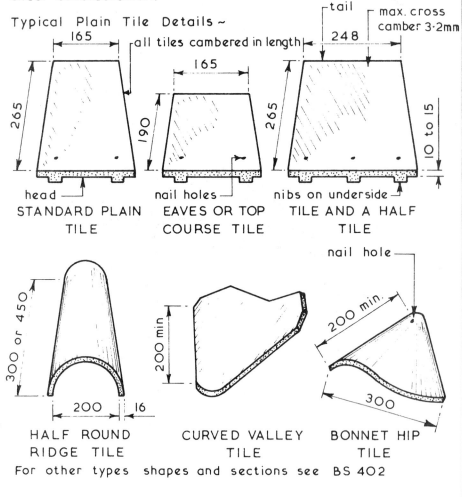

STANDARD PLAIN TILE

EAVES OR TOP COURSE TILE

TILE AND A HALF TILE

HALF ROUND RIDGE TILE

CURVED VALLEY TILE

BONNET HIP TILE

For other types shapes and sections see BS 402

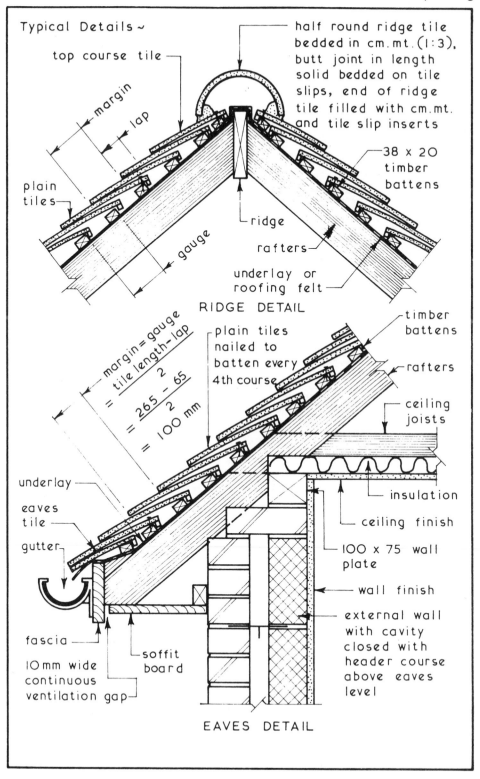

Typical Details ~

half round ridge tile
bedded in cm.mt.(1:3),
butt joint in length
solid bedded on tile
slips, end of ridge
tile filled with cm.mt.
and tile slip inserts

top course tile

margin

lap

plain
tiles

38 x 20
timber
battens

ridge

gauge

rafters

underlay or
roofing felt

RIDGE DETAIL

$$margin = gauge$$
$$= \frac{tile\ length - lap}{2}$$
$$= \frac{265 - 65}{2}$$
$$= 100\ mm$$

plain tiles
nailed to
batten every
4th course

timber
battens

rafters

ceiling
joists

underlay

insulation

eaves
tile

ceiling finish

gutter

100 x 75 wall
plate

wall finish

fascia

external wall
with cavity
closed with
header course
above eaves
level

10 mm wide
continuous
ventilation gap

soffit
board

EAVES DETAIL

329

## Double Lap Tiling

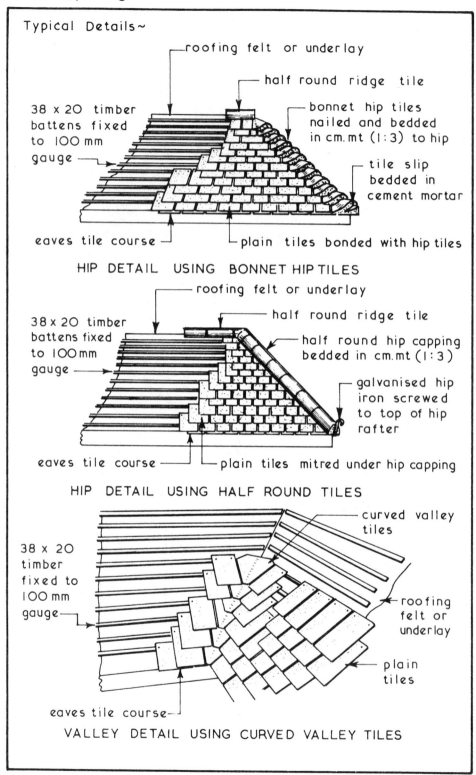

Typical Details~

roofing felt or underlay

half round ridge tile

38 x 20 timber battens fixed to 100 mm gauge

bonnet hip tiles nailed and bedded in cm. mt (1:3) to hip

tile slip bedded in cement mortar

eaves tile course

plain tiles bonded with hip tiles

**HIP DETAIL USING BONNET HIP TILES**

roofing felt or underlay

half round ridge tile

38 x 20 timber battens fixed to 100mm gauge

half round hip capping bedded in cm.mt (1:3)

galvanised hip iron screwed to top of hip rafter

eaves tile course

plain tiles mitred under hip capping

**HIP DETAIL USING HALF ROUND TILES**

curved valley tiles

38 x 20 timber fixed to 100 mm gauge

roofing felt or underlay

plain tiles

eaves tile course

**VALLEY DETAIL USING CURVED VALLEY TILES**

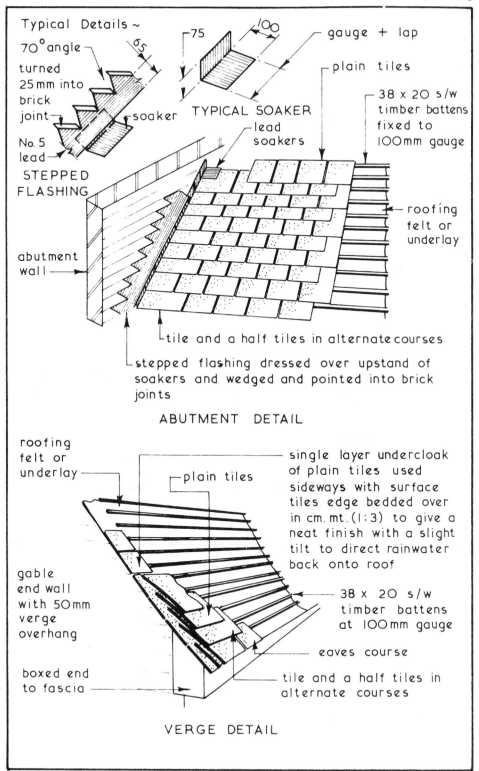

Typical Details ~

70° angle turned 25mm into brick joint

65

soaker

No. 5 lead

STEPPED FLASHING

75

100

TYPICAL SOAKER

gauge + lap

plain tiles

38 x 20 s/w timber battens fixed to 100mm gauge

lead soakers

abutment wall

roofing felt or underlay

tile and a half tiles in alternate courses

stepped flashing dressed over upstand of soakers and wedged and pointed into brick joints

ABUTMENT DETAIL

roofing felt or underlay

plain tiles

single layer undercloak of plain tiles used sideways with surface tiles edge bedded over in cm.mt.(1:3) to give a neat finish with a slight tilt to direct rainwater back onto roof

gable end wall with 50mm verge overhang

38 x 20 s/w timber battens at 100mm gauge

eaves course

boxed end to fascia

tile and a half tiles in alternate courses

VERGE DETAIL

331

Single Lap Tiling ~ so called because the single lap of one tile over another provides the weather tightness as opposed to the two layers of tiles used in double lap tiling. Most of the single lap tiles produced in clay and concrete have a tongue and groove joint along their side edges and in some patterns on all four edges which forms a series of interlocking joints and therefore these tiles are called single lap interlocking tiles. Generally there will be an overall reduction in the weight of the roof covering when compared with double lap tiling but the batten size is larger than that used for plain tiles and as a minimum every tile in alternate courses should be twice nailed although a good specification will require every tile to be twice nailed. The gauge or batten spacing for single lap tiling is found by subtracting the end lap from the length of the tile.

Typical  Single Lap  Tiles ~

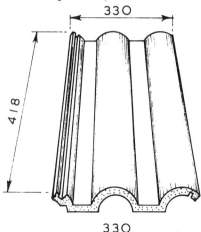

ROLL  TYPE  TILE

minimum pitch  $30°$

head lap  75mm

side lap  30mm

gauge  343mm

linear coverage  300mm

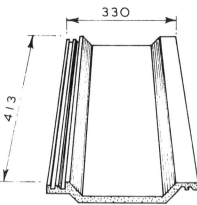

TROUGH  TYPE  TILE

minimum pitch  $15°$

head lap  75mm

side lap  38mm

gauge  338mm

linear coverage  292mm

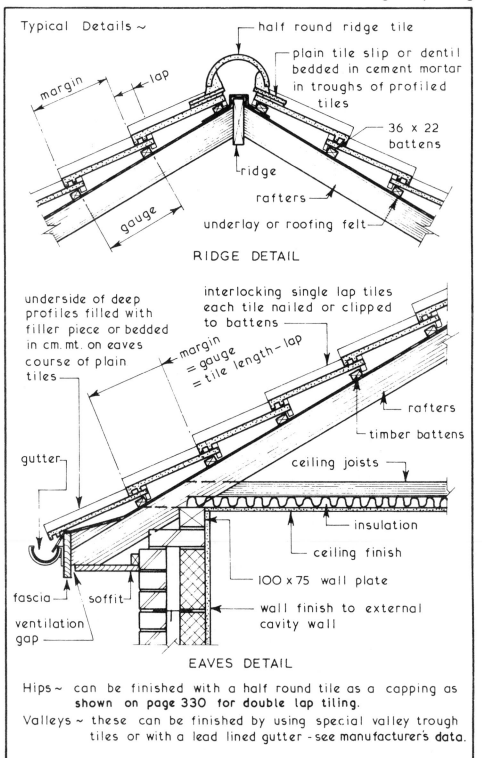

Typical Details ~

half round ridge tile

plain tile slip or dentil bedded in cement mortar in troughs of profiled tiles

margin

lap

36 x 22 battens

gauge

ridge

rafters

underlay or roofing felt

**RIDGE DETAIL**

underside of deep profiles filled with filler piece or bedded in cm. mt. on eaves course of plain tiles

interlocking single lap tiles each tile nailed or clipped to battens

margin
= gauge
= tile length - lap

rafters

timber battens

gutter

ceiling joists

insulation

ceiling finish

100 x 75 wall plate

fascia

soffit

wall finish to external cavity wall

ventilation gap

**EAVES DETAIL**

Hips ~ can be finished with a half round tile as a capping as **shown on page 330 for double lap tiling.**

Valleys ~ these can be finished by using special valley trough tiles or with a lead lined gutter - see manufacturer's data.

## Roof Slating

Slates ~ slate is a natural dense material which can be split into thin sheets and cut to form a small unit covering suitable for pitched roofs in excess of 25° pitch. Slates are graded according to thickness and texture the thinnest being known as 'Bests'. Slates are laid to the same double lap principles as plain tiles. Ridges and hips are normally covered with half round or angular tiles whereas valley junctions are usually of mitred slates over soakers. Unlike plain tiles every slate in every course is fixed to the battens by head nailing or centre nailing, the latter being used on long slates and on pitches below 35° to overcome the problem of vibration caused by the wind which can break head nailed long slates. For full range of sizes and gradings see BS 680.

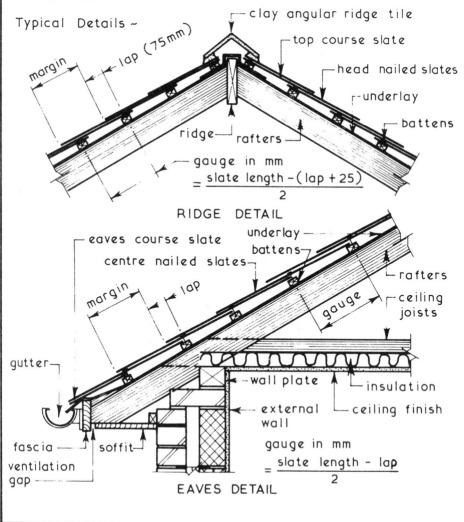

Typical Details ~

margin — lap (75mm) — clay angular ridge tile — top course slate — head nailed slates — underlay — battens — ridge — rafters

gauge in mm
$$= \frac{\text{slate length} - (\text{lap} + 25)}{2}$$

RIDGE DETAIL

eaves course slate — centre nailed slates — underlay — battens — rafters — ceiling joists — margin — lap — gauge — gutter — wall plate — insulation — external wall — ceiling finish — fascia — soffit — ventilation gap

gauge in mm
$$= \frac{\text{slate length} - \text{lap}}{2}$$

EAVES DETAIL

334

Flat Roofs ~ these roofs are very seldom flat with a pitch of 0° but are considered to be flat if the pitch does not exceed 10.° The actual pitch chosen can be governed by the roof covering selected and /or by the required rate of rainwater discharge off the roof. As a general rule the minimum pitch for smooth surfaces such as asphalt should be 1:80 or 0°-43' and for sheet coverings with laps 1:60 or 0°-57.'

Methods of Obtaining Falls ~

1. Joists cut to falls

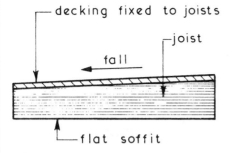

Simple to fix but could be wasteful in terms of timber unless two joists are cut from one piece of timber

2. Joists laid to falls

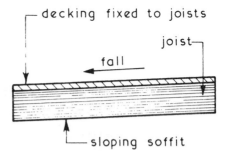

Economic and simple but sloping soffit may not be acceptable but this could be hidden by a flat suspended ceiling

3. Firrings with joist run

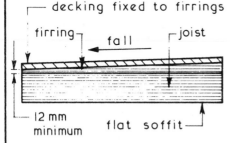

Simple and effective but does not provide a means of natural cross ventilation. Usual method employed.

4. Firrings against joist run

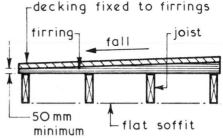

Simple and effective but uses more timber than 3 but does provide a means of natural cross ventilation

Wherever possible joists should span the shortest distance of the roof plan

## Timber Flat Roofs up to 4m Span

Timber Roof Joists ~ the spacing and sizes of joists is related to the loadings and span actual dimensions for **domestic loadings can be taken direct from Tables B21 to B24 of Approved Document A or they can be calculated from** first principles in the same manner as used for timber upper floors. Strutting between joists should be used if the span exceeds 2·400 to restrict joist movements and twisting.

Typical Eaves Details ~

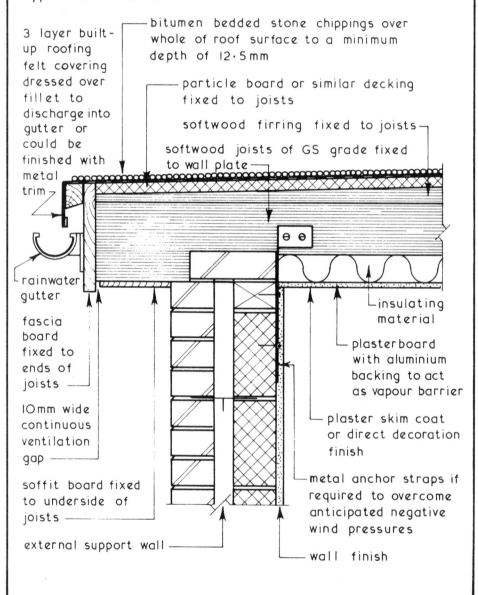

3 layer built-up roofing felt covering dressed over fillet to discharge into gutter or could be finished with metal trim

bitumen bedded stone chippings over whole of roof surface to a minimum depth of 12·5mm

particle board or similar decking fixed to joists

softwood firring fixed to joists

softwood joists of GS grade fixed to wall plate

rainwater gutter

fascia board fixed to ends of joists

10mm wide continuous ventilation gap

soffit board fixed to underside of joists

external support wall

insulating material

plasterboard with aluminium backing to act as vapour barrier

plaster skim coat or direct decoration finish

metal anchor straps if required to overcome anticipated negative wind pressures

wall finish

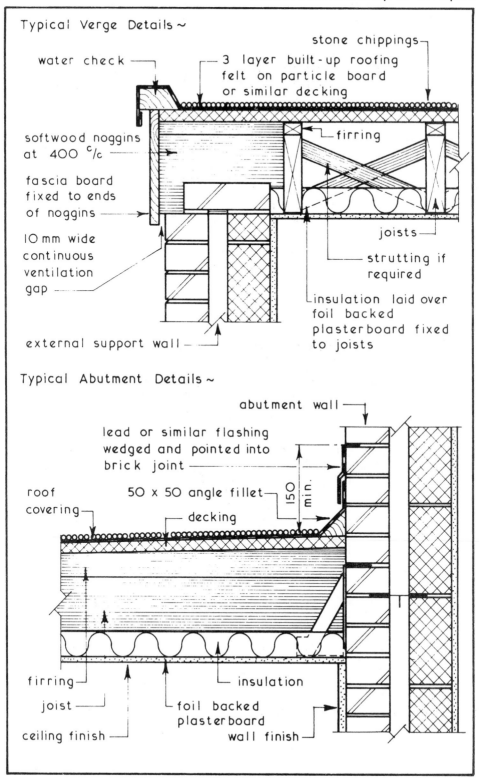

Typical Verge Details ~

water check

3 layer built-up roofing felt on particle board or similar decking

stone chippings

softwood noggins at 400 c/c

firring

fascia board fixed to ends of noggins

10 mm wide continuous ventilation gap

joists

strutting if required

insulation laid over foil backed plasterboard fixed to joists

external support wall

Typical Abutment Details ~

abutment wall

lead or similar flashing wedged and pointed into brick joint

roof covering

50 x 50 angle fillet

decking

150 min.

firring

joist

ceiling finish

insulation

foil backed plasterboard

wall finish

# Typical Timber Flat Roof Coverings

Built-up Roofing Felt ~ this consists of three layers of bitumen roofing felt to BS 747 and should be laid to the recommendations of CP 144 Part 3. The layers of felt are bonded together with hot bitumen and should have staggered laps of 50mm minimum for side laps and 75mm minimum for **end laps - for typical details see pages 336 & 337.** Other felt materials which could be used are the two layer polyester based roofing felts which use a non-woven polyester base instead of the woven base used in the BS 747 felts.

Mastic Asphalt ~ this consists of two layers of mastic asphalt laid breaking the joints and built up to a minimum thickness of 20mm and should be laid to the recommendations of CP 144 Part 4. The mastic asphalt is laid over an isolating membrane of black sheathing felt complying with BS 747A (i) which should be laid loose with 50mm minimum overlaps.

Typical  Details ~

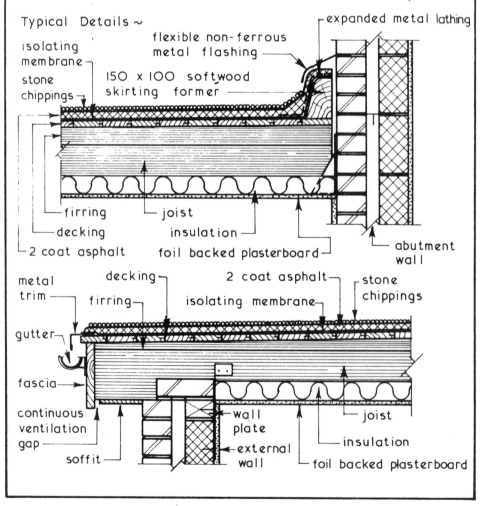

Steel Roof Trusses ~ these are triangulated plane frames which carry purlins to which the roof coverings can be fixed. Steel is stronger than timber and will not spread fire over its surface and for these reasons it is often preferred to timber for medium and long span roofs. The rafters are restrained from spreading by being connected securely at their feet by a tie member, struts and ties are provided within the basic triangle to give adequate bracing. Angle sections are usually employed for steel truss members since they are economic and accept both tensile and compressive stresses. The members of a steel roof truss are connected together with bolts or by welding to shaped plates called gussets. Steel trusses are usually placed at 3·000 to 4·500 centres which gives an economic purlin size.

Typical Steel Roof Truss Formats ~

tie – tension member
strut – compression member

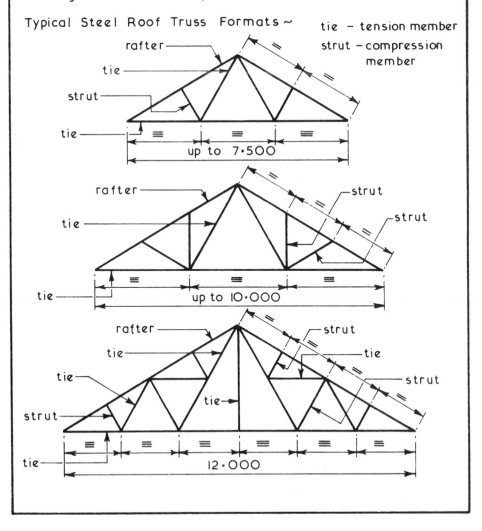

rafter
tie
strut
tie
up to 7·500

rafter
tie
strut
strut
tie
up to 10·000

rafter
tie
tie
strut
tie
strut
tie
12·000

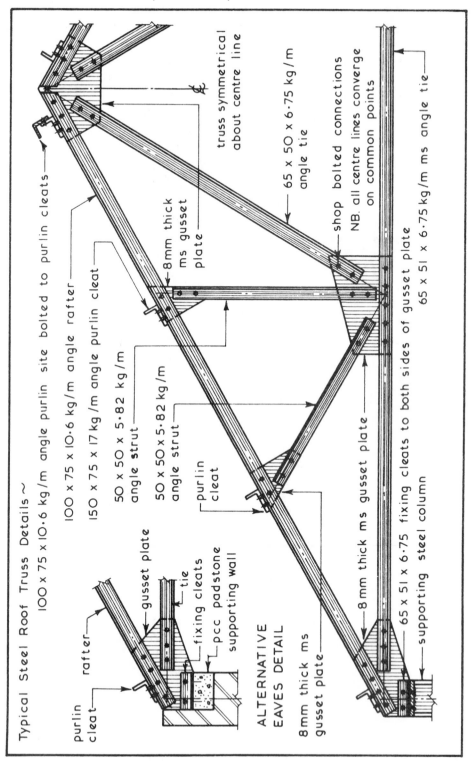

Typical Steel Roof Truss Details ~

100 × 75 × 10·6 kg/m angle purlin site bolted to purlin cleats

100 × 75 × 10·6 kg/m angle rafter

150 × 75 × 17 kg/m angle purlin cleat

50 × 50 × 5·82 kg/m angle strut

50 × 50 × 5·82 kg/m angle strut

purlin cleat

8mm thick ms gusset plate

truss symmetrical about centre line

65 × 50 × 6·75 kg/m angle tie

shop bolted connections NB. all centre lines converge on common points

65 × 51 × 6·75 kg/m ms angle tie

65 × 51 × 6·75 fixing cleats to both sides of gusset plate

8mm thick ms gusset plate

65 × 51 × 6·75 fixing cleats to both sides of gusset plate

supporting steel column

ALTERNATIVE EAVES DETAIL

8mm thick ms gusset plate

purlin cleat

rafter

gusset plate

tie

fixing cleats

pcc padstone

supporting wall

340

Sheet Coverings ~ the basic functions of sheet coverings used in conjunction with steel roof trusses are to :—

1. Provide resistance to penetration by the elements.
2. Provide restraint to wind and snow loads.
3. Provide a degree of thermal insulation of not less than that set out in Part L of the Building Regulations.
4. Provide resistance to surface spread of flame as set out in Part B of the Building Regulations.
5. Provide any natural daylight required through the roof in accordance with the maximum permitted areas set out in Part L of the Building Regulations.
6. Be of low self weight to give overall design economy.
7. Be durable to keep maintenance needs to a minimum.

Suitable Materials ~

Hot-dip galvanised corrugated steel sheets — BS 3083

Asbestos cement profiled sheets — BS 690

Aluminium profiled sheets - BS 4868.

Asbestos free profiled sheets — various manufacturers whose products are usually based on a mixture of Portland cement, mineral fibres and density modifiers.

Typical Profiles ~

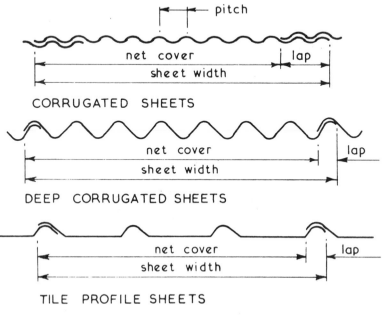

CORRUGATED SHEETS

DEEP CORRUGATED SHEETS

TILE PROFILE SHEETS

## Roof Sheet Coverings

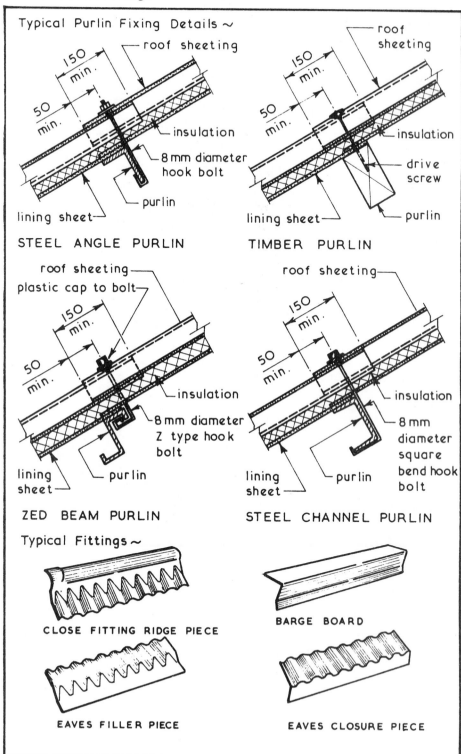

Typical Purlin Fixing Details ~

roof sheeting

150 min.

50 min.

insulation

8 mm diameter hook bolt

purlin

lining sheet

**STEEL ANGLE PURLIN**

roof sheeting

150 min.

50 min.

insulation

drive screw

purlin

lining sheet

**TIMBER PURLIN**

roof sheeting

plastic cap to bolt

150 min.

50 min.

insulation

8 mm diameter Z type hook bolt

lining sheet

purlin

**ZED BEAM PURLIN**

roof sheeting

150 min.

50 min.

insulation

8 mm diameter square bend hook bolt

lining sheet

purlin

**STEEL CHANNEL PURLIN**

Typical Fittings ~

**CLOSE FITTING RIDGE PIECE**

**BARGE BOARD**

**EAVES FILLER PIECE**

**EAVES CLOSURE PIECE**

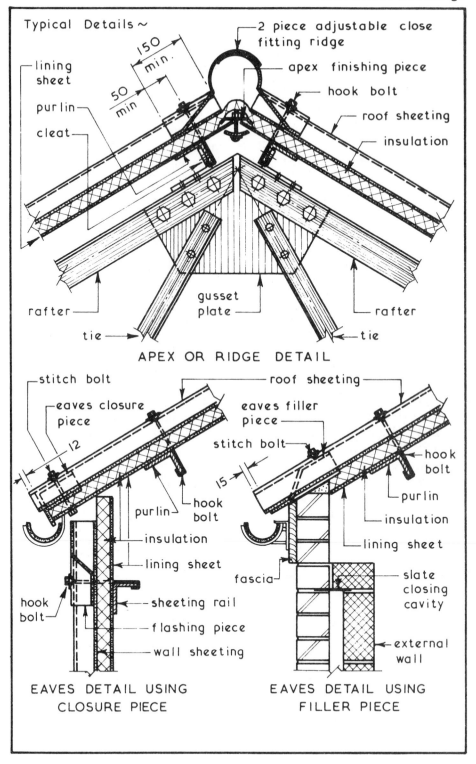

Typical Details ~

**APEX OR RIDGE DETAIL**

- 2 piece adjustable close fitting ridge
- apex finishing piece
- hook bolt
- roof sheeting
- insulation
- lining sheet
- purlin
- cleat
- 150 min.
- 50 min
- rafter
- gusset plate
- rafter
- tie
- tie

**EAVES DETAIL USING CLOSURE PIECE**

- stitch bolt
- eaves closure piece
- 12
- roof sheeting
- hook bolt
- purlin
- insulation
- lining sheet
- sheeting rail
- flashing piece
- wall sheeting
- hook bolt

**EAVES DETAIL USING FILLER PIECE**

- eaves filler piece
- stitch bolt
- 15
- hook bolt
- purlin
- insulation
- lining sheet
- fascia
- slate closing cavity
- external wall

Long Span Roofs ~ these can be defined as those exceeding 12·000 in span. They can be fabricated in steel, aluminium alloy, timber, reinforced concrete and prestressed concrete. Long span roofs can be used for buildings such as factories, large public halls and gymnasiums which require a large floor area free of roof support columns. The primary roof functions of providing weather protection, thermal insulation, sound insulation and restricting spread of fire over the roof surface are common to all roof types but these roofs may also have to provide strength sufficient to carry services, lifting equipment and provide for natural daylight to the interior by means of rooflights.

Basic Roof Forms ~

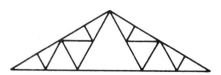

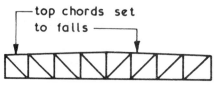

top chords set to falls

Pitched Trusses - spaced at suitable centres to carry purlins to which the roof coverings are fixed. Good rainwater run off - reasonable daylight spread from rooflights - high roof volume due to the triangulated format - on long spans roof volume can be reduced by using a series of short span trusses.

Flat Top Girders - spaced at suitable centres to carry purlins to which the roof coverings are fixed. Low pitch to give acceptable rainwater run off - reasonable daylight spread from rooflights - can be designed for very long spans but depth and hence roof volume increases with span.

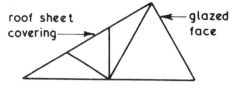

roof sheet covering — glazed face

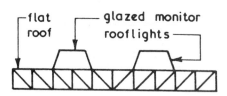

flat roof — glazed monitor rooflights

Northlight - spaced at suitable centres to carry purlins to which roof sheeting is fixed. Good rainwater run off - if correctly orientated solar glare is eliminated - long spans can be covered by a series of short span frames

Monitor - girders or cranked beams at centres to suit low pitch decking used. Good even daylight spread from monitor lights which is not affected by orientation of building.

Pitched Trusses ~ these can be constructed with a symmetrical outline (as shown on pages 339 to 343) or with an asymmetrical outline (Northlight – see detail below). They are usually made from standard steel sections with shop welded or bolted connections, alternatively they can be fabricated using timber members joined together with bolts and timber connectors or formed as a precast concrete portal frame.

Typical Multi-span Northlight Roof Details ~

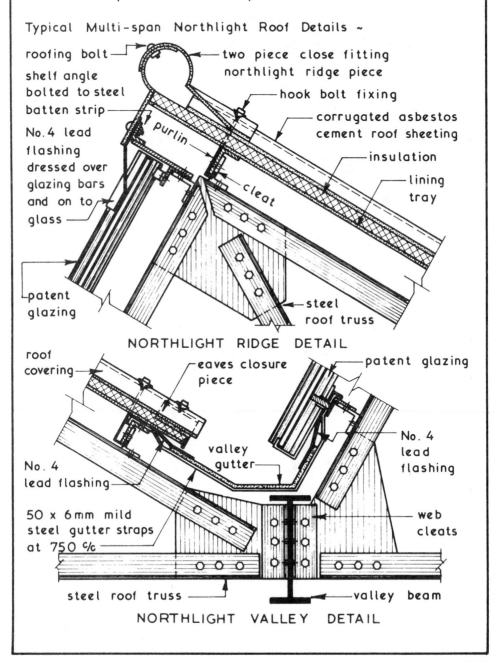

roofing bolt

shelf angle bolted to steel batten strip

No. 4 lead flashing dressed over glazing bars and on to glass

patent glazing

purlin

cleat

two piece close fitting northlight ridge piece

hook bolt fixing

corrugated asbestos cement roof sheeting

insulation

lining tray

steel roof truss

NORTHLIGHT RIDGE DETAIL

roof covering

eaves closure piece

patent glazing

No. 4 lead flashing

valley gutter

No. 4 lead flashing

50 x 6mm mild steel gutter straps at 750 c/c

web cleats

steel roof truss

valley beam

NORTHLIGHT VALLEY DETAIL

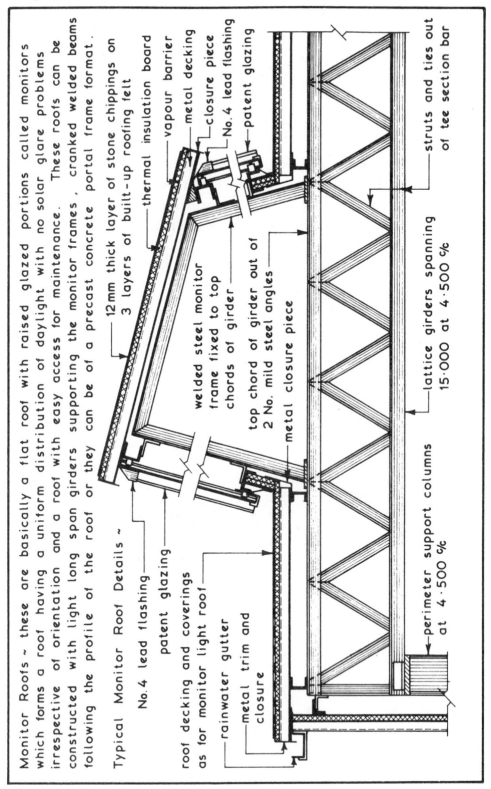

Monitor Roofs ~ these are basically a flat roof with raised glazed portions called monitors which forms a roof having a uniform distribution of daylight with no solar glare problems irrespective of orientation and a roof with easy access for maintenance. These roofs can be constructed with light long span girders supporting the monitor frames, cranked welded beams following the profile of the roof or they can be of a precast concrete portal frame format.

Typical Monitor Roof Details ~

No.4 lead flashing

patent glazing

roof decking and coverings as for monitor light roof

rainwater gutter

metal trim and closure

12mm thick layer of stone chippings on 3 layers of built-up roofing felt

thermal insulation board

vapour barrier

metal decking

closure piece

No.4 lead flashing

patent glazing

welded steel monitor frame fixed to top chords of girder

top chord of girder out of 2 No. mild steel angles

metal closure piece

struts and ties out of tee section bar

lattice girders spanning 15·000 at 4·500 c/c

perimeter support columns at 4·500 c/c

Flat Top Girders ~ these are suitable for roof spans ranging from 15·000 to 45·000 and are basically low pitched lattice beams used to carry purlins which support the roof coverings. One of the main advantages of this form of roof is the reduction in roof volume. The usual materials employed in the fabrication of flat top girders are timber and steel.

Typical Flat Top Girder Details ~

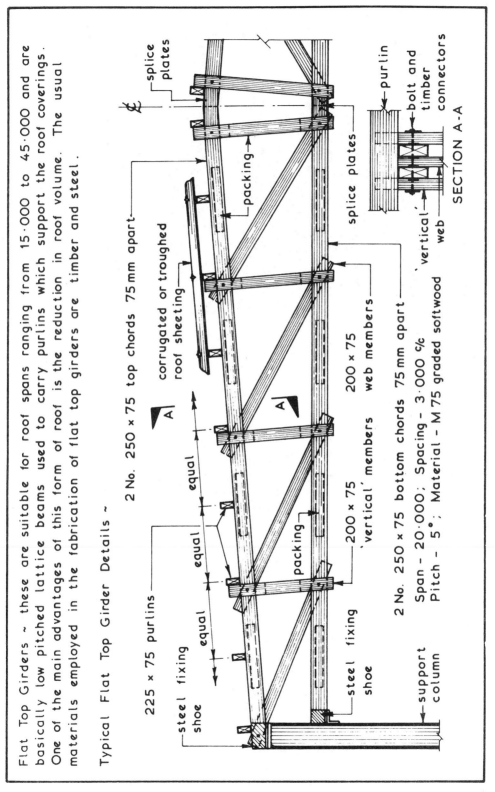

347

Connections~ nails, screws and bolts have their limitations when used to join structural timber members. The low efficiency of joints made with a rigid bar such as a bolt is caused by the usual low shear strength of timber parallel to the grain and the non-uniform distribution of bearing stress along the shank of the bolt -

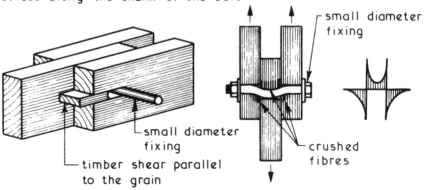

small diameter
fixing

small diameter
fixing

timber shear parallel
to the grain

crushed
fibres

SHEARING  EFFECT                STRESS  DISTRIBUTION

Timber Connectors~ these are designed to overcome the problems of structural timber connections outlined above by increasing the effective bearing area of the bolts.

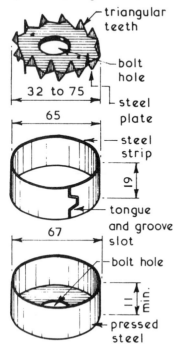

triangular
teeth

bolt
hole

32 to 75

steel
plate

65

steel
strip

19

tongue
and groove
slot

67

bolt hole

pressed
steel

Toothed Plate Connector – provides an efficient joint without special tools or equipment - suitable for all connections especially small sections - bolt holes are drilled 2mm larger than the bolt diameter, the timbers forming the joint being held together whilst being drilled.

Split Ring Connector – very efficient and develops a high joint strength – suitable for all connections – split ring connectors are inserted into a precut groove formed with a special tool making the connector independent from the bolt.

Shear Plate Connector – counterpart of a split ring connector – housed flush into timber – used for temporary joints.

Space Deck ~ this is a structural roofing system based on a simple repetitive pyramidal unit to give large clear spans of up to 22·000 for single spanning designs and up to 33·000 for two way spanning designs. The steel units are easily transported to site before assembly into beams and the complete space deck at ground level before being hoisted into position on top of the perimeter supports. The recommended roof covering is 50mm thick wood wool slabs with built-up roofing felt although any suitable lightweight decking could be used. Rooflights can be mounted directly onto the square top space deck units.

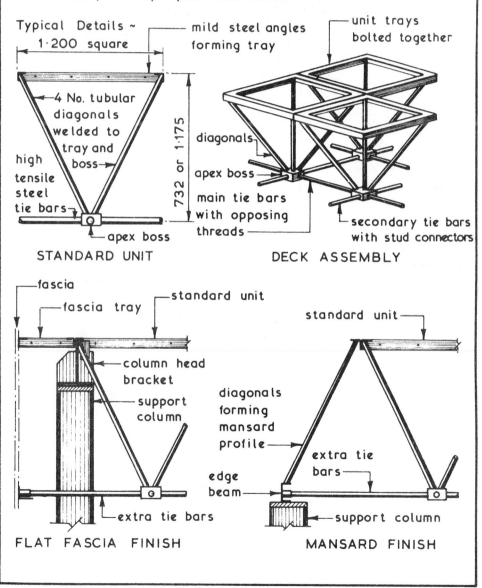

Typical Details ~

1·200 square

mild steel angles forming tray

4 No. tubular diagonals welded to tray and boss

high tensile steel tie bars

apex boss

732 or 1·175

STANDARD UNIT

unit trays bolted together

diagonals

apex boss

main tie bars with opposing threads

secondary tie bars with stud connectors

DECK ASSEMBLY

fascia

fascia tray

standard unit

column head bracket

support column

extra tie bars

FLAT FASCIA FINISH

standard unit

diagonals forming mansard profile

extra tie bars

edge beam

support column

MANSARD FINISH

Space Frames ~ these are roofing systems which consist of basically a series of connectors which joins together the chords and bracing members of the system. Single or double layer grids are possible, the former usually employed in connection with small domes or curved roofs. Space frames are similar in concept to space decks but they have greater flexibility in design and layout possibilities. Most space frames are fabricated from structural steel tubes or tubes of aluminium alloy although any suitable structural material could be used.

Typical Examples ~

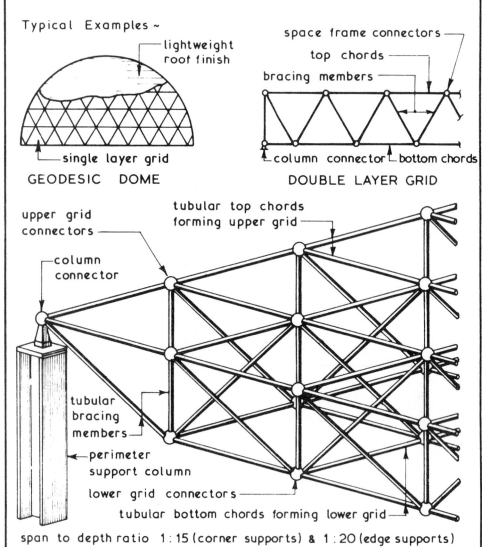

lightweight roof finish

single layer grid

GEODESIC DOME

space frame connectors

top chords

bracing members

column connector — bottom chords

DOUBLE LAYER GRID

upper grid connectors

column connector

tubular top chords forming upper grid

tubular bracing members

perimeter support column

lower grid connectors

tubular bottom chords forming lower grid

span to depth ratio 1:15 (corner supports) & 1:20 (edge supports)

TYPICAL DOUBLE LAYER GRID FORMAT

Shell Roofs ~ these can be defined as a structural curved skin covering a given plan shape and area where the forces in the shell or membrane are compressive and in the restraining edge beams are tensile. The usual materials employed in shell roof construction are insitu reinforced concrete and timber. Concrete shell roofs are constructed over formwork which in itself is very often a shell roof making this format expensive since the principle of use and reuse of formwork can not normally be applied. The main factors of shell roofs are :-

1. The entire roof is primarily a structural element.

2. Basic strength of any particular shell is inherent in its geometrical shape and form.

3. Comparatively less material is required for shell roofs than other forms of roof construction.

Domes - these are double curvature shells which can be rotationally formed by any curved geometrical plane figure rotating about a central vertical axis. Translation domes are formed by a curved line moving over another curved line whereas pendentive domes are formed by inscribing within the base circle a regular polygon and vertical planes through the true hemispherical dome.

Typical Examples ~

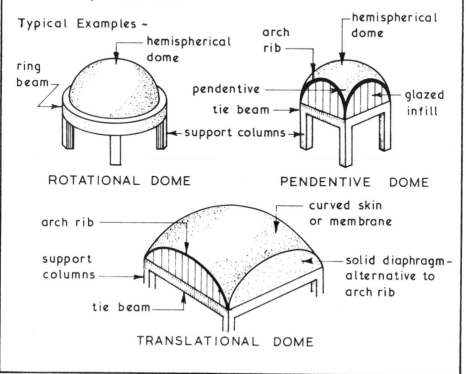

ROTATIONAL DOME

PENDENTIVE DOME

TRANSLATIONAL DOME

Barrel Vaults ~ these are single curvature shells which are essentially a cut cylinder which must be restrained at both ends to overcome the tendency to flatten. A barrel vault acts as a beam whose span is equal to the length of the roof. Long span barrel vaults are those whose span is longer than its width or chord length and conversely short barrel vaults are those whose span is shorter than its width or chord length. In very long span barrel vaults thermal expansion joints will be required at 30·000 centres which will create a series of abutting barrel vault roofs weather sealed together (see page 353 ).

Typical Single Barrel Vault Principles ~

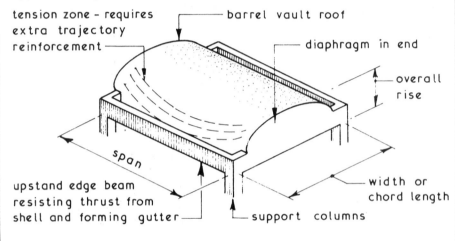

tension zone - requires extra trajectory reinforcement — barrel vault roof — diaphragm in end — overall rise

span

upstand edge beam resisting thrust from shell and forming gutter — width or chord length — support columns

economic design ratios - width : span 1 : 2 to 1 : 5
rise : span 1 : 10 to 1 : 15

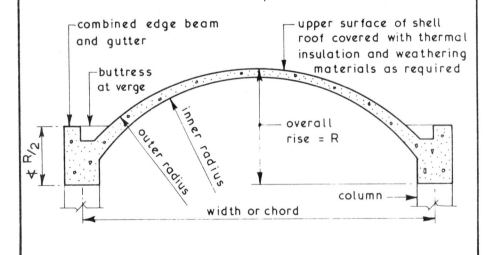

combined edge beam and gutter

upper surface of shell roof covered with thermal insulation and weathering materials as required

buttress at verge

inner radius

outer radius

overall rise = R

R/2

column

width or chord

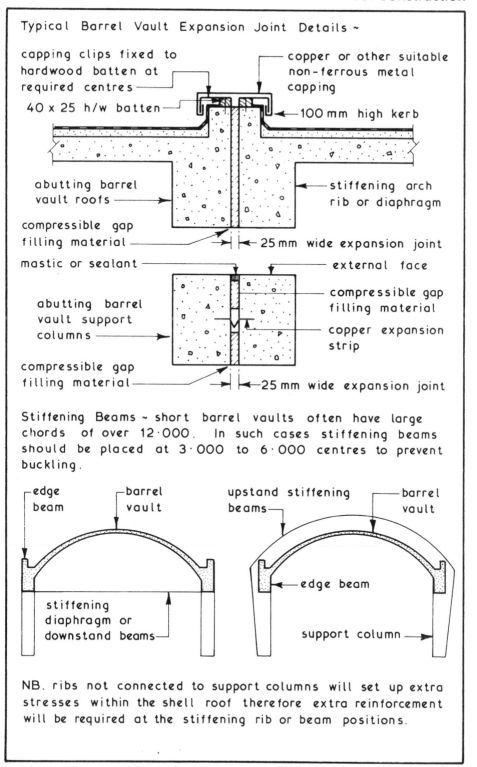

Typical Barrel Vault Expansion Joint Details ~

capping clips fixed to hardwood batten at required centres

copper or other suitable non-ferrous metal capping

40 x 25 h/w batten

100 mm high kerb

abutting barrel vault roofs

stiffening arch rib or diaphragm

compressible gap filling material

25 mm wide expansion joint

mastic or sealant

external face

abutting barrel vault support columns

compressible gap filling material

copper expansion strip

compressible gap filling material

25 mm wide expansion joint

Stiffening Beams ~ short barrel vaults often have large chords of over 12·000. In such cases stiffening beams should be placed at 3·000 to 6·000 centres to prevent buckling.

edge beam

barrel vault

upstand stiffening beams

barrel vault

stiffening diaphragm or downstand beams

edge beam

support column

NB. ribs not connected to support columns will set up extra stresses within the shell roof therefore extra reinforcement will be required at the stiffening rib or beam positions.

Other Forms of Barrel Vault ~ by cutting intersecting and placing at different levels the basic barrel vault roof can be formed into a groin or northlight barrel vault roof :-

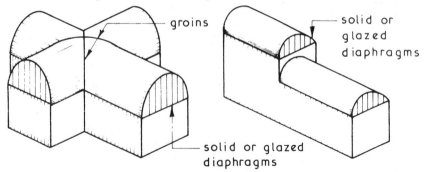

INTERSECTING BARREL VAULTS    STEPPED BARREL VAULTS

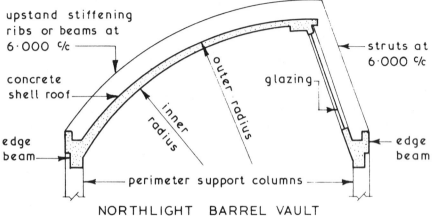

NORTHLIGHT BARREL VAULT

Conoids ~ these are double curvature shell roofs which can be considered as an alternative to barrel vaults. Spans up to 12·000 with chord lengths up to 24·000 are possible. Typical chord to span ratio 2 : 1.

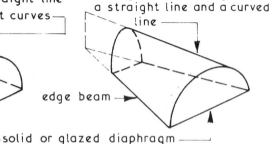

Hyperbolic Paraboloids ~ the true hyperbolic paraboloid shell roof shape is generated by moving a vertical parabola (the generator) over another vertical parabola (the directrix) set at right angles to the moving parabola. This forms a saddle shape where horizontal sections taken through the roof are hyperbolic in format and vertical sections are parabolic. The resultant shape is not very suitable for roofing purposes therefore only part of the saddle shape is used and this is formed by joining the centre points thus :-

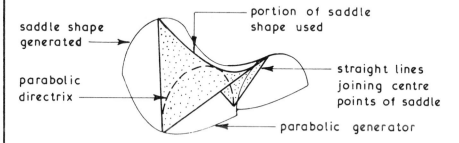

saddle shape generated ⟶

parabolic directrix ⟶

portion of saddle shape used

straight lines joining centre points of saddle

parabolic generator

To obtain a more practical shape than the true saddle a straight line limited hyperbolic paraboloid is used. This is formed by raising or lowering one or more corners of a square forming a warped parallelogram thus :-

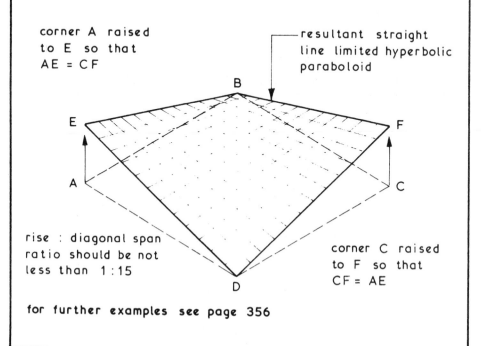

corner A raised to E so that AE = CF

resultant straight line limited hyperbolic paraboloid

rise : diagonal span ratio should be not less than 1 : 15

corner C raised to F so that CF = AE

for further examples see page 356

# Shell Roof Construction

Typical Straight Line Limited Hyperbolic Paraboloid Formats~

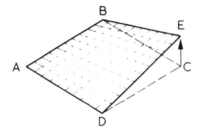

corner C raised to E

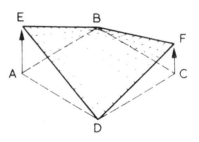

corner A raised to E and
corner C raised to F so that
AE ≠ CF

resultant hyperbolic
paraboloid ────────┐

original
square ───┐

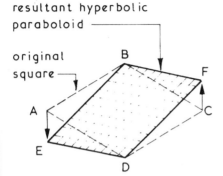

corner A lowered to E and
corner C raised to F so that
AE = CF

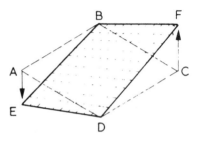

corner A lowered to E and
corner C raised to F so that
AE ≠ CF

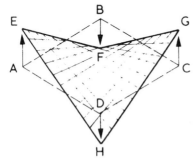

corners A & C raised to E & G
corners B & D lowered to F & H
so that AE = CG & BF = DH

Combination of Hyperbolic
Paraboloid Shell Roofs ~

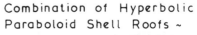
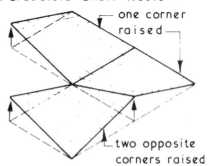

one corner
raised ───┐

two opposite
corners raised

NB. any combination possible

Concrete Hyperbolic Paraboloid Shell Roofs ~ these can be constructed in reinforced concrete (characteristic strength 25 or 30 N/mm² ) with a minimum shell thickness of 50 mm with diagonal spans up to 35·000. These shells are cast over a timber form in the shape of the required hyperbolic paraboloid format. In practice therefore two roofs are constructed and it is one of the reasons for the popularity of timber versions of this form of shell roof.

Timber Hyperbolic Paraboloid Shell Roofs ~ these are usually constructed using laminated edge beams and layers of t & g boarding to form the shell membrane. For roofs with a plan size of up to 6·000 × 6·000 only 2 layers of boards are required and these are laid parallel to the diagonals with both layers running in opposite directions. Roofs with a plan size of over 6·000 × 6·000 require 3 layers of board as shown below. The weather protective cover can be of any suitable flexible material such as built-up roofing felt, copper and lead. During construction the relatively lightweight roof is tied down to a framework of scaffolding until the anchorages and wall infilling have been completed. This is to overcome any negative and positive wind pressures due to the open sides.

Typical Details ~

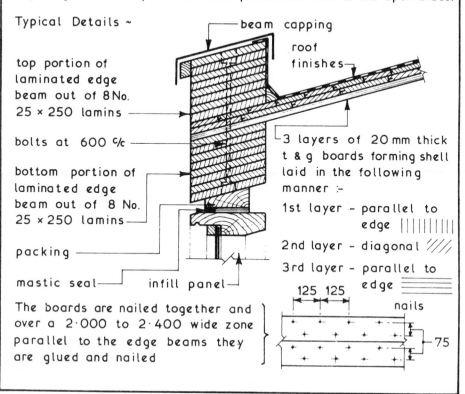

top portion of laminated edge beam out of 8 No. 25 × 250 lamins

bolts at 600 ℃

bottom portion of laminated edge beam out of 8 No. 25 × 250 lamins

packing

mastic seal        infill panel

beam capping

roof finishes

3 layers of 20 mm thick t & g boards forming shell laid in the following manner :-

1st layer - parallel to edge ||||||||||||
2nd layer - diagonal ////
3rd layer - parallel to edge ≡

125  125

nails

75

The boards are nailed together and over a 2·000 to 2·400 wide zone parallel to the edge beams they are glued and nailed

## Shell Roof Construction

Support Considerations ~ in timber hyperbolic paraboloid shell roofs only two supports are required :-

Edge beams are in compression forces P are transmitted to B and D resulting in a vertical force V and a horizontal force H at both positions therefore support columns are required at B and D.

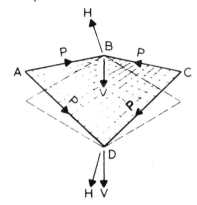

Vertical force V is transmitted directly down the columns to a suitable foundation. The outward or horizontal force H can be accommodated in one of two ways :-

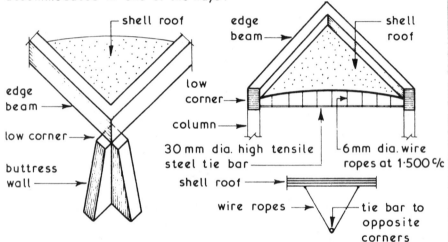

If shell roof is to be supported at high corners the edge beams will be in tension and horizontal force will be inwards. This can be resisted by a diagonal strut between the high corners.

Combination Roof Support Example ~

4 No. roof shells of equal loading joined together

Supports required at A; C; G and E. with ties between AC; CE; EG and GA. Forces at J cancel each other therefore no support required at J.

**Fig 83 Shell Roof Construction 8**

358

Rooflights ~ the useful penetration of daylight through the windows in external walls of buildings is from 6·000 to 9·000 depending on the height and size of the window. In buildings with spans over 18·000 side wall daylighting needs to be supplemented by artificial lighting or in the case of top floors or single storey buildings by rooflights. The total maximum area of wall window openings and rooflights for the various purpose groups is set out in the Building Regulations with allowances for increased areas if double or triple glazing is used. In pitched roofs such as northlight and monitor roofs the rooflights are usually in the form of patent glazing (see **Long Span Roofs** on pages 345 and 346) In flat roof construction natural daylighting can be provided by one or more of the following methods :-

1. **Lantern lights - see page 361**

2. **Lens lights - see page 361**

3. **Dome, pyramid and similar rooflights - see page 362**

Patent Glazing ~ these are systems of steel or aluminium alloy glazing bars which span the distance to be glazed whilst giving continuous edge support to the glass. They can be used in the roof forms noted above as well as in pitched roofs with profiled coverings where the patent glazing bars **are fixed above and below the profiled sheets - see page 360** .

Typical Patent Glazing Bar Sections ~

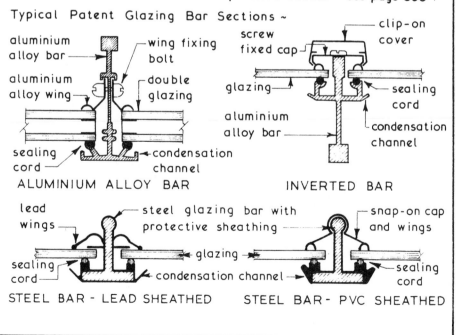

ALUMINIUM ALLOY BAR    INVERTED BAR

STEEL BAR - LEAD SHEATHED    STEEL BAR - PVC SHEATHED

# Rooflights

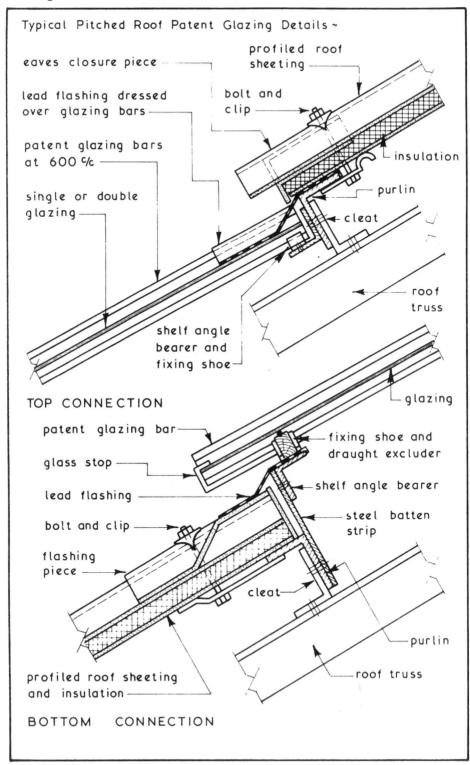

Typical Pitched Roof Patent Glazing Details ~

**TOP CONNECTION**

- profiled roof sheeting
- eaves closure piece
- bolt and clip
- lead flashing dressed over glazing bars
- insulation
- patent glazing bars at 600 c/c
- purlin
- cleat
- single or double glazing
- roof truss
- shelf angle bearer and fixing shoe

**BOTTOM CONNECTION**

- glazing
- patent glazing bar
- fixing shoe and draught excluder
- glass stop
- shelf angle bearer
- lead flashing
- steel batten strip
- bolt and clip
- flashing piece
- cleat
- purlin
- profiled roof sheeting and insulation
- roof truss

Lantern Lights ~ these are a form of rooflight used in conjunction with flat roofs. They consist of glazed vertical sides and fully glazed pitched roof which is usually hipped at both ends. Part of the glazed upstand sides is usually formed as an opening light or alternatively glazed with louvres to provide a degree of controllable ventilation. They can be constructed of timber, metal or a combination of these two materials. Lantern lights in the context of new buildings have been generally superseded by the various **forms of dome light (see page 362).**

Typical Lantern Light Details ~

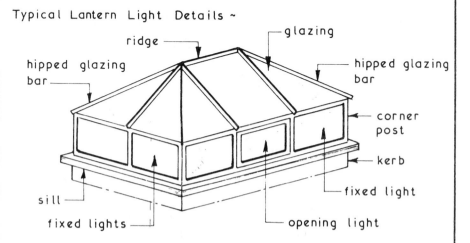

Lens Lights ~ these are small square or round blocks of translucent toughened glass especially designed for casting into concrete and are suitable for use in flat roofs and curved roofs such as barrel vaults. They can also be incorporated in precast concrete frames for inclusion into a cast insitu roof.

Typical Detail ~

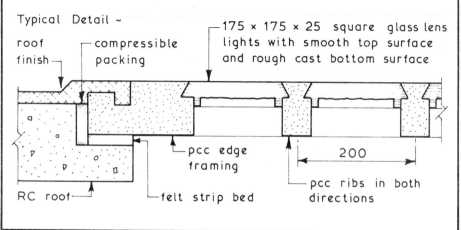

# Rooflights

Dome, Pyramid and Similar Rooflights ~ these are used in conjunction with flat roofs and may be framed or unframed. The glazing can be of glass or plastics such as polycarbonate, acrylic, PVC and glass fibre reinforced polyester resin (grp). The whole component is fixed to a kerb and may have a raising piece containing hit and miss ventilators, louvres or flaps for controllable ventilation purposes.

Typical Details ~

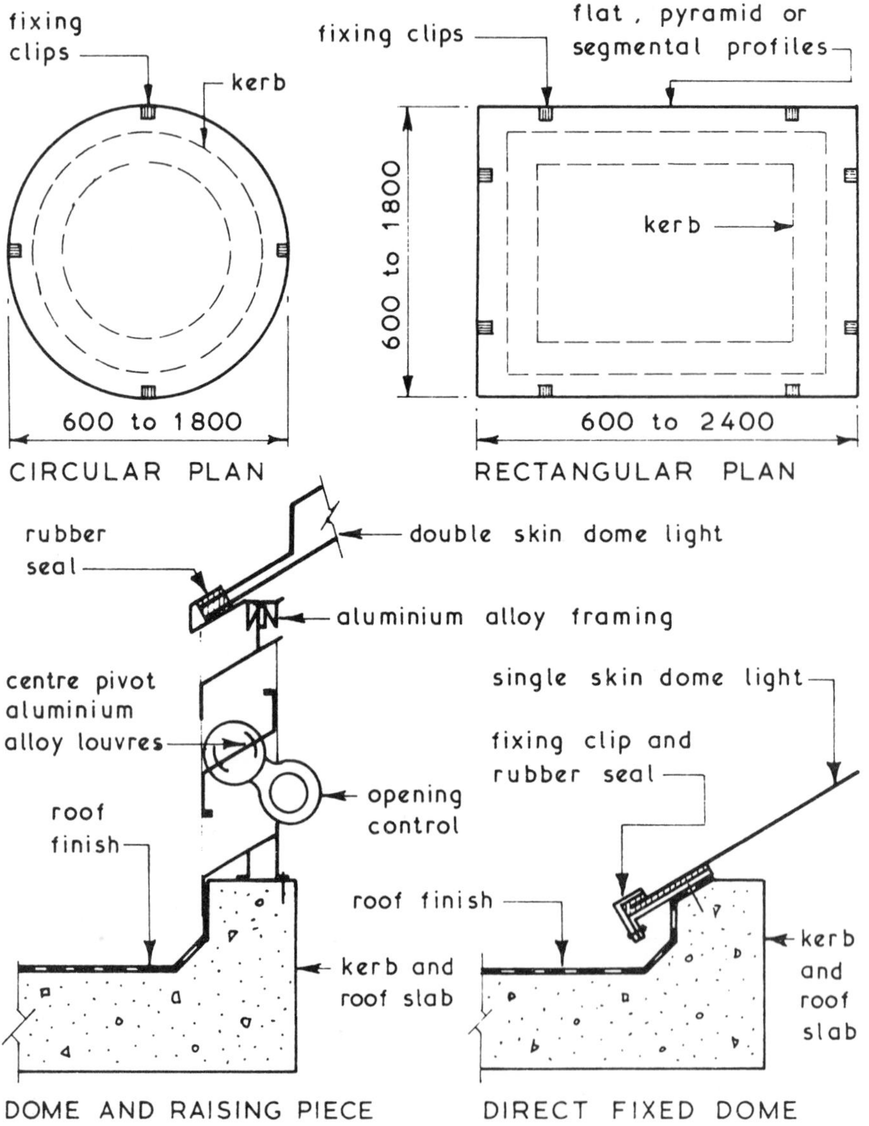

fixing clips
kerb

600 to 1800

CIRCULAR PLAN

fixing clips
flat, pyramid or segmental profiles
kerb

600 to 1800

600 to 2400

RECTANGULAR PLAN

rubber seal
double skin dome light
aluminium alloy framing
centre pivot aluminium alloy louvres
opening control
roof finish
kerb and roof slab

DOME AND RAISING PIECE

single skin dome light
fixing clip and rubber seal
roof finish
kerb and roof slab

DIRECT FIXED DOME

**Fig 87 Rooflights 4**

362

Non-load Bearing Brick Panel Walls ~ these are used in conjunction with framed structures as an infill between the beams and columns. They are constructed in the same manner as ordinary brick walls with the openings being formed by traditional methods.

Basic Requirements ~

1. To be adequately supported by and tied to the structural frame.
2. Have sufficient strength to support own self weight plus any attached finishes and imposed loads such as wind pressures.
3. Provide the necessary resistance to penetration by the natural elements.
4. Provide the required degree of thermal insulation, sound insulation and fire resistance.
5. Have sufficient durability to reduce maintenance costs to a minimum.
6. Provide for movements due to moisture and thermal expansion of the panel and for contraction of the frame.

Typical Details ~

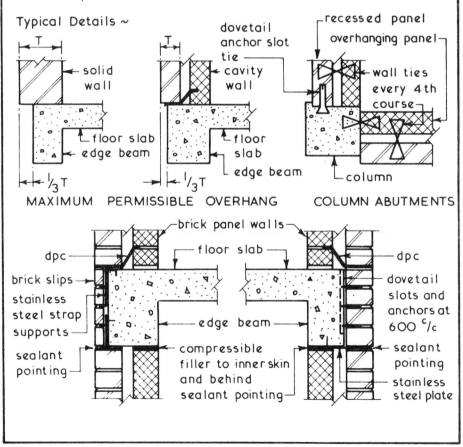

MAXIMUM  PERMISSIBLE  OVERHANG          COLUMN ABUTMENTS

Infill Panel Walls ~ these can be used between the framing members of a building to provide the cladding and division between the internal and external environments and are distinct from claddings and facings :-

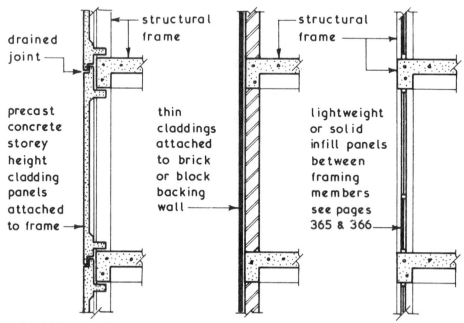

drained joint

precast concrete storey height cladding panels attached to frame →

structural frame

thin claddings attached to brick or block backing wall →

structural frame

lightweight or solid infill panels between framing members see pages 365 & 366 →

CLADDING PANELS     PANEL & FACINGS     INFILL PANELS

Functional Requirements ~ all forms of infill panel should be designed and constructed to fulfil the following functional requirements :-

1. Self supporting between structural framing members.

2. Provide resistance to the penetration of the elements.

3. Provide resistance to positive and negative wind pressures.

4. Give the required degree of thermal insulation.

5. Give the required degree of sound insulation.

6. Give the required degree of fire resistance.

7. Have sufficient openings to provide the required amount of natural ventilation.

8. Have sufficient glazed area to fulfil the natural daylight and vision out requirements.

9. Be economic in the context of construction and maintenance.

10. Provide for any differential movements between panel and structural frame.

Brick Infill Panels ~ these can be constructed in a solid or cavity format the latter usually having an inner skin of blockwork to increase the thermal insulation properties of the panel. All the fundamental construction processes and detail of solid and cavity walls ( bonding, lintels over openings, wall ties, damp-proof courses etc.,) apply equally to infill panel walls. The infill panel walls can be tied to the columns by means of wall ties cast into the columns at 300 mm centres or located in cast-in dovetail anchor slots. The head of every infill panel should have a compressible joint to allow for any differential movements between the frame and panel.

Typical Details ~

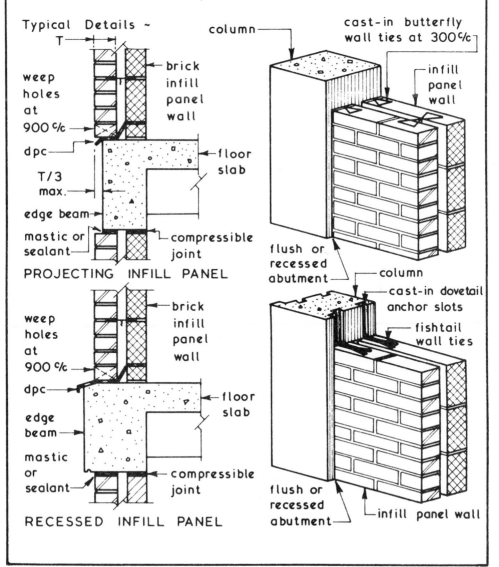

PROJECTING INFILL PANEL

RECESSED INFILL PANEL

365

## Infill Panel Walls

Lightweight Infill Panels ~ these can be constructed from a wide variety or combination of materials such as timber, metals and plastics into which single or double glazing can be fitted. If solid panels are to be used below a transom they are usually of a composite or sandwich construction to provide the required sound insulation, thermal insulation and fire resistance properties.

Typical Example ~

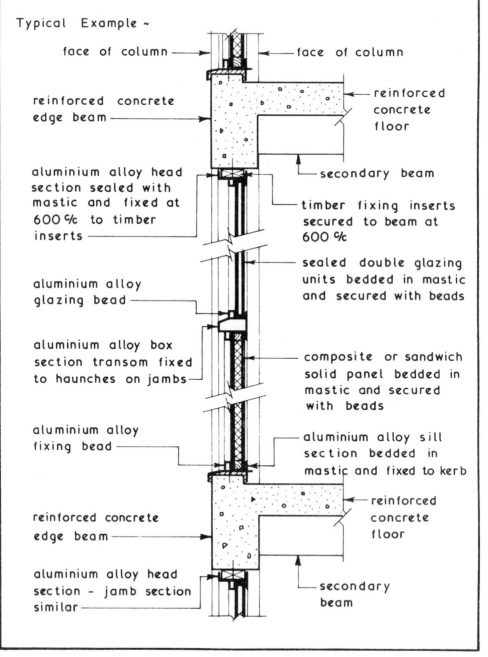

face of column
face of column

reinforced concrete edge beam

reinforced concrete floor

secondary beam

aluminium alloy head section sealed with mastic and fixed at 600 c/c to timber inserts

timber fixing inserts secured to beam at 600 c/c

sealed double glazing units bedded in mastic and secured with beads

aluminium alloy glazing bead

aluminium alloy box section transom fixed to haunches on jambs

composite or sandwich solid panel bedded in mastic and secured with beads

aluminium alloy fixing bead

aluminium alloy sill section bedded in mastic and fixed to kerb

reinforced concrete edge beam

reinforced concrete floor

aluminium alloy head section - jamb section similar

secondary beam

Lightweight Infill Panels ~ these can be fixed between the structural horizontal and vertical members of the frame or fixed to the face of either the columns or beams to give a grid, horizontal or vertical emphasis to the façade thus –

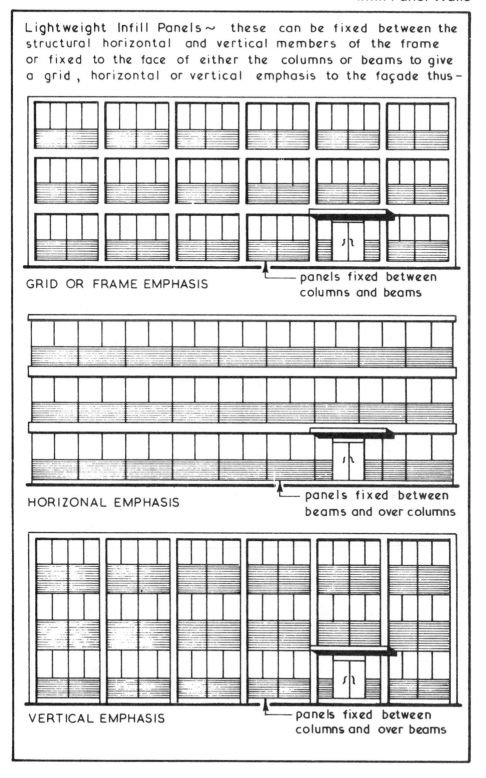

GRID OR FRAME EMPHASIS — panels fixed between columns and beams

HORIZONAL EMPHASIS — panels fixed between beams and over columns

VERTICAL EMPHASIS — panels fixed between columns and over beams

Curtain Walling ~ this is a form of lightweight non-load bearing external cladding which forms a complete envelope or sheath around the structural frame. In low rise structures the curtain wall framing could be of timber or patent glazing but in the usual high rise context box or solid members of steel or aluminium alloy are normally employed.

Basic Requirements for Curtain Walls ~

1. Provide the necessary resistance to penetration by the elements.

2. Have sufficient strength to carry own self weight and provide resistance to both positive and negative wind pressures.

3. Provide required degree of fire resistance – glazed areas are classified in the Building Regulations as unprotected areas therefore any required fire resistance must be obtained from the infill or undersill panels and any backing wall or beam.

4. Be easy to assemble, fix and maintain.

5. Provide the required degree of sound and thermal insulation.

6. Provide for thermal and structural movements.

Typical Curtain Walling Arrangement ~

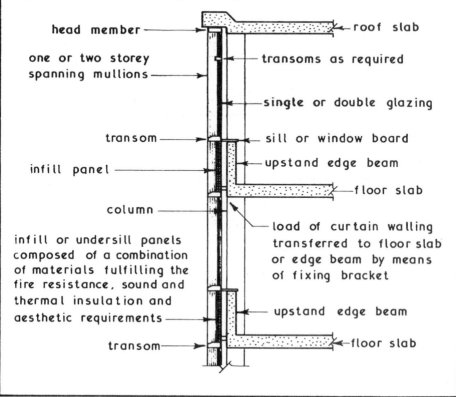

head member ———— roof slab

one or two storey spanning mullions ———— transoms as required

single or double glazing

transom ———— sill or window board

infill panel ———— upstand edge beam

floor slab

column ———— load of curtain walling transferred to floor slab or edge beam by means of fixing bracket

infill or undersill panels composed of a combination of materials fulfilling the fire resistance, sound and thermal insulation and aesthetic requirements ———— upstand edge beam

transom ———— floor slab

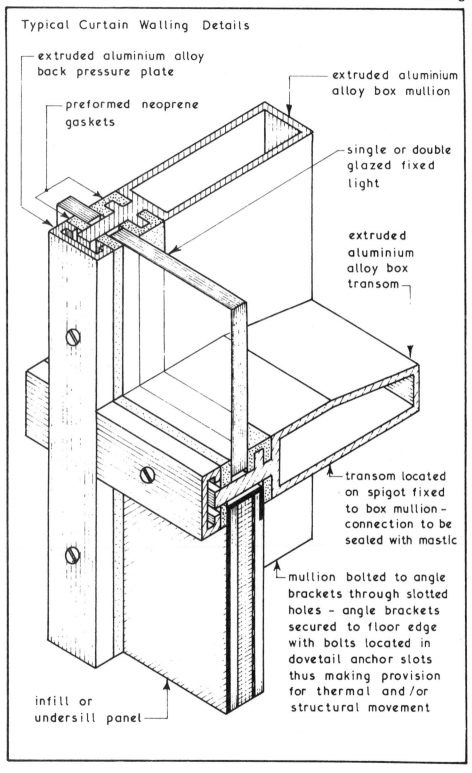

Typical Curtain Walling Details

extruded aluminium alloy
back pressure plate

preformed neoprene
gaskets

extruded aluminium
alloy box mullion

single or double
glazed fixed
light

extruded
aluminium
alloy box
transom

transom located
on spigot fixed
to box mullion -
connection to be
sealed with mastic

mullion bolted to angle
brackets through slotted
holes - angle brackets
secured to floor edge
with bolts located in
dovetail anchor slots
thus making provision
for thermal and/or
structural movement

infill or
undersill panel

Fixing Curtain Walling to the Structure ~ in curtain walling systems it is the main vertical component or mullion which carries the loads and transfers them to the structural frame at every or alternate floor levels depending on the spanning ability of the mullion. At each fixing point the load must be transferred and an allowance made for thermal expansion and differential movement between the structural frame and curtain walling. The usual method employed is slotted bolt fixings.

Typical Examples ~

fixing bracket

mullion

mullion joint

edge of beam or floor

packing as required

fixing bolt with plastic washer

threaded cast-in socket

ASSEMBLY

sliding joint fixing bracket

slotted bolt holes to both sides of stem

box mullion

COMPONENTS

fixing bracket

mullion

mullion joint

edge of beam or floor

slotted bolt holes

packing as required

anchor fixing bolt with plastic washer

cast-in channel insert

ASSEMBLY

fixing bracket

slotted bolt holes to both sides of stem

slotted bolt holes for through bolts

mullion

COMPONENTS

Loadbearing Concrete Panels ~ this form of construction uses storey height loadbearing precast reinforced concrete perimeter panels. The width and depth of the panels is governed by the load(s) to be carried, the height and exposure of the building. Panel can be plain or fenestrated proving the latter leaves sufficient concrete to transmit the load(s) around the opening. The cladding panels, being structural, eliminate the need for perimeter columns and beams and provides an internal surface ready to receive insulation, attached services and decorations. In the context of design these structures must be formed in such a manner that should a single member be removed by an internal explosion, wind pressure or similar force progressive or structural collapse will not occur, the minimum requirements being set out in Part A of the Building Regulations. Loadbearing concrete panel construction can be a cost effective method of building.

Typical Details ~

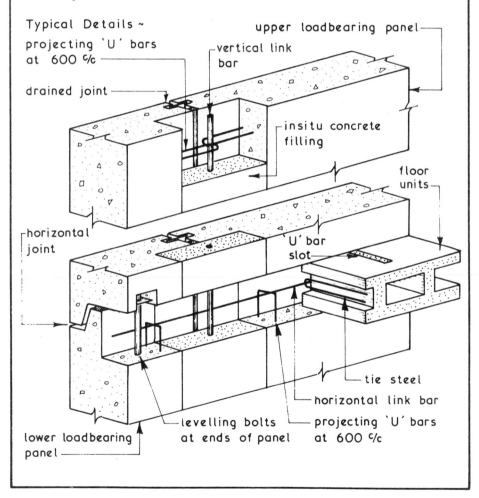

projecting 'U' bars at 600 c/c

drained joint

vertical link bar

upper loadbearing panel

insitu concrete filling

floor units

horizontal joint

'U' bar slot

tie steel

horizontal link bar

levelling bolts at ends of panel

projecting 'U' bars at 600 c/c

lower loadbearing panel

371

Concrete Cladding Panels ~ these are usually of reinforced precast concrete to an undersill or storey height format, the former being sometimes called apron panels. All precast concrete cladding panels should be designed and installed to fulfil the following functions :-

1. Self supporting between framing members.

2. Provide resistance to penetration by the natural elements.

3. Resist both positive and negative wind pressures.

4. Provide required degree of fire resistance.

5. Provide required degree of thermal insulation by having the insulating material incorporated within the body of the cladding or alternatively allow the cladding to act as the outer leaf of cavity wall panel.

6. Provide required degree of sound insulation.

Undersill or Apron Cladding Panels ~ these are designed to span from column to column and provide a seating for the windows located above. Levelling is usually carried out by wedging and packing from the lower edge before being fixed with grouted dowels.

Typical Details ~

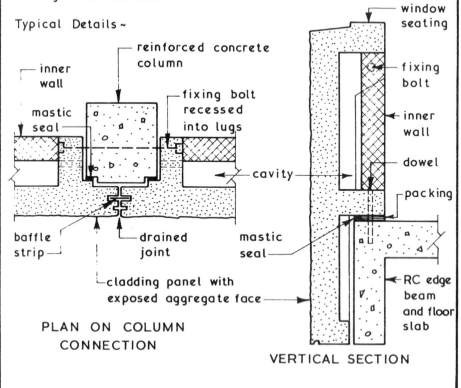

PLAN ON COLUMN
CONNECTION

VERTICAL SECTION

Storey Height Cladding Panels ~ these are designed to span vertically from beam to beam and can be fenestrated if required. Levelling is usually carried out by wedging and packing from floor level before being fixed by bolts or grouted dowels.

Typical Details ~

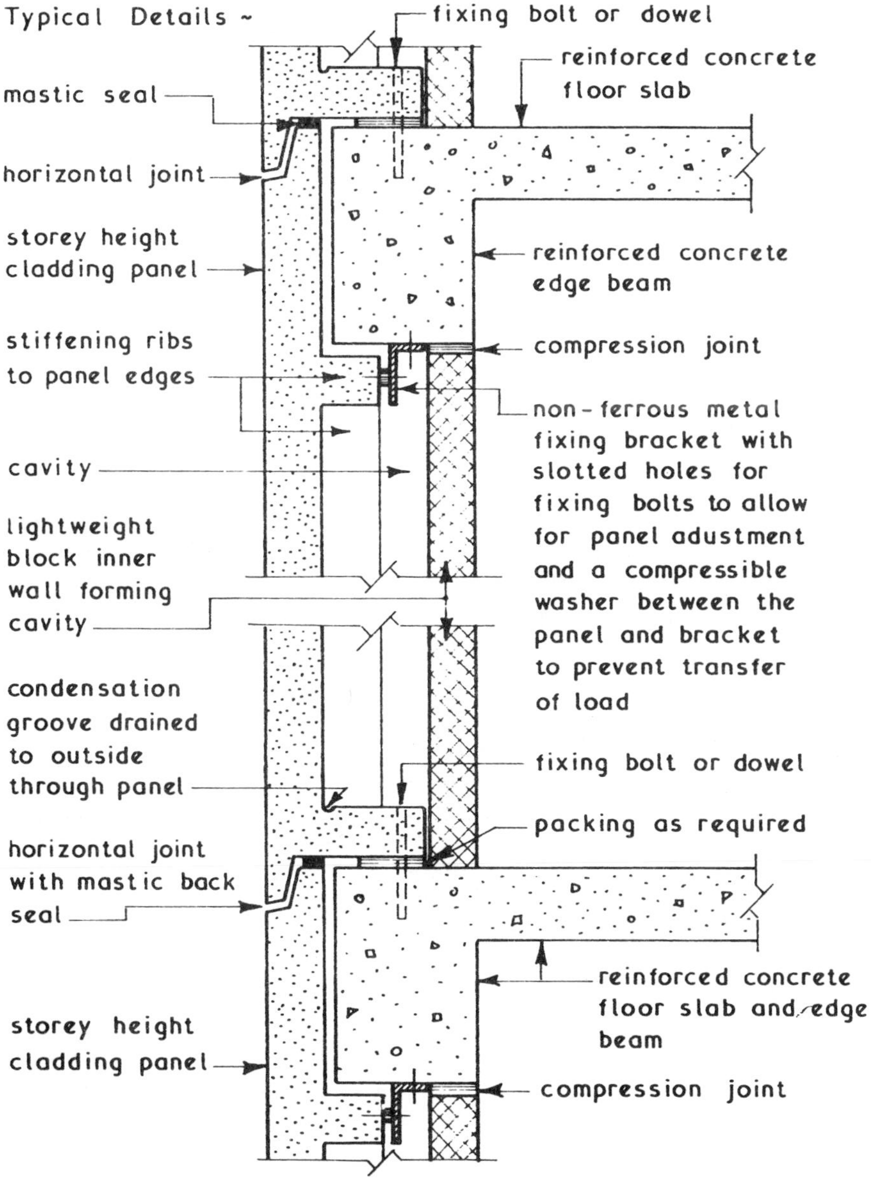

mastic seal

horizontal joint

storey height cladding panel

stiffening ribs to panel edges

cavity

lightweight block inner wall forming cavity

condensation groove drained to outside through panel

horizontal joint with mastic back seal

storey height cladding panel

fixing bolt or dowel

reinforced concrete floor slab

reinforced concrete edge beam

compression joint

non-ferrous metal fixing bracket with slotted holes for fixing bolts to allow for panel adustment and a compressible washer between the panel and bracket to prevent transfer of load

fixing bolt or dowel

packing as required

reinforced concrete floor slab and edge beam

compression joint

**VERTICAL SECTION**

# Prestressed Concrete

Principles ~ the well known properties of concrete are that it has high compressive strength and low tensile strength. The basic concept of reinforced concrete is to include a designed amount of steel bars in a predetermined pattern to give the concrete a reasonable amount of tensile strength. In prestressed concrete a precompression is induced into the member to make full use of its own inherent compressive strength when loaded. The design aim is to achieve a balance of tensile and compressive forces so that the end result is a concrete member which is resisting only stresses which are compressive. In practice a small amount of tension may be present but providing this does not exceed the tensile strength of the concrete being used tensile failure will not occur.

Comparison of Reinforced and Prestressed Concrete ~

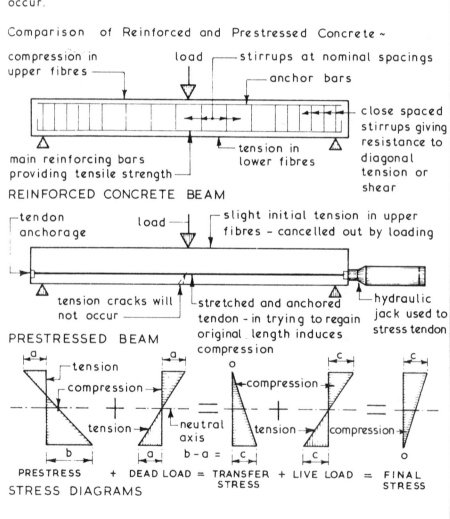

REINFORCED CONCRETE BEAM

PRESTRESSED BEAM

STRESS DIAGRAMS

Materials ~ concrete will shrink whilst curing and it can also suffer sectional losses due to creep when subjected to pressure. The amount of shrinkage and creep likely to occur can be controlled by designing the strength and workability of the concrete; high strength and low workability giving the greatest reduction in both shrinkage and creep. Mild steel will suffer from relaxation losses which is where the stresses in steel under load decrease to a minimum value after a period of time and this can be overcome by increasing the initial stress in the steel. If mild steel is used for prestressing the summation of shrinkage, creep and relaxation losses will cancel out any induced compression, therefore special alloy steels must be used to form tendons for prestressed work.

Tendons - these can be of small diameter wires (2 to 7 mm) in a plain round, crimped or indented format, these wires may be individual or grouped to form cables. Another form of tendon is strand which consists of a straight core wire around which is helically wound further wires to give formats such as 7 wire (6 over 1) and 19 wire (9 over 9 over 1) and like wire tendons strand can be used individually or in groups to form cables. The two main advantages of strand are :-

1. A large prestressing force can be provided over a restricted area.

2. Strand can be supplied in long flexible lengths capable of being stored on drums thus saving site storage and site fabrication space.

Typical Tendon Formats ~

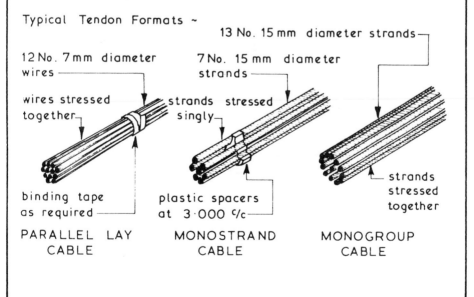

13 No. 15 mm diameter strands

12 No. 7 mm diameter wires

7 No. 15 mm diameter strands

wires stressed together

strands stressed singly

binding tape as required

plastic spacers at 3·000 c/c

strands stressed together

PARALLEL LAY CABLE

MONOSTRAND CABLE

MONOGROUP CABLE

Pre-tensioning ~ this method is used mainly in the factory production of precast concrete components such as lintels, floor units and small beams. Many of these units are formed by the long line method where precision steel moulds up to 120·000 long are used with spacer or dividing plates to form the various lengths required. In pre-tensioning the wires are stressed within the mould before the concrete is placed around them. Steam curing is often used to accelerate this process to achieve a 24 hour characteristic strength of $28N/mm^2$ with a typical 28 day cube strength of $40N/mm^2$. Stressing of the wires is carried out by using hydraulic jacks operating from one or both ends of the mould to achieve an initial 10% overstress to counteract expected looses. After curing the wires are released or cut and the bond between the stressed wires and the concrete prevents the tendons from regaining their original length thus maintaining the precompression or prestress.

At the extreme ends of the members the bond between the stressed wires and concrete is not fully developed due to low frictional resistance. This results in a small contraction and swelling at the ends of the wire forming in effect a cone shape anchorage. The distance over which this contraction occurs is called the transfer length and is equal to 80 to 120 times the wire diameter. To achieve a greater total surface contact area it is common practice to use a larger number of small diameter wires rather than a smaller number of large diameter wires giving the same total cross sectional area.

Typical Pre-tensioning Arrangement ~

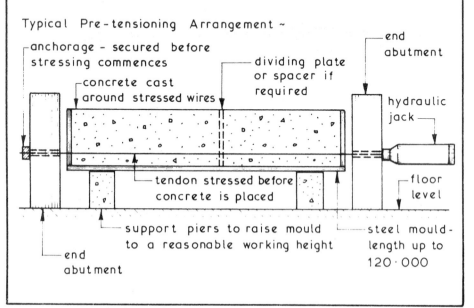

anchorage - secured before stressing commences

end abutment

concrete cast around stressed wires

dividing plate or spacer if required

hydraulic jack

tendon stressed before concrete is placed

floor level

support piers to raise mould to a reasonable working height

steel mould - length up to 120·000

end abutment

Post-tensioning ~ this method is usually employed where stressing is to be carried out on site after casting an insitu component or where a series of precast concrete units are to be joined together to form the required member. It can also be used where curved tendons are to be used to overcome negative bending moments. In post-tensioning the concrete is cast around ducts or sheathing in which the tendons are to be housed. Stressing is carried out after the concrete has cured by means of hydraulic jacks operating from one or both ends of the member. The anchorages (see page 378) which form part of the complete component prevent the stressed tendon from regaining its original length thus maintaining the precompression or prestress. After stressing the annular space in the tendon ducts should be filled with grout to prevent corrosion of the tendons due to any entrapped moisture and to assist in stress distribution. Due to the high local stresses at the anchorage positions it is usual for a reinforcing spiral to be included in the design.

Typical Post-tensioning Arrangement ~

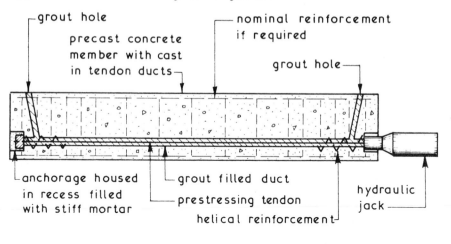

Curved Tendons for Negative Bending Moments ~

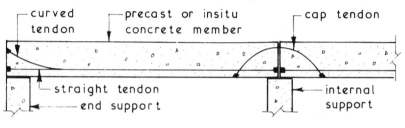

Anchorages ~ the formats for anchorages used in conjunction with post-tensioned prestressed concrete works depends mainly on whether the tendons are to be stressed individually or as a group, but most systems use a form of split cone wedges or jaws acting against a form of bearing or pressure plate.

Typical Anchorage Details ~

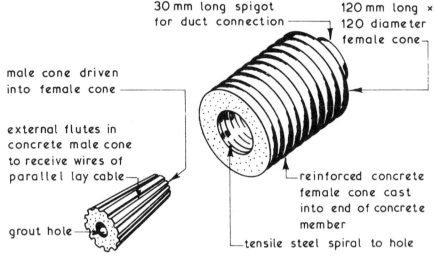

30 mm long spigot for duct connection

120 mm long × 120 diameter female cone

male cone driven into female cone

external flutes in concrete male cone to receive wires of parallel lay cable

grout hole

reinforced concrete female cone cast into end of concrete member

tensile steel spiral to hole

FREYSSINET ANCHORAGE

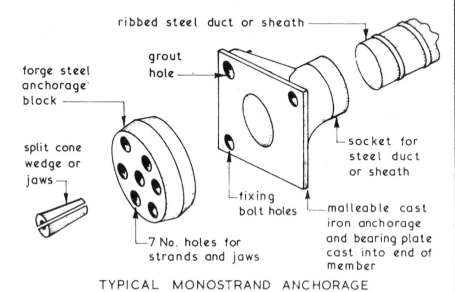

ribbed steel duct or sheath

grout hole

forge steel anchorage block

split cone wedge or jaws

7 No. holes for strands and jaws

fixing bolt holes

socket for steel duct or sheath

malleable cast iron anchorage and bearing plate cast into end of member

TYPICAL MONOSTRAND ANCHORAGE

Comparison with Reinforced Concrete ~ when comparing prestressed concrete with conventional reinforced concrete the main advantages and disadvantages can be enumerated but in the final analysis each structure and/or component must be decided on its own merit.

Main Advantages :-

1. Makes full use of the inherent compressive strength of concrete.

2. Makes full use of the special alloy steels used to form the prestressing tendons.

3. Eliminates tension cracks thus reducing the risk of corrosion of steel components.

4. Reduces shear stresses.

5. For any given span and loading condition a component with a smaller cross section can be used thus giving a reduction in weight.

6. Individual precast concrete units can be joined together to form a composite member.

Main Disadvantages :-

1. High degree of control over materials, design and quality of workmanship is required.

2. Special alloy steels are dearer than most traditional steels used in reinforced concrete.

3. Extra cost of special equipment required to carry out the prestressing activities.

4. Cost of extra safety requirements needed whilst stressing tendons.

As a general comparison between the two structural options under consideration it is usually found that :-

1. Up to 6·000 span traditional reinforced concrete is the most economic method.

2. Spans between 6·000 and 9·000 the two options are compatible.

3. Over 9·000 span prestressed concrete is more economical than reinforced concrete.

It should be noted that generally columns and walls do not need prestressing but in tall columns and high retaining walls where the bending stresses are high prestressing techniques can sometimes be economically applied.

# Prestressed Concrete

Ground Anchors ~ these are a particular application of post-tensioning prestressing techniques and can be used to form ground tie backs to cofferdams, retaining walls and basement walls. They can also be used as vertical tie downs to basement and similar slabs to prevent floatation during and after construction. Ground anchors can be of a solid bar format (rock anchors) or of a wire or cable format for granular and cohesive soils. A lined or unlined bore hole must be drilled into the soil to the design depth and at the required angle to house the ground anchor. In clay soils the bore hole needs to be underreamed over the anchorage length to provide adequate bond. The tail end of the anchor is pressure grouted to form a bond with the surrounding soil, the remaining length being unbonded so that it can be stressed and anchored at head thus inducing the prestress. The void around the unbonded or elastic length is gravity grouted after completion of the stressing operation.

Typical Ground Anchor Details ~

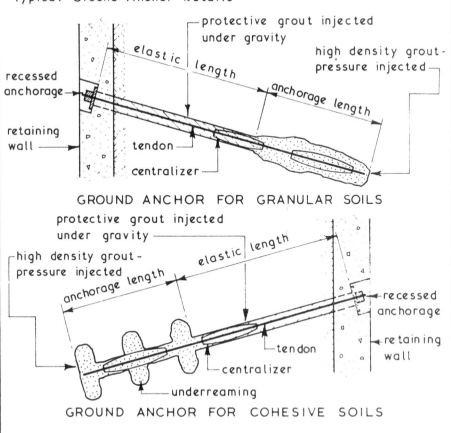

GROUND ANCHOR FOR GRANULAR SOILS

GROUND ANCHOR FOR COHESIVE SOILS

Concrete Surface Finishes ~ it is not easy to produce a concrete surface with a smooth finish of uniform colour direct from the mould or formwork since the colour of the concrete can be affected by the cement and fine aggregate used. The concrete surface texture can be affected by the aggregate grading, cement content, water content, degree of compaction, pin holes caused by entrapped air and rough patches caused by adhesion to parts of the formworks. Complete control over the above mentioned causes is difficult under ideal factory conditions and almost impossible under normal site conditions. The use of textured and applied finishes has therefore the primary function of improving the appearance of the concrete surface and in some cases it will help to restrict the amount of water which reaches a vertical joint.

Casting ~ concrete components can usually be cast insitu or precast in moulds. Obtaining a surface finish to concrete cast insitu is usually carried out against a vertical face, whereas precast concrete components can be cast horizontally and treated on either upper or lower mould face. Apart from a plain surface concrete the other main options are :-

1. Textured and profiled surfaces.

2. Tooled finishes.

3. Cast-on finishes. (see page 382)

4. Exposed aggregate finishes. (see page 382)

Textured and Profiled Surfaces ~ these can be produced on the upper surface of a horizontal casting by rolling, tamping, brushing and sawing techniques but variations in colour are difficult to avoid. Textured and profiled surfaces can be produced on the lower face of a horizontal casting by using suitable mould linings.

Tooled Finishes ~ the surface of hardened concrete can be tooled by bush hammering, point tooling and grinding. Bush hammering and point tooling can be carried out by using an electric or pneumatic hammer on concrete which is at least three weeks old provided gravel aggregates have not been used since these tend to shatter leaving surface pits. Tooling up to the arris could cause spalling therefore a 10mm wide edge margin should be left untooled. Grinding the hardened concrete consists of smoothing the surface with a rotary carborundum disc which may have an integral water feed. Grinding is a suitable treatment for concrete containing the softer aggregates such as limestone.

# Concrete Surface Finishes

Cast-on Finishes ~ these finishes include split blocks, bricks, stone, tiles and mosaic. Cast-on finishes to the upper surface of a horizontal casting are not recommended although such finishes could be bedded onto the fresh concrete. Lower face treatment is by laying the materials with sealed or grouted joints onto the base of mould or alternatively the materials to be cast-on may be located in a sand bed spread over the base of the mould.

Exposed Aggregate Finishes ~ attractive effects can be obtained by removing the skin of hardened cement paste or surface matrix, which forms on the surface of concrete, to expose the aggregate. The methods which can be employed differ with the casting position.

Horizontal Casting – treatment to the upper face can consist of spraying with water and brushing some two hours after casting, trowelling aggregate into the fresh concrete surface or by using the felt-float method. This method consists of trowelling 10mm of dry mix fine concrete onto the fresh concrete surface and using the felt pad to pick up the cement and fine particles from the surface leaving a clean exposed aggregate finish.

Treatment to the lower face can consist of applying a retarder to the base of the mould so that the partially set surface matrix can be removed by water and/or brushing as soon as the castings are removed from the moulds. When special face aggregates are used the sand bed method could be employed.

Vertical Casting – exposed aggregate finishes to the vertical faces can be obtained by tooling the hardened concrete or they can be cast-on by the aggregate transfer process. This consists of sticking the selected aggregate onto the rough side of pegboard sheets with a mixture of water soluble cellulose compounds and sand fillers. The cream like mixture is spread evenly over the surface of the pegboard to a depth of one third the aggregate size and the aggregate sprinkled or placed evenly over the surface before being lightly tamped into the adhesive. The prepared board is then set aside for 36 hours to set before being used as a liner to the formwork or mould. The liner is used in conjunction with a loose plywood or hardboard baffle placed against the face of the aggregate. The baffle board is removed as the concrete is being placed.

aggregate exposed by washing away adhesive

Thermal Insulation ~ the primary objective of thermal insulation is to reduce the rate of heat loss through the external fabric of the building. This reduction in the rate of heat loss will result in a lower total heat input into the building thus saving fuel and power. It will also give rise to higher thermal comfort for the occupants of the building.

Part L of the Building Regulations is concerned with the conservation of fuel and power and applies to dwellings and other buildings whose floor area exceeds $30\,m^2$. Approved Document L gives guidance in the form of maximum U values for the elements of a building together with limitations for windows and rooflights. Alternative U values for use with double glazing or half double glazing in dwellings are given together with caculation procedures to comply with the performance standards required by Part L.

Typical Examples ~

single glazing [SG] $U = 5\cdot7\,W/m^2K$ ; double glazing [DG] $U = 2\cdot8\,W/m^2K$

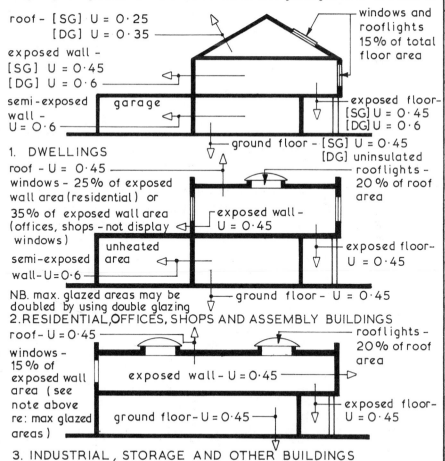

roof - [SG] U = 0·25
      [DG] U = 0·35

exposed wall -
[SG] U = 0·45
[DG] U = 0·6

semi-exposed
wall -
U = 0·6

garage

windows and rooflights 15% of total floor area

exposed floor-
[SG] U = 0·45
[DG] U = 0·6

ground floor - [SG] U = 0·45
               [DG] uninsulated

1. DWELLINGS

roof - U = 0·45
windows - 25% of exposed wall area (residential) or 35% of exposed wall area (offices, shops - not display windows)

semi-exposed wall-U=0·6

exposed wall -
U = 0·45

unheated area

rooflights - 20% of roof area

exposed floor-
U = 0·45

NB. max. glazed areas may be doubled by using double glazing

ground floor - U = 0·45

2. RESIDENTIAL, OFFICES, SHOPS AND ASSEMBLY BUILDINGS

roof- U = 0·45
windows - 15% of exposed wall area (see note above re: max glazed areas)

exposed wall - U = 0·45

ground floor - U = 0·45

rooflights - 20% of roof area

exposed floor-
U = 0·45

3. INDUSTRIAL, STORAGE AND OTHER BUILDINGS

# Thermal Insulation

Thermal Insulation ~ this is required in most roofs to reduce the heat loss from the interior of the building which will create a better internal environment reducing the risk of condensation and give a saving on heating costs.

Part L of the Building Regulations when dealing with dwellings gives the need to make reasonable provision for the conservation of fuel and power in buildings. To satisfy this requirement Approved Document L gives a maximum allowable thermal transmittance coefficient or U value of $0.25 \, W/m^2 K$ for single glazing and $0.35 \, W/m \, K$ for total double glazing for the roofs of dwellings. This is usually achieved by placing thermal insulating material(s) at ceiling level creating a cold roof void. Alternatively the insulation can be placed at rafter level thus creating a warm roof void.

Typical Detail ~

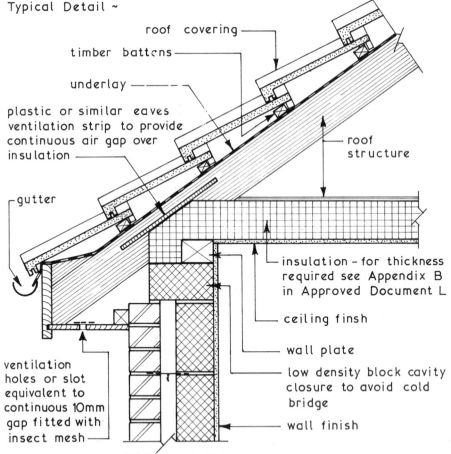

- roof covering
- timber battens
- underlay
- plastic or similar eaves ventilation strip to provide continuous air gap over insulation
- gutter
- roof structure
- insulation – for thickness required see Appendix B in Approved Document L
- ceiling finsh
- wall plate
- low density block cavity closure to avoid cold bridge
- wall finish
- ventilation holes or slot equivalent to continuous 10mm gap fitted with insect mesh

NB. all pipework in roof space should insulated to prevent frost attack (see BS 6700). The sides and top of cold water storage cisterns should be insulated to prevent freezing.

Thermal Insulation to Walls ~ the minimum performance standards for exposed walls set out in Approved Document L to meet the requirements of Part L of the Building Regulations (see page 383) can be achieved in several ways. The usual methods require careful specification, detail and construction of the wall fabric, insulating material(s) and/or applied finishes.

Typical Examples ~

20mm thick external cement rendering

215mm thick aerated concrete blocks – density 475 kg/m³

13mm thick plasterboard with vapour barrier on 20mm thick battens

SOLID BLOCK WALL ( U = 0·39 W/m²K )

50mm wide cavity

50mm thick mineral wool/fibre cavity bats

external brick outer leaf

100mm thick lightweight concrete block inner leaf – density 600 kg/m³

13mm thick lightweight plaster

CAVITY WALL WITH CAVITY INSULATING BATS ( U = 0·39W/m²K )

50mm wide cavity

external brick outer leaf

150mm thick lightweight concrete block inner leaf – density 1100 kg/m³

13mm thick lightweight plaster

TRADITIONAL CAVITY WALL ( U = 0·52W/m²K )

Sound Insulation ~ sound can be defined as vibrations of air which are registered by the human ear. All sounds are produced by a vibrating object which causes tiny particles of air around it to move in unison. These displaced air particles collide with adjacent air particles setting them in motion and in unison with the vibrating object. This continuous chain reaction creates a sound wave which travels through the air until at some distance the air particle movement is so small that it is inaudible to the human ear. Sounds are defined as either impact or airborne sound, the definition being determined by the source producing the sound. Impact sounds are created when the fabric of structure is vibrated by direct contact whereas airborne sound only sets the structural fabric vibrating in unison when the emitted sound wave reaches the enclosing structural fabric. The vibrations set up by the structural fabric can therefore transmit the sound to adjacent rooms which can cause annoyance, disturbance of sleep and of the ability to hold a normal conversation. The objective of sound insulation is therefore to reduce this transmitted sound to an acceptable level.

Part E of the Building Regulations makes it a mandatory requirement to reduce airborne sound to a reasonable level for walls separating dwellings, or a dwelling from another building, or a habitable room within a dwelling from another part of the same building which is not used exclusively with the dwelling. Part E also makes similar requirements for impact and airborne sounds in the context of floors.

External sources of sound such as traffic and aircraft noise can cause annoyance to the occupants of a building and this is usually overcome by a system of double glazing. (see page 278)

Typical Sources and Transmission of Sound ~

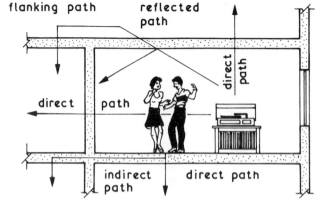

flanking path    reflected
                 path

direct    path

indirect    direct path
path

sound reduction
through :-
closed single
glazed window -
20 to 25 dB.

open window -
10 dB

cavity wall -
40 to 45 dB

tiled roof -
35 dB

direct path

Sound Insulation Between Dwellings ~ to reduce the sound transmitted between dwellings a fabric constructed of high mass could be used since heavy and stiff elements are not easily set into vibration. Aternatively lighter materials can be used provided there is a degree of isolation between the materials incorporated within the design of the element. Obviously if the source of sound can be fixed or constructed so as to reduce the noise emitted this would help to solve the problem of designing for sound insulation.

Typical Details ~

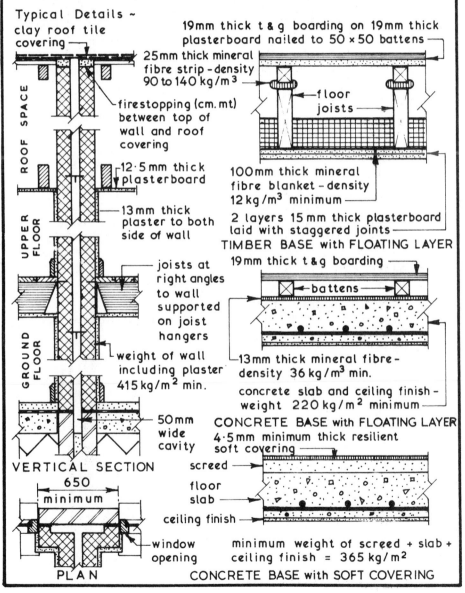

clay roof tile covering

19mm thick t & g boarding on 19mm thick plasterboard nailed to 50 × 50 battens

25mm thick mineral fibre strip – density 90 to 140 kg/m³

firestopping (cm. mt) between top of wall and roof covering

floor joists

12·5mm thick plasterboard

13mm thick plaster to both side of wall

100mm thick mineral fibre blanket – density 12 kg/m³ minimum

2 layers 15mm thick plasterboard laid with staggered joints

TIMBER BASE with FLOATING LAYER

19mm thick t & g boarding

battens

joists at right angles to wall supported on joist hangers

13mm thick mineral fibre – density 36 kg/m³ min.

concrete slab and ceiling finish – weight 220 kg/m² minimum

weight of wall including plaster 415 kg/m² min.

CONCRETE BASE with FLOATING LAYER

50mm wide cavity

4·5mm minimum thick resilient soft covering

screed

floor slab

ceiling finish

VERTICAL SECTION

650 minimum

window opening

PLAN

minimum weight of screed + slab + ceiling finish = 365 kg/m²

CONCRETE BASE with SOFT COVERING

ROOF SPACE

UPPER FLOOR

GROUND FLOOR

Claddings to External Walls ~ external walls of block or timber frame construction can be clad with tiles, timber boards or plastic board sections. The tiles used are plain roofing tiles with either a straight or patterned bottom edge. They are applied to the vertical surface in the same manner as tiles laid on a sloping surface (see pages 328 to 331) except that the gauge can be wider and each tile is twice nailed. External and internal angles can be formed using special tiles or they can be mitred. Timber boards such as matchboarding and shiplap can be fixed vertically to horizontal battens or horizontally to vertical battens. Plastic moulded board claddings can be applied in a similar manner. The battens to which the claddings are fixed should be treated with a preservative against fungi and beetle attack and should be fixed with corrosion resistant nails.

Typical Details ~

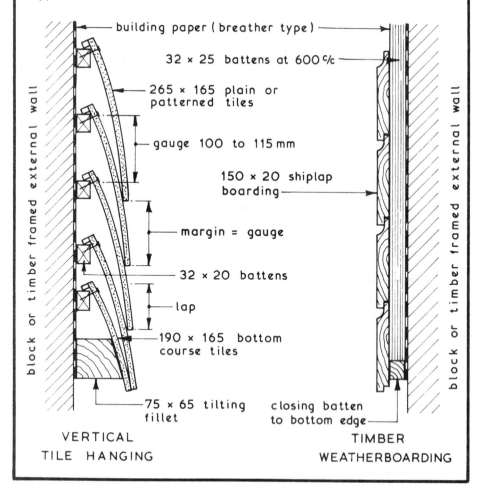

building paper (breather type)

32 × 25 battens at 600 ℅

265 × 165 plain or patterned tiles

gauge 100 to 115 mm

150 × 20 shiplap boarding

margin = gauge

32 × 20 battens

lap

190 × 165 bottom course tiles

75 × 65 tilting fillet

closing batten to bottom edge

block or timber framed external wall

block or timber framed external wall

VERTICAL
TILE HANGING

TIMBER
WEATHERBOARDING

# 6 INTERNAL CONSTRUCTION AND FINISHES

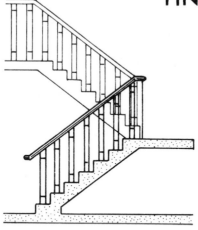

INTERNAL WALLS

PARTITIONS

PLASTERS AND PLASTERING

DRY LINING TECHNIQUES

WALL TILING

DOMESTIC FLOORS AND FINISHES

LARGE CAST INSITU GROUND FLOORS

CONCRETE FLOOR SCREEDS

TIMBER SUSPENDED FLOORS

REINFORCED CONCRETE SUSPENDED FLOORS

PRECAST CONCRETE FLOORS

TIMBER, CONCRETE AND METAL STAIRS

INTERNAL DOORS

PLASTERBOARD CEILINGS

SUSPENDED CEILINGS

PAINTS AND PAINTING

JOINERY PRODUCTION

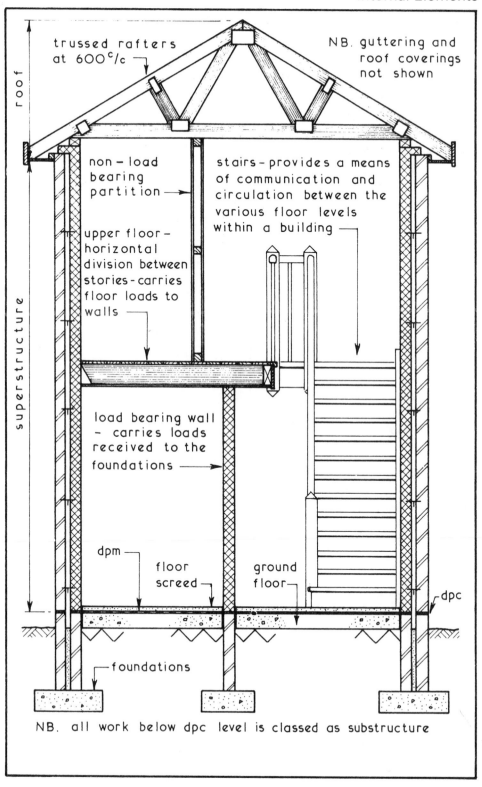

trussed rafters
at 600°/c

NB. guttering and
roof coverings
not shown

roof

non-load
bearing
partition

stairs-provides a means
of communication and
circulation between the
various floor levels
within a building

upper floor-
horizontal
division between
stories-carries
floor loads to
walls

superstructure

load bearing wall
- carries loads
received to the
foundations

dpm

floor
screed

ground
floor

dpc

foundations

NB. all work below dpc level is classed as substructure

Internal Walls ~ their primary function is to act as a vertical divider of floor space and in so doing form a storey height enclosing element.

Other Possible Functions :-

provide a degree of fire resistance

provide a degree of sound insulation

provide a degree of thermal insulation

provide the required degree of durability

provide for low future maintenance

constructed as a load bearing or non-load bearing wall ( see page 393 )

provide borrowed light facilities

provide a good fixing medium for trims

provide a means of access to , egress from and communication between internal spaces

provide a suitable background for fixtures and fittings

provide a suitable medium for housing services

provide the necessary resistance to impact damage and vibration due to doors being slammed

provide the actual or suitable background for finishes and decorations

Internal Walls ~ there are two basic design concepts for internal walls those which accept and transmit structural loads to the foundations are called Load Bearing Walls and those which support only their own self-weight and do not accept any structural loads are called Non-load Bearing Walls or Partitions.

Typical Examples ~

roof spans from external wall to external wall

NB. guttering and roof coverings not shown

external wall

upper floor

non-load bearing wall built off upper floor surface

external wall

non-load bearing wall built off ground floor surface

ground floor

floor spans from external wall to external wall

foundations

roof spans between external walls and internal load bearing wall

NB. guttering and roof coverings not shown

floor spans from external wall to internal load bearing wall

load bearing wall transmits part of roof load to fnds.

external wall

non-load bearing wall

load bearing wall transmits part of roof and floor load to foundations

external wall

foundations

393

Internal Brick Walls ~ these can be load bearing or non-load bearing (see page 393) and for most two storey dwellings are built in half brick thickness in stretcher bond.

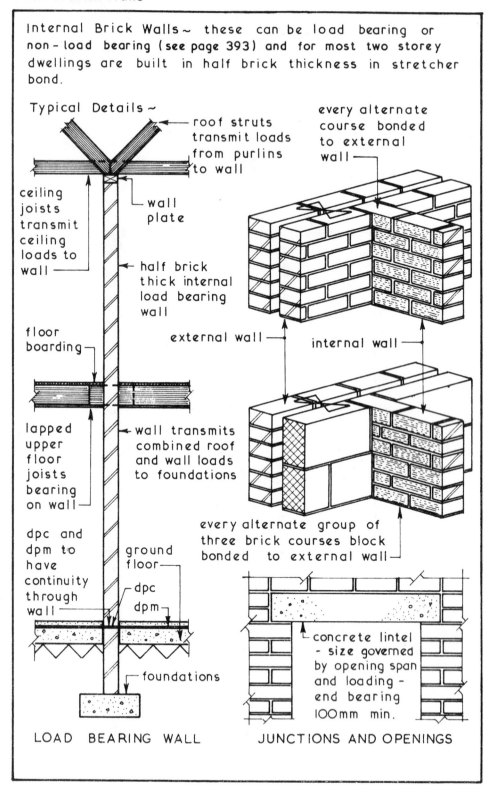

Typical Details ~

roof struts transmit loads from purlins to wall

wall plate

ceiling joists transmit ceiling loads to wall

half brick thick internal load bearing wall

floor boarding

lapped upper floor joists bearing on wall

wall transmits combined roof and wall loads to foundations

dpc and dpm to have continuity through wall

ground floor

dpc

dpm

foundations

LOAD BEARING WALL

every alternate course bonded to external wall

external wall

internal wall

every alternate group of three brick courses block bonded to external wall

concrete lintel - size governed by opening span and loading - end bearing 100mm min.

JUNCTIONS AND OPENINGS

Internal Block Walls ~ these can be load bearing or non-load bearing (see page 393) the thickness and type of block to be used will depend upon the loadings it has to carry.

Typical Details ~

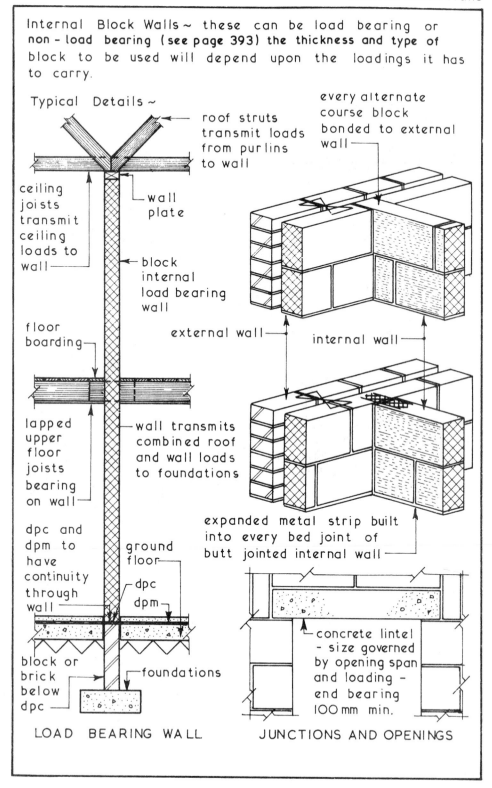

roof struts transmit loads from purlins to wall

every alternate course block bonded to external wall

ceiling joists transmit ceiling loads to wall

wall plate

block internal load bearing wall

external wall

internal wall

floor boarding

lapped upper floor joists bearing on wall

wall transmits combined roof and wall loads to foundations

expanded metal strip built into every bed joint of butt jointed internal wall

dpc and dpm to have continuity through wall

ground floor

dpc

dpm

concrete lintel - size governed by opening span and loading - end bearing 100 mm min.

block or brick below dpc

foundations

LOAD BEARING WALL

JUNCTIONS AND OPENINGS

## Partitions

Internal Partitions ~ these are vertical dividers which are used to separate the internal space of a building into rooms and circulation areas such as corridors. Partitions which give support to a floor or roof are classified as load bearing whereas those which give no such support are called non-load bearing.

Load Bearing Partitions – these walls can be constructed of bricks, blocks or insitu concrete by traditional methods and have the design advantages of being capable of having good fire resistance and/or high sound insulation. Their main disadvantage is permanence giving rise to an inflexible internal layout.

Non-load Bearing Partitions – the wide variety of methods available makes it difficult to classify the form of partition but most can be placed into one of three groups :-

1. Masonry partitions.

2 Stud partitions – **see page 398.**

3. Demountable partitions – **see page 400.**

Masonry Partitions ~ these are usually built with blocks of clay or lightweight concrete which are readily available and easy to construct thus making them popular. These masonry partitions should be adequately tied to the structure or load bearing walls to provide continuity as a sound barrier, provide edge restraint and to reduce the shrinkage cracking which inevitably occurs at abutments. Wherever possible openings for doors should be in the form of storey height frames to provide extra stiffness at these positions.

Typical Details ~

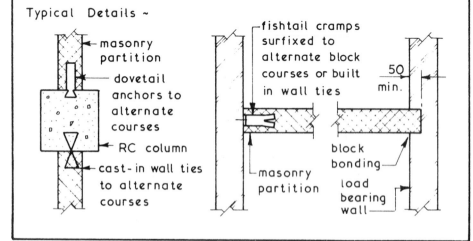

masonry partition

dovetail anchors to alternate courses

RC column

cast-in wall ties to alternate courses

fishtail cramps surfixed to alternate block courses or built in wall ties

masonry partition

block bonding

load bearing wall

50 min.

Timber Stud Partitions ~ these are non-load bearing internal dividing walls which are easy to construct, lightweight, adaptable and can be clad and infilled with various materials to give different finishes and properties. The timber studs should be of prepared or planed material to ensure that the wall is of constant thickness with parallel faces. Stud spacings will be governed by the size and spanning ability of the facing or cladding material.

Typical Details ~

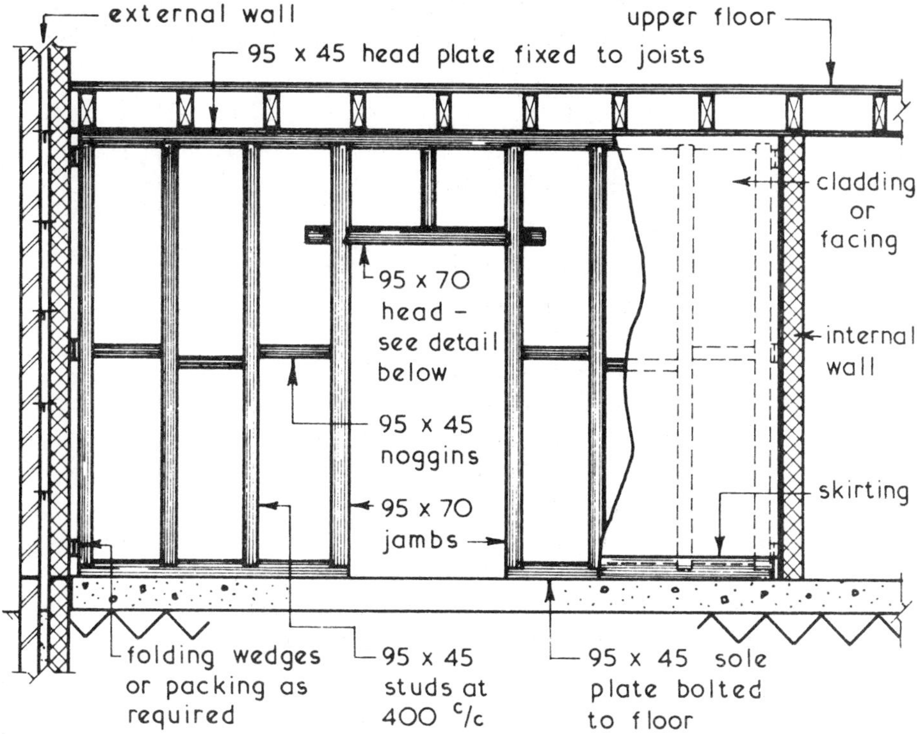

external wall

upper floor

95 x 45 head plate fixed to joists

cladding or facing

95 x 70 head – see detail below

internal wall

95 x 45 noggins

95 x 70 jambs

skirting

folding wedges or packing as required

95 x 45 studs at 400 $^c$/$_c$

95 x 45 sole plate bolted to floor

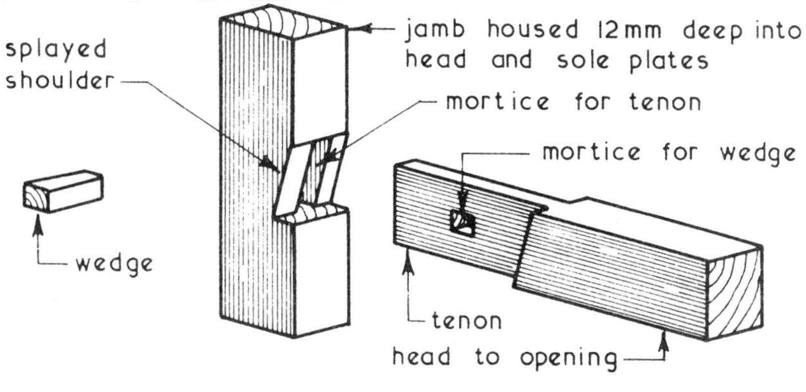

splayed shoulder

wedge

jamb housed 12mm deep into head and sole plates

mortice for tenon

mortice for wedge

tenon

head to opening

# Metal Stud Partitions

Stud Partitions ~ these non-load bearing partitions consist of a framework of vertical studs to which the facing material can be attached. The void between the studs created by the two faces can be infilled to meet specific design needs. The traditional material for stud partitions is timber [see **(Timber Stud Partitions** on page 397 **)** but a similar arrangement can be constructed using metal studs faced on both sides with plasterboard.

Typical Metal Stud Partition Details ~

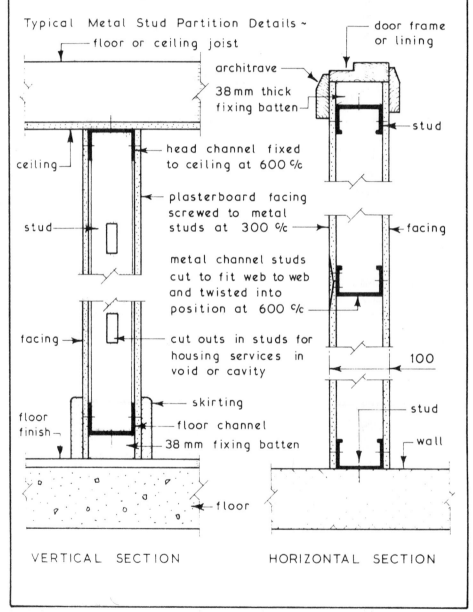

floor or ceiling joist

door frame or lining

architrave

38 mm thick fixing batten

stud

ceiling

head channel fixed to ceiling at 600 %

plasterboard facing screwed to metal studs at 300 %

facing

stud

metal channel studs cut to fit web to web and twisted into position at 600 %

facing

cut outs in studs for housing services in void or cavity

100

skirting

stud

floor finish

floor channel

38 mm fixing batten

wall

floor

VERTICAL SECTION

HORIZONTAL SECTION

Partitions ~ these can be defined as vertical internal space dividers and are usually non-loadbearing. They can be permanent, constructed of materials such as bricks or blocks or they can be demountable constructed using lightweight materials and capable of being taken down and moved to a new location incurring little or no damage to the structure or finishes. There is a wide range of demountable partitions available constructed from a variety of materials giving a range that will be suitable for most situations. Many of these partitions have a permanent finish which requires no decoration and only periodic cleaning in the context of planned maintenance.

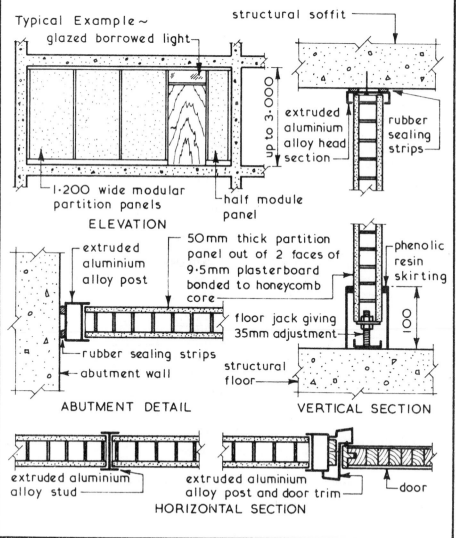

Typical Example ~
glazed borrowed light
structural soffit
up to 3·000
extruded aluminium alloy head section
rubber sealing strips
1·200 wide modular partition panels
half module panel

**ELEVATION**

extruded aluminium alloy post
50mm thick partition panel out of 2 faces of 9·5mm plasterboard bonded to honeycomb core
phenolic resin skirting
floor jack giving 35mm adjustment
rubber sealing strips
abutment wall
structural floor
100

**ABUTMENT DETAIL**

**VERTICAL SECTION**

extruded aluminium alloy stud
extruded aluminium alloy post and door trim
door

**HORIZONTAL SECTION**

Demountable Partitions ~ it can be argued that all internal
non-load bearing partitions are demountable and therefore
the major problem is the amount of demountability required
in the context of ease of moving and the possible frequency
anticipated. The range of partitions available is very wide
including **stud partitions, framed panel partitions**
(**see Demountable Partitions on page 399**) panel to
panel partitions and sliding/folding partitions which are
similar in concept to industrial doors (**see Industrial
Doors on pages 284 and 286**). The latter type is often
used where movement of the partition is required frequently.
The choice is therefore based on the above stated factors
taking into account finish and glazing requirements together
with any personal preference for a particular system but in
all cases the same basic problems will have to be considered:-

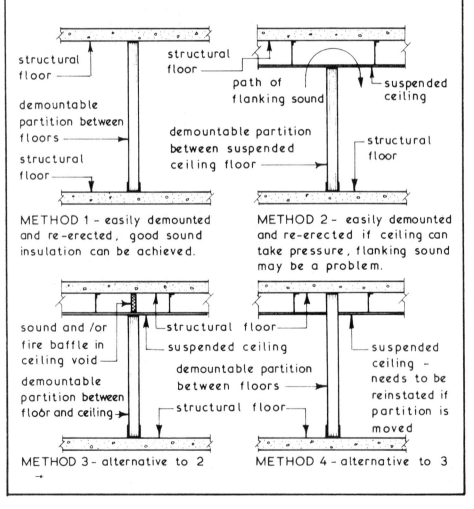

METHOD 1 - easily demounted
and re-erected, good sound
insulation can be achieved.

METHOD 2 - easily demounted
and re-erected if ceiling can
take pressure, flanking sound
may be a problem.

METHOD 3 - alternative to 2

METHOD 4 - alternative to 3

Plaster ~ this is a wet mixed material applied to internal walls as a finish to fill in any irregularities in the wall surface and to provide a smooth continuous surface suitable for direct decoration. The plaster finish also needs to have a good resistance to impact damage. The material used to fulfil these requirements is gypsum plaster. Gypsum is a crystalline combination of calcium sulphate and water. The raw material is crushed, screened and heated to dehydrate the gypsum and this process together with various additives defines its type as set out in  BS 1191 :-

Raw material (gypsum) is heated to 150 to 170°C to drive off 75% of the combined water

Hemi-hydrate Plaster of Paris Class A to BS1191: Part 1

Retarder added giving - Retarded hemi-hydrate Class B finish plaster types b1 and b2 to BS1191: Part 1

Expanded perlite and other additives added giving - One coat plaster; Renovating grade plaster; Spray plaster.

Lightweight aggregates added giving - Premixed lightweight plaster to BS 1191 : Part 2

Usual lightweight additives :- Expanded perlite giving - Browning plaster type a1; Exfoliated vermiculite + perlite + rust inhibitor giving- Metal lath plaster type a2; Exfoliated vermiculite giving - Bonding plaster type a3 or Finish plaster type b

Plaster of Paris is quick setting plaster ( 5 to 10 minutes) and is therefore not suitable for walls but can be used for filling cracks and cast or run mouldings.

Browning and Bonding plasters are used as undercoats to Premixed lightweight plasters

All plaster should be stored in dry conditions since any absorption of moisture before mixing may shorten the normal setting time of about one and a half hours which can reduce the strength of the set plaster.   Gypsum plasters are not suitable for use in temperatures exceeding 43°C and should not be applied to frozen backgrounds.

A good key to the background and between successive coats is essential for successful plastering.  Generally brick and block walls provide the key whereas concrete unless cast against rough formwork will need to be treated to provide the key.

## Plaster Finish to Internal Walls

Internal Wall Finishes~ these can be classified as wet or dry. The traditional wet finish is plaster which is mixed and applied to the wall in layers to achieve a smooth and durable finish suitable for decorative treatments such as paint and wallpaper.

Most plasters are supplied in 50 kg. paper sacks and require only the addition of clean water or sand and clean water according to the type of plaster being used.

Typical Method of Application ~

surface well brushed with hard broom to remove loose material and dust¬

chases cut before plastering¬

thin coats of undercoat plaster applied and built up to required thickness

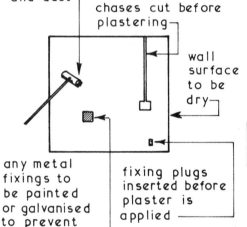

wall surface to be dry¬

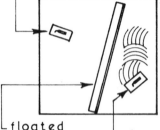

any metal fixings to be painted or galvanised to prevent staining

fixing plugs inserted before plaster is applied

floated undercoat brought to a true and level surface with a rule or straightedge

fine wooden scratcher used to form key for finishing coat

### 1. PREPARATION      2. UNDERCOATING

finishing coat of plaster applied with steel trowel to give a smooth finish

trims and decorative finishes applied after plaster has set and cured

TYPICAL DATA FOR BRICK AND BLOCK BACKGROUNDS

textured surfaces can be obtained by using a sponge, hair brush, felt float or steel combs

### 3. FINISHING

Undercoat - 12 mm thick
Finishing coat - 2 mm thick
Setting times ~
Undercoat - 2 hours
Finishing coat - 1 hour

Plasterboard ~ a board material made of two sheets of thin mill-board with gypsum plaster between - three edge profiles are available :-

Tapered Edge - joint filling

a flush seamless surface is obtained by filling the joint with a special filling plaster, applying a joint tape over the filling and finishing with a thin layer of joint filling plaster the edge of which is feathered out using a slightly damp jointing sponge.

Square Edge - edges are close butted and finished with a cover fillet or the joint is covered with a jute scrim before being plastered.

Bevelled Edge - edges are close butted forming a vee-joint which becomes a feature of the lining.

Typical Details ~

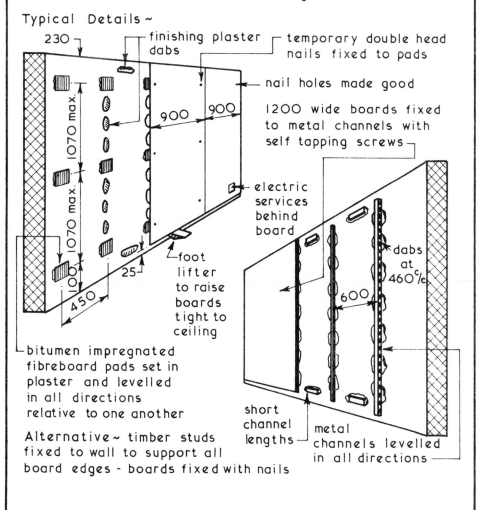

finishing plaster dabs

temporary double head nails fixed to pads

nail holes made good

230

900  900

1070 max.

1070 max.

100

25

450

1200 wide boards fixed to metal channels with self tapping screws

electric services behind board

foot lifter to raise boards tight to ceiling

dabs at 460% c

600

bitumen impregnated fibreboard pads set in plaster and levelled in all directions relative to one another

short channel lengths

metal channels levelled in all directions

Alternative~ timber studs fixed to wall to support all board edges - boards fixed with nails

403

Dry Linings ~ the internal surfaces of walls and partitions are usually covered with a wet finish (plaster or rendering) or with a dry lining such as plasterboard, insulating fibre board, hardboard, timber boards, and plywood all of which can be supplied with a permanent finish or they can be supplied to accept an applied finish such as paint or wallpaper. The main purpose of any applied covering to an internal wall surface is to provide an acceptable but not necessarily an elegant or expensive wall finish, it is also very difficult and expensive to build a brick or block wall which has a fair face to both sides since this would involve the hand selection of bricks and blocks to ensure a constant thickness together with a high degree of skill to construct a satisfactory wall. The main advantage of dry lining walls is that the drying out period required with wet finishes is eliminated. By careful selection and fixing of some dry lining materials it is possible to improve the thermal insulation properties of a wall. Dry linings can be fixed direct to the backing by means of a recommended adhesive or they can be fixed to a suitable arrangement of wall battens.

Typical Example ~

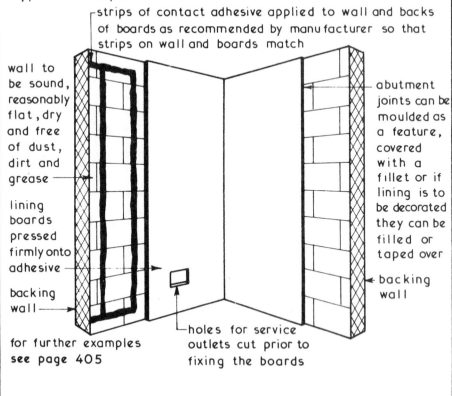

strips of contact adhesive applied to wall and backs of boards as recommended by manufacturer so that strips on wall and boards match

wall to be sound, reasonably flat, dry and free of dust, dirt and grease

lining boards pressed firmly onto adhesive

backing wall

abutment joints can be moulded as a feature, covered with a fillet or if lining is to be decorated they can be filled or taped over

backing wall

holes for service outlets cut prior to fixing the boards

for further examples **see** page 405

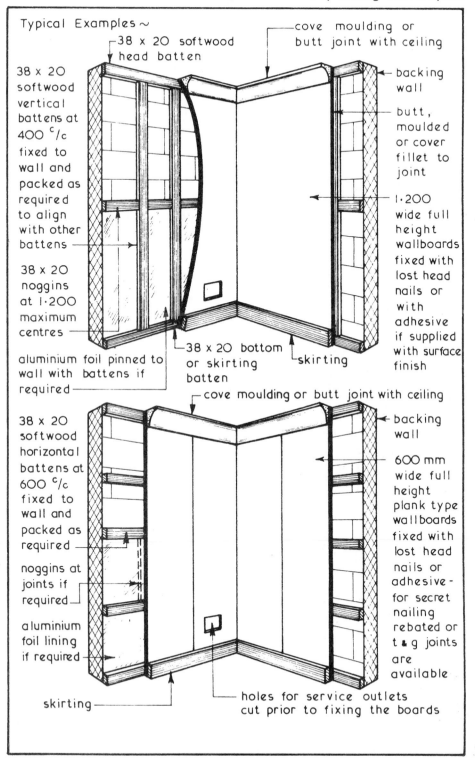

Typical Examples ~

38 x 20 softwood head batten

cove moulding or butt joint with ceiling

38 x 20 softwood vertical battens at 400 c/c fixed to wall and packed as required to align with other battens

backing wall

butt, moulded or cover fillet to joint

1·200 wide full height wallboards fixed with lost head nails or with adhesive if supplied with surface finish

38 x 20 noggins at 1·200 maximum centres

aluminium foil pinned to wall with battens if required

38 x 20 bottom or skirting batten

skirting

cove moulding or butt joint with ceiling

38 x 20 softwood horizontal battens at 600 c/c fixed to wall and packed as required

backing wall

600 mm wide full height plank type wallboards fixed with lost head nails or adhesive – for secret nailing rebated or t & g joints are available

noggins at joints if required

aluminium foil lining if required

skirting

holes for service outlets cut prior to fixing the boards

405

Glazed Wall Tiles ~ internal glazed wall tiles are usually made to the recommendations of BS 6431. External glazed wall tiles made from clay or clay/ceramic mixtures are manufactured but there is no British Standard available.

Internal Glazed Wall Tiles ~ the body of the tile can be made from ball-clay, china clay, china stone, flint and limestone. The material is usually mixed with water to the desired consistency, shaped and then fired in a tunnel oven at a high temperature ($1150^\circ$C) for several days to form the unglazed biscuit tile. The glaze, pattern and colour can now be imparted onto to the biscuit tile before the final firing process at a temperature slightly lower than that of the first firing ($1050^\circ$C) for about two days.

Typical Internal Glazed Wall Tiles and Fittings ~

Sizes - Modular - 100 x 100 x 5mm thick and 200 x 100 x 6·5mm thick.

Non-modular - 152 x 152 x 5 to 8mm thick and 108 x 108 x 4 and 6·5mm thick.

Fittings - wide range available particularly in the non-modular format.

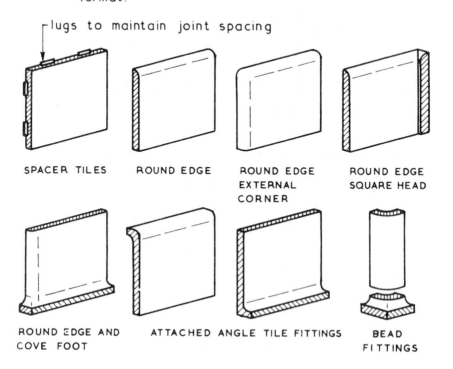

lugs to maintain joint spacing

SPACER TILES    ROUND EDGE    ROUND EDGE EXTERNAL CORNER    ROUND EDGE SQUARE HEAD

ROUND EDGE AND COVE FOOT    ATTACHED ANGLE TILE FITTINGS    BEAD FITTINGS

Bedding of Internal Wall Tiles ~ generally glazed internal wall tiles are considered to be inert in the context of moisture and thermal movement, therefore if movement of the applied wall tile finish is to be avoided attention must be given to the background and the method of fixing the tiles.

Backgrounds ~ these are usually of a cement rendered or plastered surface and should be flat, dry, stable, firmly attached to the substrate and sufficiently old enough for any initial shrinkage to have taken place. The flatness of the background should be not more than 3mm in 2·000 for the thin bedding of tiles and not more than 6mm in 2·000 for thick bedded tiles.

Fixing Wall Tiles ~ two methods are in general use:-

1. Thin Bedding - lightweight internal glazed wall tiles fixed dry using a recommended adhesive which is applied to wall in small areas (1 m²) at a time with a notched trowel, the tile being pressed or tapped into the adhesive.

2. Thick Bedding - cement mortar within the mix range of 1:3 to 1:4 is used as the adhesive either by buttering the backs of the tiles which are then pressed or tapped into position or by rendering the wall surface to a thickness of approximately 10mm and then applying the lightly buttered tiles (1:2 mix) to the rendered wall surface within two hours. It is usually necessary to soak the wall tiles in water to reduce suction before they are placed in position.

Grouting ~ when the wall tiles have set the joints can be grouted by rubbing into the joints a grout paste either using a sponge or brush. Most grouting materials are based on cement with inert fillers and are used neat.

Typical Example ~

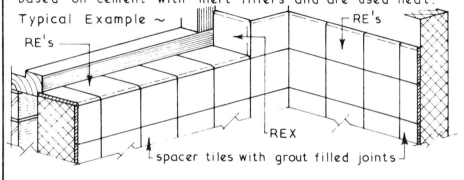

RE's

RE's

REX

spacer tiles with grout filled joints

Primary Functions ~
1. Provide a level surface with sufficient strength to support the imposed loads of people and furniture.
2. Exclude the passage of water and water vapour from the interior of the building.
3. Provide resistance to unacceptable heat loss through the floor.
4. Provide the correct type of surface to receive the chosen finish.

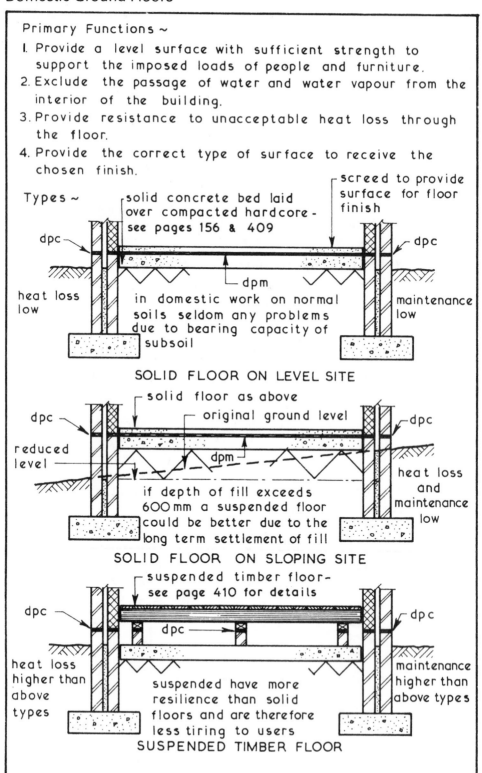

Types ~

screed to provide surface for floor finish

solid concrete bed laid over compacted hardcore - see pages 156 & 409

dpc

dpc

dpm

heat loss low

in domestic work on normal soils seldom any problems due to bearing capacity of subsoil

maintenance low

SOLID FLOOR ON LEVEL SITE

solid floor as above

original ground level

dpc

dpc

reduced level

dpm

heat loss and maintenance low

if depth of fill exceeds 600mm a suspended floor could be better due to the long term settlement of fill

SOLID FLOOR ON SLOPING SITE

suspended timber floor - see page 410 for details

dpc

dpc

dpc

heat loss higher than above types

suspended have more resilience than solid floors and are therefore less tiring to users

maintenance higher than above types

SUSPENDED TIMBER FLOOR

This drawing should be read in conjunction with page 156 - Foundation beds.

A domestic solid ground floor consists of three components

1. Hardcore – a suitable filling material to make up the reduced level and top soil removal excavations. It should have a top surface which can be rolled out to ensure that cement grout is not lost from the concrete It may be necessary to blind the top surface with a layer of sand or fine ash especially if the damp-proof ·membrane is to be placed under the concrete bed.

2. Damp-proof Membrane – an impervious layer such as heavy duty polythene sheet to prevent moisture passing through the floor to the interior of the building.

3. Concrete Bed– the component providing the solid level surface to which screeds and finishes can be applied.

Typical Details ~

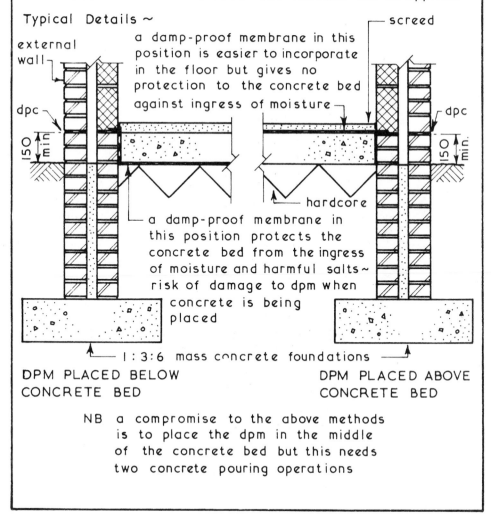

a damp-proof membrane in this position is easier to incorporate in the floor but gives no protection to the concrete bed against ingress of moisture

a damp-proof membrane in this position protects the concrete bed from the ingress of moisture and harmful salts~ risk of damage to dpm when concrete is being placed

1 : 3:6 mass concrete foundations

**DPM PLACED BELOW CONCRETE BED**

**DPM PLACED ABOVE CONCRETE BED**

NB a compromise to the above methods is to place the dpm in the middle of the concrete bed but this needs two concrete pouring operations

Suspended Timber Ground Floors ~ these need to have a well ventilated space beneath the floor construction to prevent the moisture content of the timber rising above an unacceptable level ( i.e. ≯ 20% ) which would create the conditions for possible fungal attack.

Typical Details ~

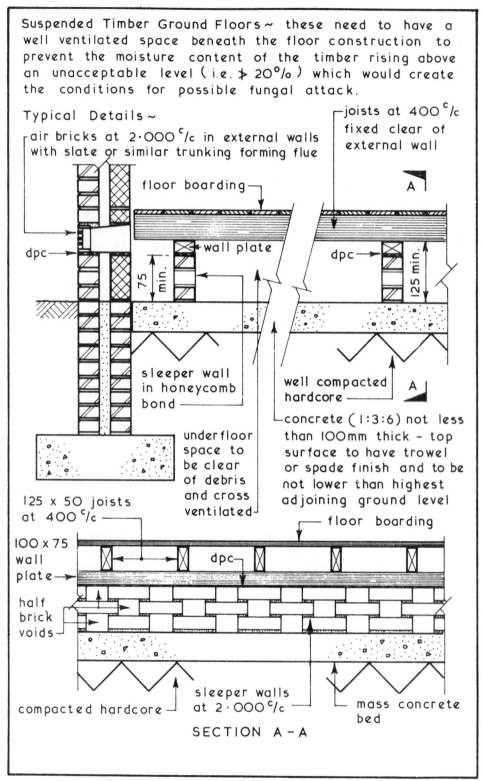

joists at 400 $^c$/c fixed clear of external wall

air bricks at 2·000 $^c$/c in external walls with slate or similar trunking forming flue

floor boarding

A

dpc

wall plate

75 min.

dpc

125 min.

sleeper wall in honeycomb bond

well compacted hardcore

A

concrete ( 1:3:6 ) not less than 100mm thick – top surface to have trowel or spade finish and to be not lower than highest adjoining ground level

underfloor space to be clear of debris and cross ventilated

125 x 50 joists at 400 $^c$/c

floor boarding

100 x 75 wall plate

dpc

half brick voids

compacted hardcore

sleeper walls at 2·000 $^c$/c

mass concrete bed

SECTION A – A

Floor Finishes ~ these are usually applied to a structural base but may form part of the floor structure as in the case of floor boards. Most finishes are chosen to fulfil a particular function such as :-

1. Appearance – chosen mainly for their aesthetic appeal or effect but should however have reasonable wearing properties. Examples are carpets; carpet tiles and wood blocks.

2. High Resistance – chosen mainly for their wearing and impact resistance properties and for high usuage areas such as kitchens. Examples are quarry tiles and granolithic pavings.

3. Hygiene – chosen to provide an impervious easy to clean surface with reasonable aesthetic appeal. Examples are quarry tiles and polyvinyl chloride (PVC) sheet and tiles.

Carpets and Carpet Tiles ~ made from animal hair, mineral fibres and man made fibres such as nylon and acrylic. They are also available in mixtures of the above. A wide range of patterns ; sizes and colours are available. Carpets and carpet tiles can be laid loose, stuck with a suitable adhesive or in the case of carpets edge fixed using special strips.

PVC Tiles ~ made from a blended mix of thermoplastic binders; fillers and pigments in a wide variety of colours and patterns to the recommendations of BS 3261. PVC tiles are usually 305 x 305 x 1·6mm thick and are stuck to a suitable base with special adhesives as recommended by the manufacturer.

Quarry Tiles ~

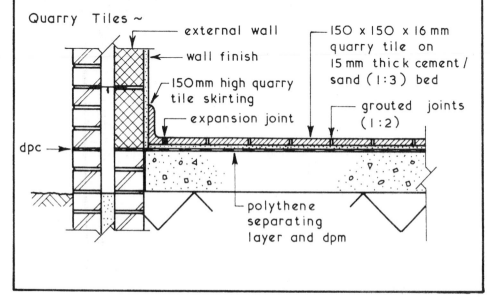

- external wall
- wall finish
- 150mm high quarry tile skirting
- expansion joint
- 150 x 150 x 16 mm quarry tile on 15 mm thick cement / sand (1:3) bed
- grouted joints (1:2)
- dpc
- polythene separating layer and dpm

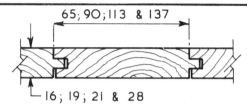

65; 90; 113 & 137

16; 19; 21 & 28

Tongue and Groove Boarding ~ prepared from softwoods to the recommendations of BS 1297. Boards are laid at right angles to the joists and are fixed with 2 No. 65mm long cut floor brads per joists. The ends of board lengths are butt jointed on the centre line of the supporting joist. Maximum board spans are:-

| | | |
|---|---|---|
| 16 mm thick | — | 505 mm |
| 19 mm thick | — | 600 mm |
| 21 mm thick | — | 635 mm |
| 28 mm thick | — | 790 mm. |

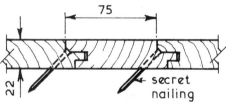

75

22

secret nailing

Timber Strip Flooring ~ strip flooring is usually considered to be boards under 100 mm face width.

In good class work hardwoods would be specified the boards being individually laid and secret nailed. Strip flooring can be obtained treated with a spirit-based fungicide. Spacing of supports depends on type of timber used and applied loading. After laying the strip flooring should be finely sanded and treated with a seal or wax. In common with all timber floorings a narrow perimeter gap should be left for moisture movement

Chipboard ~ sometimes called Particle Board is made from particles of wood bonded with a synthetic resin and / or other organic binders to the recommendations of BS 5669

It can be obtained with a rebated or tongue and groove joint in 600 mm wide boards 19mm thick. The former must be supported on all the longitudinal edges whereas the latter should be supported at all cross joints.

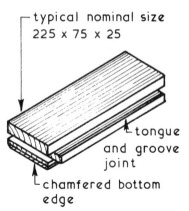

typical nominal size 225 x 75 x 25

tongue and groove joint

chamfered bottom edge

Wood Blocks ~ prepared from hardwoods and softwoods to the recommendations of BS 1187. Wood blocks can be laid to a variety of patterns, also different timbers can be used to create colour and grain effects. Laid blocks should be finely sanded and sealed or polished.

Large Cast-Insitu Ground Floors ~ these are floors designed to
carry medium to heavy loadings such as those used in factories,
warehouses, shops, garages and similar buildings. Their design
and construction is similar to that used for small roads.(see
pages 84 to 87)    Floors of this type are usually  laid in
alternate  4·500 wide strips running the length of the building
or in line with the anticipated traffic flow where applicable.
Transverse joints will be required to control the tensile
stresses due to the thermal  movement and contraction of the
slab.   The spacing of these joints will be determined by the
design and the amount of reinforcement used.  Such joints can
either be formed by using a crack inducer or by sawing a
20 to 25mm deep groove into the upper surface of the slab
within  20 to 30 hours of casting.

Typical Layout ~

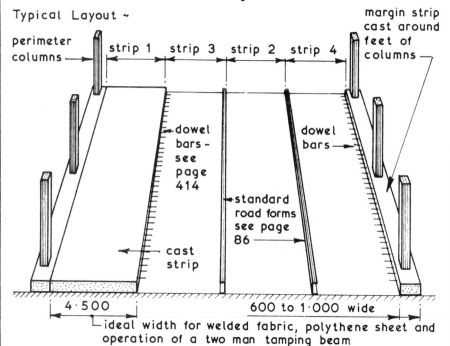

perimeter columns → strip 1  strip 3  strip 2  strip 4  margin strip cast around feet of columns

dowel bars – see page 414

dowel bars →

standard road forms see page 86 →

cast strip

4·500    600 to 1·000 wide

└ideal width for welded fabric, polythene sheet and
operation of a two man tamping beam

Surface Finishing ~ the surface of the concrete may be finished
by power floating or trowelling which is carried  out whilst the
concrete is still plastic  but with sufficient resistance to the
weight of machine and operator whose footprint should not leave
a depression of more than 3mm.    Power grinding of the surface
is an alternative  method which is carried out  within a few days
of the concrete hardening.   The wet concrete having been
surface finished with a  skip float after the initial levelling
with a tamping has been carried out.   Power grinding removes
1 to 2 mm from the surface and is intended to improve surface
texture and not to make good deficiencies in levels.

Vacuum Dewatering ~ if the specification calls for a power float surface finish vacuum dewatering could be used to shorten the time delay between tamping the concrete and power floating the surface. This method is suitable for slabs up to 300 mm thick. The vacuum should be applied for approximately 3 minutes for every 25 mm depth of concrete which will allow power floating to take place usually within 20 to 30 minutes of the tamping operation. The applied vacuum forces out the surplus water by compressing the slab and this causes a reduction in slab depth of approximately 2 % therefore packing strips should be placed on the side forms before tamping to allow for sufficient surcharge of concrete.

Typical Details ~

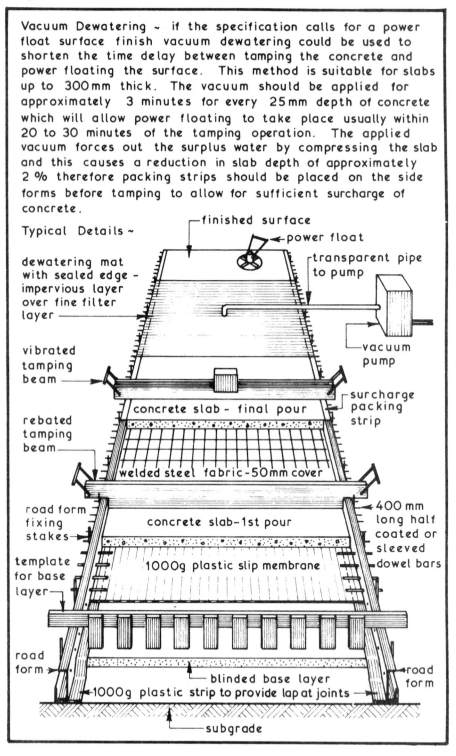

finished surface

power float

transparent pipe to pump

dewatering mat with sealed edge - impervious layer over fine filter layer

vacuum pump

vibrated tamping beam

surcharge packing strip

concrete slab - final pour

rebated tamping beam

welded steel fabric - 50 mm cover

road form fixing stakes

concrete slab - 1st pour

400 mm long half coated or sleeved dowel bars

template for base layer

1000 g plastic slip membrane

road form

road form

blinded base layer

1000 g plastic strip to provide lap at joints

subgrade

Concrete Floor Screeds~ these are used to give a concrete floor a finish suitable to receive the floor finish or covering specified. It should be noted that it is not always necessary or desirable to apply a floor screed to receive a floor covering, techniques are available to enable the concrete floor surface to be prepared at the time of casting to receive the coverings at a later stage.

Typical Screed Mixes~

| Screed Thickness | Cement | Dry Fine Agg < 5mm | Coarse Agg. >5mm<10mm |
|---|---|---|---|
| up to 40 mm | 1 | 3 to 4$\frac{1}{2}$ | — |
| 40 to 75mm | 1 | 3 to 4$\frac{1}{2}$ | — |
| | 1 | 1$\frac{1}{2}$ | 3 |

Laying Floor Screeds~ floor screeds should not be laid in bays since this can cause curling at the edges, screeds can however be laid in 3·000 wide strips to receive thin coverings. Levelling of screeds is achieved by working to levelled timber screeding batten or alternatively a 75mm wide band of levelled screed with square edges can be laid to the perimeter of the floor prior to the general screed laying operation.

Screed Types~

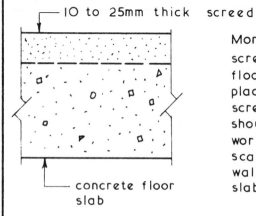

— 10 to 25mm thick screed

concrete floor slab

Monolithic Screeds — screed laid directly on concrete floor slab within three hours of placing concrete - before any screed is placed all surface water should be removed - all screeding work should be carried out from scaffold board runways to avoid walking on the 'green' concrete slab.

# Concrete Floor Screeds

Screed Types ~

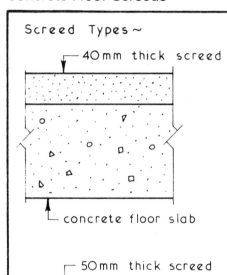

40 mm thick screed

concrete floor slab

**Separate Screeds —**
screed is laid onto the concrete floor slab after it has cured. The floor surface must be clean and rough enough to ensure an adequate bond unless the floor surface is prepared by applying a suitable bonding agent or by brushing with a cement / water grout of a thick cream like consistency just before laying the screed.

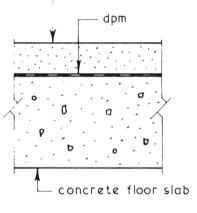

50 mm thick screed

dpm

concrete floor slab

**Unbonded Screeds —**
screed laid directly over a damp-proof membrane — care must be taken during this operation to ensure that the damp-proof membrane is not damaged.

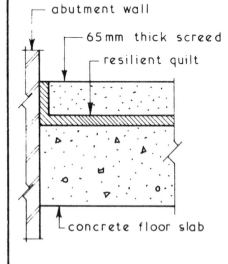

abutment wall

65 mm thick screed

resilient quilt

concrete floor slab

**Floating Screeds —**
a resilient quilt of 25 mm thickness is laid with butt joints and turned up at the edges against the abutment walls the screed being laid directly over the quilt. The main objective of this form of floor screed is to improve the sound insulation properties of the floor.

Primary Functions ~

1. Provide a level surface with sufficient strength to support the imposed loads of people and furniture plus the dead loads of flooring and ceiling.
2. Reduce heat loss from lower floor as required.
3. Provide required degree of sound insulation.
4. Provide required degree of fire resistance.

Basic Construction - a timber suspended upper floor consists of a series of beams or joists support by load bearing walls sized and spaced to carry all the dead and imposed loads.

Joist Sizing - three methods can be used:-

1. Building Regulations imposed load to be not more than $1 \cdot 5 \, kN/m^2$ Calculate dead load in $kg/m^2$ supported by joist excluding the mass of the joist and select suitable size from Tables B3 and B4 in Approved Document A

2. Calculation formula:-

$$BM = \frac{f\,b\,d^2}{6}$$

where
BM = bending moment
f = fibre stress
b = breadth
d = depth in mm
b must be assumed

3. Empirical formula:-

$$D = \frac{\text{span in mm}}{24} + 50$$

where
D = depth of joist in mm
above assumes that joists have a breadth of 50 mm and are at 400 $^c/c$ spacing

Support and Restraint ~

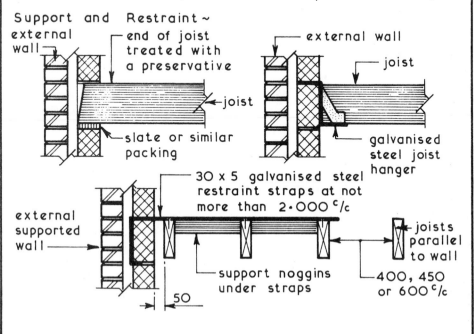

external wall — end of joist treated with a preservative — joist — slate or similar packing

external wall — joist — galvanised steel joist hanger

external supported wall — 30 x 5 galvanised steel restraint straps at not more than 2·000 $^c/c$ — support noggins under straps — 50 — joists parallel to wall — 400, 450 or 600 $^c/c$

Strutting ~ used in timber suspended floors to restrict the movements due to twisting and vibrations which could damage ceiling finishes. Strutting should be included if the span of the floor exceeds 2·400 and is positioned on the centre line.

Typical Details ~

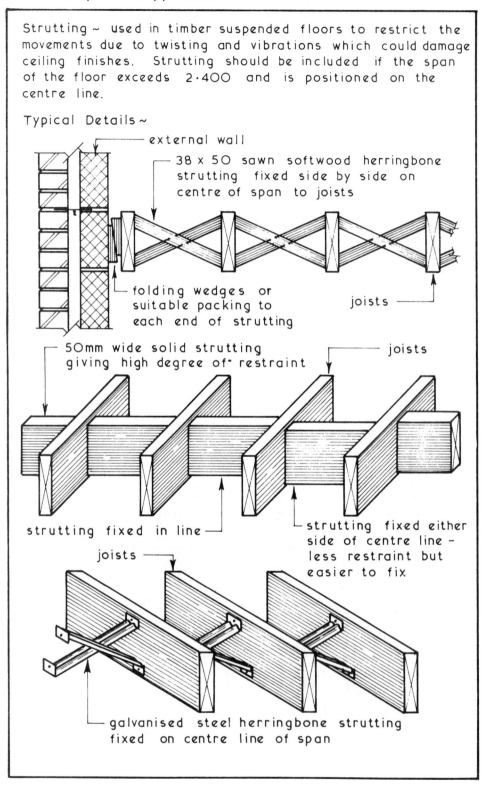

external wall

38 x 50 sawn softwood herringbone strutting fixed side by side on centre of span to joists

folding wedges or suitable packing to each end of strutting

joists

50mm wide solid strutting giving high degree of restraint

joists

strutting fixed in line

strutting fixed either side of centre line - less restraint but easier to fix

joists

galvanised steel herringbone strutting fixed on centre line of span

Trimming Members~ these are the edge members of an opening in a floor and are the same depth as common joists but are usually 25mm wider.

Typical Details ~

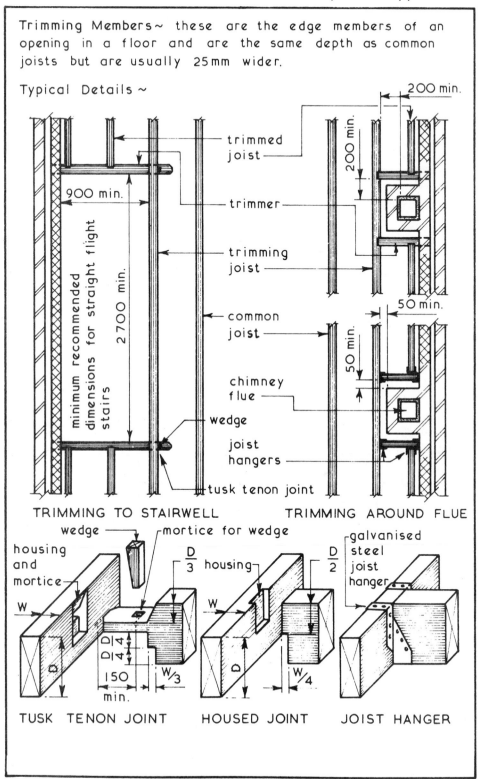

TRIMMING TO STAIRWELL     TRIMMING AROUND FLUE

TUSK TENON JOINT     HOUSED JOINT     JOIST HANGER

Reinforced Concrete Suspended Floors ~ a simple reinforced concrete flat slab cast to act as a suspended floor is not usually economical for spans over 5·000. To overcome this problem beams can be incorporated into the design to span in one or two directions. Such beams usually span between columns which transfers their loads to the foundations. The disadvantages of introducing beams are the greater overall depth of the floor construction and the increased complexity of the formwork and reinforcement. To reduce the overall depth of the floor construction flat slabs can be used where the beam is incorporated with the depth of the slab. This method usually results in a deeper slab with complex reinforcement especially at the column positions.

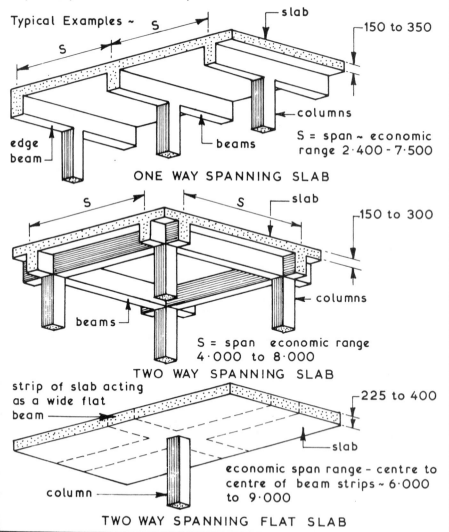

Typical Examples ~

**ONE WAY SPANNING SLAB**

slab — 150 to 350
columns
edge beam — beams
S = span ~ economic range 2·400 - 7·500

**TWO WAY SPANNING SLAB**

slab — 150 to 300
columns
beams
S = span economic range 4·000 to 8·000

**TWO WAY SPANNING FLAT SLAB**

strip of slab acting as a wide flat beam — 225 to 400
slab
economic span range - centre to centre of beam strips ~ 6·000 to 9·000
column

Ribbed Floors ~ to reduce the overall depth of a traditional cast insitu reinforced concrete beam and slab suspended floor a ribbed floor could be used. The basic concept is to replace the wide spaced deep beams with narrow spaced shallow beams or ribs which will carry only a small amount of slab loading. These floors can be designed as one or two way spanning floors. One way spanning ribbed floors are sometimes called troughed floors whereas the two way spanning ribbed floors are called coffered or waffle floors. Ribbed floors are usually cast against metal or reinforced plastic formers in the form of preformed moulds which can be left in place to provide the soffit finish.

Typical Examples ~

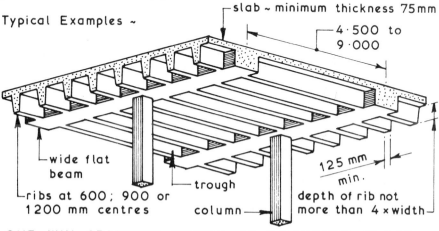

slab ~ minimum thickness 75mm

4·500 to 9·000

wide flat beam

trough

ribs at 600; 900 or 1200 mm centres

column

125 mm min.

depth of rib not more than 4 × width

ONE WAY SPANNING RIBBED OR TROUGHED FLOOR

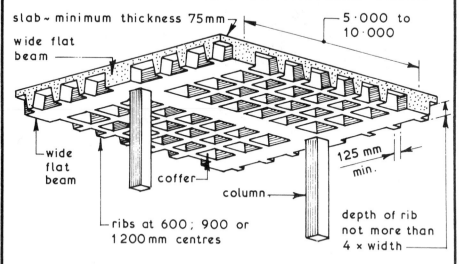

slab ~ minimum thickness 75mm

5·000 to 10·000

wide flat beam

wide flat beam

coffer

column

125 mm min.

ribs at 600; 900 or 1200 mm centres

depth of rib not more than 4 × width

TWO WAY SPANNING COFFERED OR WAFFLE FLOOR

Hollow Pot Floors ~ these are in essence a ribbed floor with permanent formwork in the form of hollow clay or concrete pots. The main advantage of this type of cast insitu floor is that it has a flat soffit which is suitable for the direct application of a plaster finish or an attached dry lining. The voids in the pots can be utilised to house small diameter services within the overall depth of the slab. These floors can be designed as one or two way spanning slabs, the common format being the one way spanning floor.

Typical Example ~

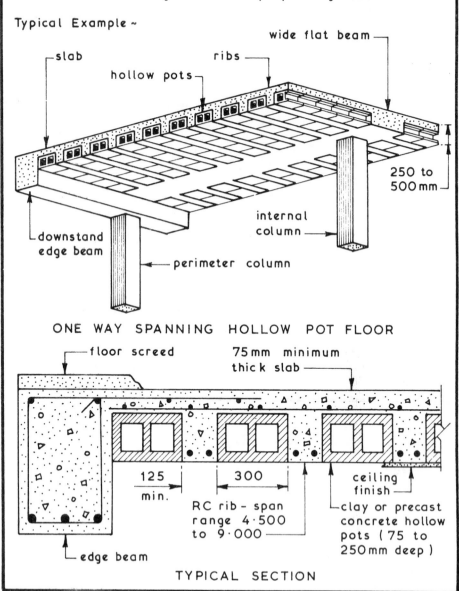

ONE WAY SPANNING HOLLOW POT FLOOR

TYPICAL SECTION

Soffit and Beam Fixings ~ concrete suspended floors can be designed to carry loads other than the direct upper surface loadings. Services can be housed within the voids created by the beams or ribs and suspended or attached ceilings can be supported by the floor. Services which run at right angles to the beams or ribs are usually housed in cast-in holes. There are many types of fixings available for use in conjunction with floor slabs, some are designed to be cast-in whilst others are fitted after the concrete has cured. All fixings must be positioned and installed so that they are not detrimental to the structural integrity of the floor.

Typical Examples ~

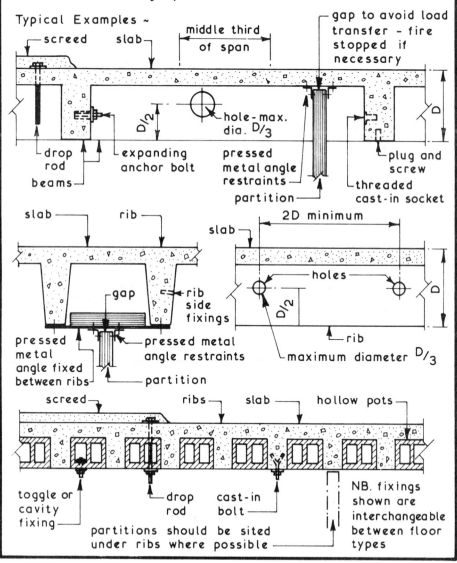

Precast Concrete Floors ~ these are available in several basic formats and provide an alternative form of floor construction to suspended timber floors and insitu reinforced concrete suspended floors. The main advantages of precast concrete floors are :—

1. Elimination of the need for formwork except for nominal propping which is required with some systems.
2. Curing time of concrete is eliminated therefore the floor is available for use as a working platform at an earlier stage.
3. Superior quality control of product is possible with factory produced componets.

The main disadvantages of precast concrete floors when compared with insitu reinforced concrete floors are :—

1. Less flexible in design terms.
2. Formation of large openings in the floor for ducts, shafts and stairwells usually have to be formed by casting an insitu reinforced concrete floor strip around the opening position.
3. Higher degree of site accuracy is required to ensure that the precast concrete floor units can be accommodated without any alterations or making good

Typical Basic Formats ~

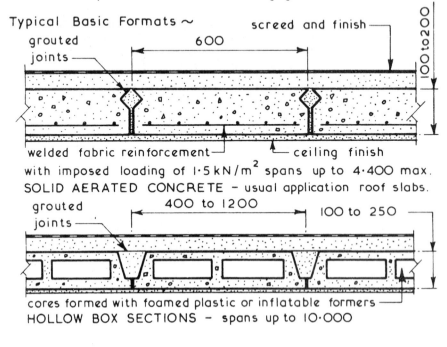

grouted joints — screed and finish — 600 — 100 to 200

welded fabric reinforcement — ceiling finish

with imposed loading of $1.5 \, kN/m^2$ spans up to 4.400 max.
SOLID AERATED CONCRETE – usual application roof slabs.

grouted joints — 400 to 1200 — 100 to 250

cores formed with foamed plastic or inflatable formers —
HOLLOW BOX SECTIONS – spans up to 10.000

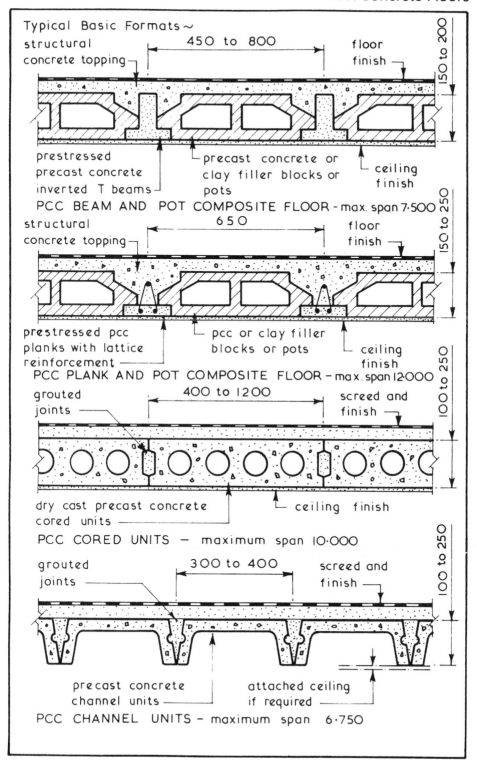

Typical Basic Formats~

structural concrete topping

450 to 800

floor finish

150 to 200

prestressed precast concrete inverted T beams

precast concrete or clay filler blocks or pots

ceiling finish

PCC BEAM AND POT COMPOSITE FLOOR - max. span 7·500

structural concrete topping

650

floor finish

150 to 250

prestressed pcc planks with lattice reinforcement

pcc or clay filler blocks or pots

ceiling finish

PCC PLANK AND POT COMPOSITE FLOOR – max. span 12·000

grouted joints

400 to 1200

screed and finish

100 to 250

dry cast precast concrete cored units

ceiling finish

PCC CORED UNITS — maximum span 10·000

grouted joints

300 to 400

screed and finish

100 to 250

precast concrete channel units

attached ceiling if required

PCC CHANNEL UNITS - maximum span 6·750

425

# Precast Concrete Floors

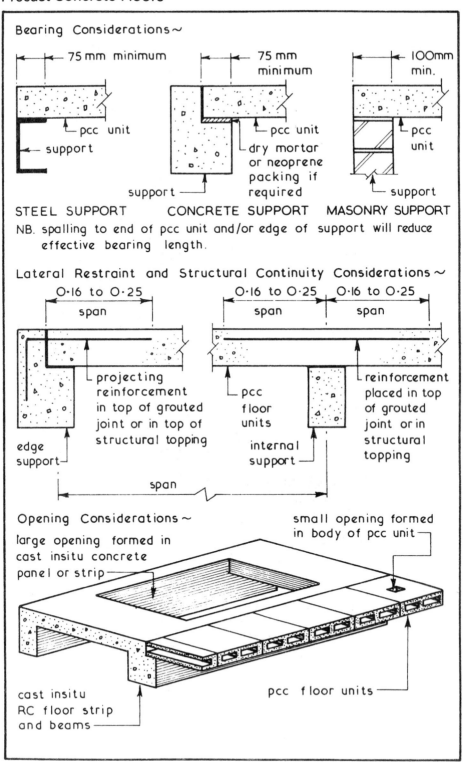

Bearing Considerations ~

75 mm minimum

pcc unit

support

STEEL SUPPORT

75 mm minimum

pcc unit

dry mortar or neoprene packing if required

support

CONCRETE SUPPORT

100mm min.

pcc unit

support

MASONRY SUPPORT

NB. spalling to end of pcc unit and/or edge of support will reduce effective bearing length.

Lateral Restraint and Structural Continuity Considerations ~

O·16 to O·25 span

projecting reinforcement in top of grouted joint or in top of structural topping

edge support

O·16 to O·25 span    O·16 to O·25 span

pcc floor units

internal support

reinforcement placed in top of grouted joint or in structural topping

span

Opening Considerations ~

large opening formed in cast insitu concrete panel or strip

small opening formed in body of pcc unit

cast insitu RC floor strip and beams

pcc floor units

Primary Functions ~

1. Provide a means of circulation between floor levels.
2. Establish a safe means of travel between floor levels.
3. Provide an easy means of travel between floor levels.
4. Provide a means of conveying fittings and furniture between floor levels.

Constituent Parts ~

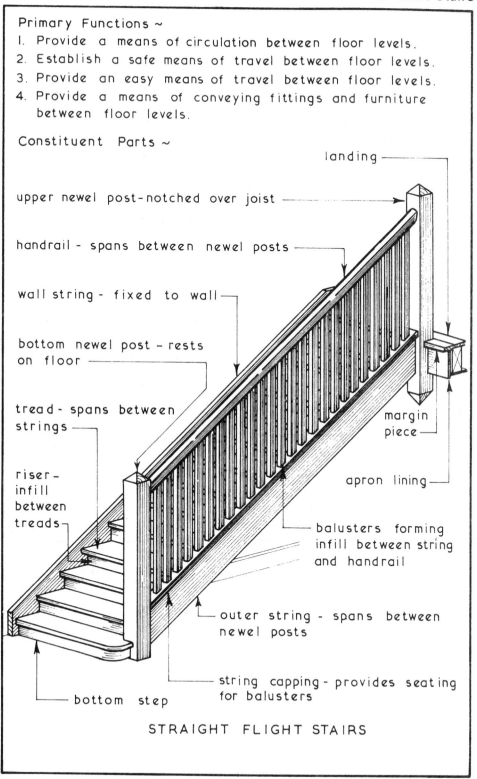

landing

upper newel post-notched over joist

handrail - spans between newel posts

wall string - fixed to wall

bottom newel post - rests on floor

tread - spans between strings

riser - infill between treads

margin piece

apron lining

balusters forming infill between string and handrail

outer string - spans between newel posts

bottom step

string capping - provides seating for balusters

STRAIGHT FLIGHT STAIRS

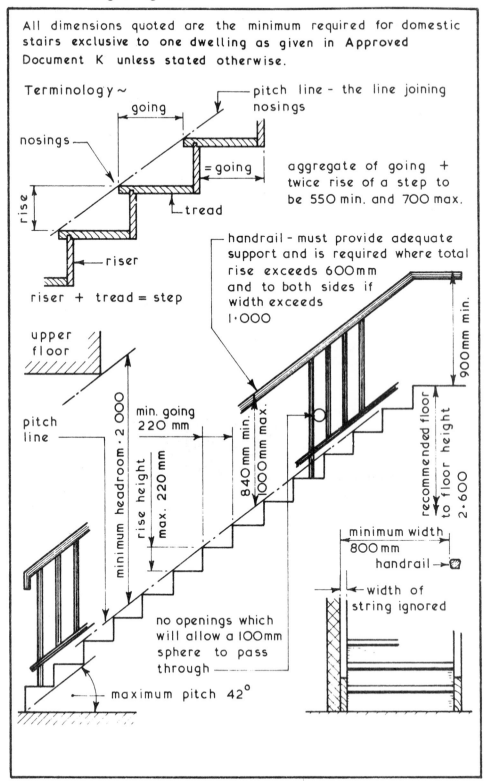

All dimensions quoted are the minimum required for domestic stairs exclusive to one dwelling as given in Approved Document K unless stated otherwise.

Terminology ~

pitch line - the line joining nosings

going

nosings

= going

tread

rise

riser

riser + tread = step

aggregate of going + twice rise of a step to be 550 min. and 700 max.

handrail - must provide adequate support and is required where total rise exceeds 600mm and to both sides if width exceeds 1·000

upper floor

pitch line

minimum headroom·2·000

min. going 220 mm

rise height max. 220 mm

840 mm min. 1000mm max.

900mm min.

recommended floor to floor height 2·600

minimum width 800 mm

handrail

width of string ignored

no openings which will allow a 100mm sphere to pass through

maximum pitch 42°

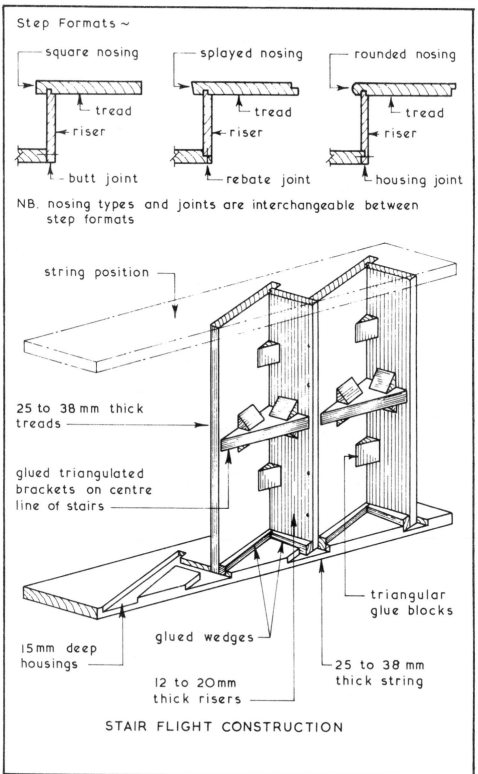

Step Formats ~

square nosing
tread
riser
butt joint

splayed nosing
tread
riser
rebate joint

rounded nosing
tread
riser
housing joint

NB. nosing types and joints are interchangeable between step formats

string position

25 to 38 mm thick treads

glued triangulated brackets on centre line of stairs

triangular glue blocks

glued wedges

15mm deep housings

12 to 20mm thick risers

25 to 38 mm thick string

STAIR FLIGHT CONSTRUCTION

Bottom Step Arrangements ~

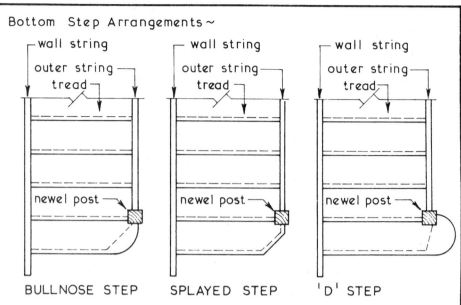

BULLNOSE STEP   SPLAYED STEP   'D' STEP

Projecting bottom steps are usually included to enable the outer string to be securely jointed to the back face of the newel post and to provide an easy line of travel when ascending or descending at the foot of the stairs.

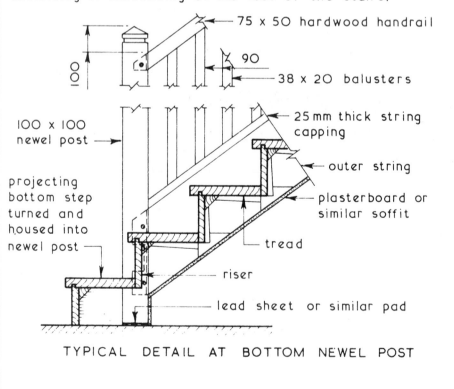

TYPICAL DETAIL AT BOTTOM NEWEL POST

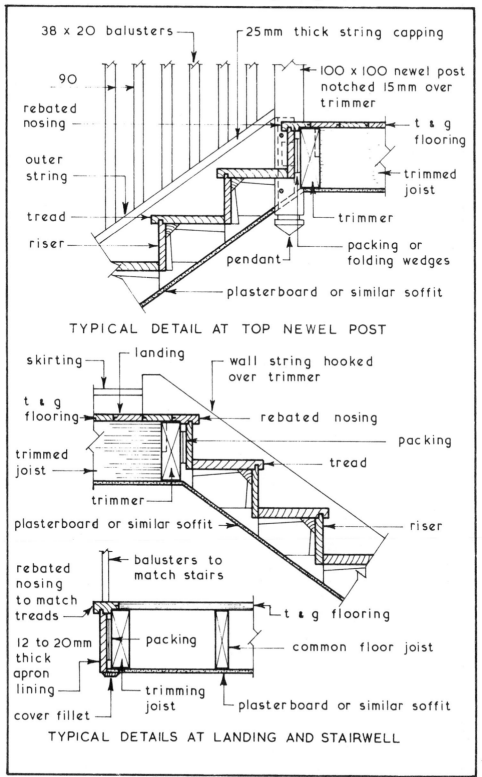

38 x 20 balusters

90

rebated nosing

outer string

tread

riser

25mm thick string capping

100 x 100 newel post notched 15mm over trimmer

t & g flooring

trimmed joist

trimmer

packing or folding wedges

pendant

plasterboard or similar soffit

TYPICAL DETAIL AT TOP NEWEL POST

skirting

landing

wall string hooked over trimmer

t & g flooring

rebated nosing

packing

trimmed joist

tread

trimmer

plasterboard or similar soffit

riser

rebated nosing to match treads

balusters to match stairs

t & g flooring

12 to 20mm thick apron lining

packing

common floor joist

trimming joist

cover fillet

plasterboard or similar soffit

TYPICAL DETAILS AT LANDING AND STAIRWELL

# Timber Open Riser Stairs

Open Riser Timber Stairs ~ these are timber stairs constructed to the same basic principles as standard timber stairs excluding the use of a riser. They have no real advantage over traditional stairs except for the generally accepted aesthetic appeal of elegance. Like the traditional timber stairs they must comply **with the minimum requirements set out** in Part K of the Building Regulations.

Typical Requirements for Stairs in a Small Residential Building ~

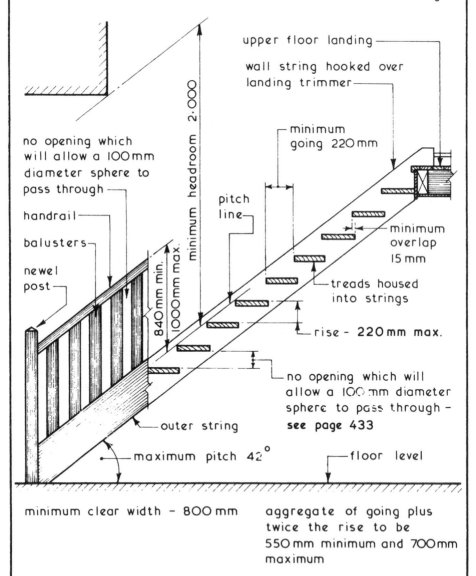

no opening which will allow a 100mm diameter sphere to pass through

handrail

balusters

newel post

minimum headroom 2·000

840mm min.
1000mm max.

pitch line

outer string

maximum pitch 42°

minimum clear width – 800 mm

upper floor landing

wall string hooked over landing trimmer

minimum going 220 mm

minimum overlap 15 mm

treads housed into strings

rise – 220mm max.

no opening which will allow a 100 mm diameter sphere to pass through – **see page 433**

floor level

aggregate of going plus twice the rise to be 550 mm minimum and 700mm maximum

Design and Construction ~ because of the legal requirement of not having a gap between any two consecutive treads through which a 100mm diameter sphere can pass and the limitation **relating to the going and rise (see page 432)** it is generally not practicable to have a completely riserless stair for residential buildings since by using minimum dimensions a very low pitch of approximately $27\frac{1}{2}°$ would result and by choosing an acceptable pitch a very thick tread would have to be used to restrict the gap to 100mm.

Possible Solutions ~

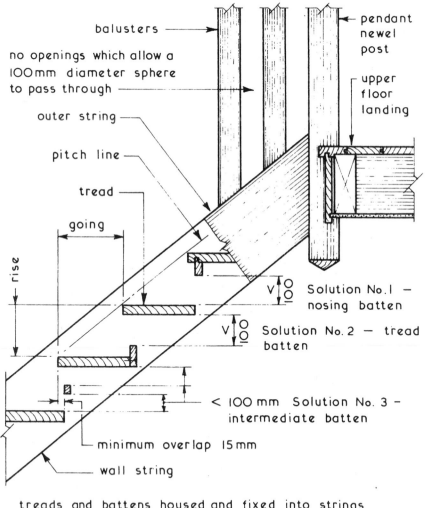

balusters

pendant newel post

no openings which allow a 100mm diameter sphere to pass through

upper floor landing

outer string

pitch line

tread

going

rise

Solution No.1 — nosing batten

Solution No. 2 — tread batten

< 100 mm   Solution No. 3 — intermediate batten

minimum overlap 15mm

wall string

treads and battens housed and fixed into strings

Timber Stairs ~ these must comply with the minimum requirements set out in Part K of the Building Regulations. Straight flight stairs are simple, easy to construct and install but by the introduction of intermediate landings stairs can be designed to change direction of travel and be more compact in plan than the straight flight stairs.

Landings ~ these are designed and constructed in the same manner as timber upper floors but due to the shorter spans they require smaller joist sections. Landings can be detailed for a 90° change of direction (quarter space landing) or a 180° change of direction (half space landing) and can be introduced at any position between the two floors being served by the stairs.

Typical Layouts ~

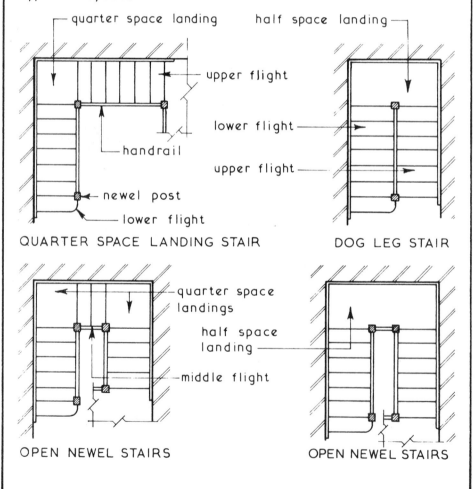

QUARTER SPACE LANDING STAIR     DOG LEG STAIR

OPEN NEWEL STAIRS     OPEN NEWEL STAIRS

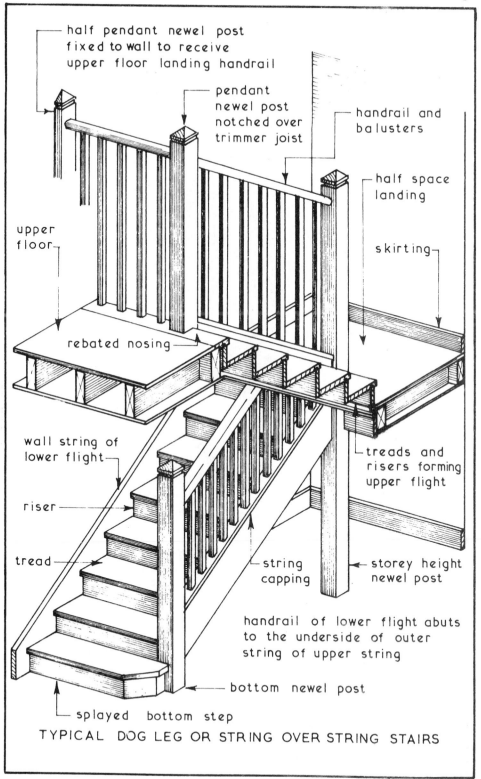

half pendant newel post
fixed to wall to receive
upper floor landing handrail

pendant
newel post
notched over
trimmer joist

handrail and
balusters

half space
landing

upper
floor

skirting

rebated nosing

wall string of
lower flight

treads and
risers forming
upper flight

riser

tread

string
capping

storey height
newel post

handrail of lower flight abuts
to the underside of outer
string of upper string

bottom newel post

splayed bottom step

TYPICAL DOG LEG OR STRING OVER STRING STAIRS

435

Insitu Reinforced Concrete Stairs ~ a variety of stair types and arrangements are possible each having its own appearance and design characteristics. In all cases these stairs must **comply with the minimum requirements** set out in Part K of the Building Regulations in accordance with the purpose group of the building in which the stairs are situated.

Typical Examples ~

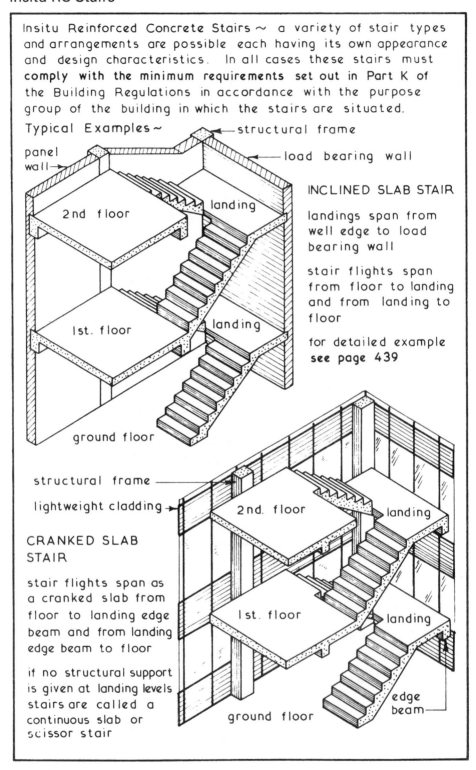

structural frame

panel wall →

load bearing wall

**INCLINED SLAB STAIR**

landings span from well edge to load bearing wall

stair flights span from floor to landing and from landing to floor

for detailed example **see page 439**

2nd floor

landing

1st. floor

landing

ground floor

structural frame

lightweight cladding →

**CRANKED SLAB STAIR**

stair flights span as a cranked slab from floor to landing edge beam and from landing edge beam to floor

if no structural support is given at landing levels stairs are called a continuous slab or scissor stair

2nd. floor

landing

1st. floor

landing

edge beam

ground floor

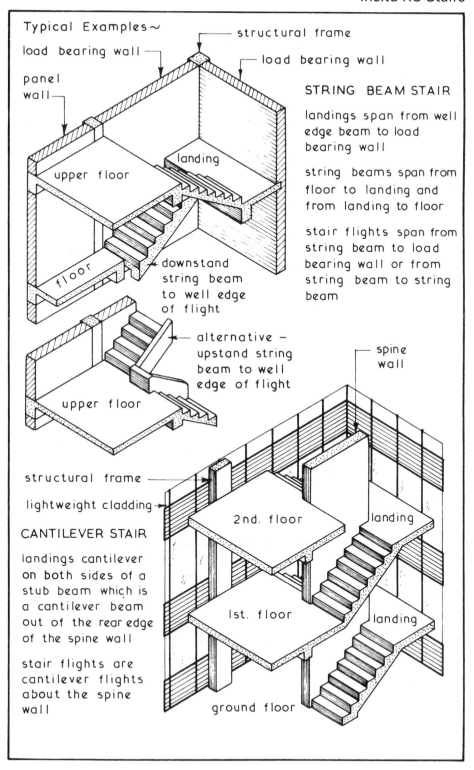

Typical Examples~

load bearing wall

panel wall

structural frame

load bearing wall

**STRING BEAM STAIR**

landings span from well edge beam to load bearing wall

string beams span from floor to landing and from landing to floor

stair flights span from string beam to load bearing wall or from string beam to string beam

upper floor

landing

floor

downstand string beam to well edge of flight

alternative – upstand string beam to well edge of flight

upper floor

spine wall

structural frame

lightweight cladding

**CANTILEVER STAIR**

landings cantilever on both sides of a stub beam which is a cantilever beam out of the rear edge of the spine wall

stair flights are cantilever flights about the spine wall

2nd. floor

landing

1st. floor

landing

ground floor

437

Spiral and Helical Stairs ~ these stairs constructed in insitu reinforced concrete are considered to be aesthetically pleasing but are expensive to construct. They are therefore mainly confined to prestige buildings usually as accommodation stairs linking floors within the same compartment. Like all other forms of stair they must conform to the requirements of Part K of the Building Regulations and if used as a means of escape in case of fire with the requirements of Part B. Spiral stairs can be defined as those describing a helix around a central column whereas a helical stair has an open well. The open well of a helical stair is usually circular or elliptical in plan and the formwork is built up around a vertical timber core.

Typical Example of a Helical Stair ~

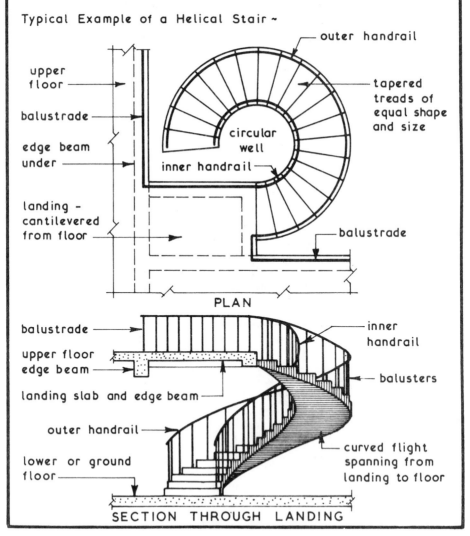

PLAN

SECTION THROUGH LANDING

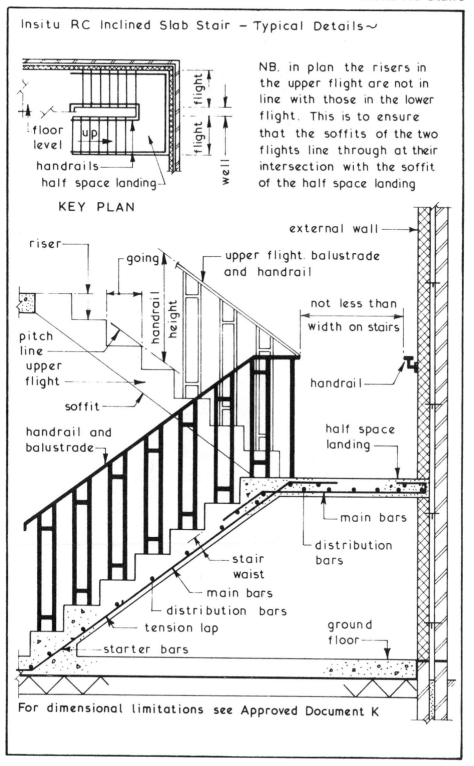

Insitu RC Inclined Slab Stair – Typical Details ～

NB. in plan the risers in the upper flight are not in line with those in the lower flight. This is to ensure that the soffits of the two flights line through at their intersection with the soffit of the half space landing

floor level

up

handrails

half space landing

KEY PLAN

flight

flight

flight

well

riser

going

upper flight. balustrade and handrail

external wall

handrail height

not less than

width on stairs

pitch line

upper flight

handrail

soffit

handrail and balustrade

half space landing

main bars

distribution bars

stair waist

main bars

distribution bars

tension lap

starter bars

ground floor

For dimensional limitations see Approved Document K

Insitu Reinforced Concrete Stair Formwork ~ in specific detail the formwork will vary for the different types of reinforced concrete stair but the basic principles for each format will remain constant.

## Typical RC Stair Formwork Details ~ (see page 439 for Key Plan)

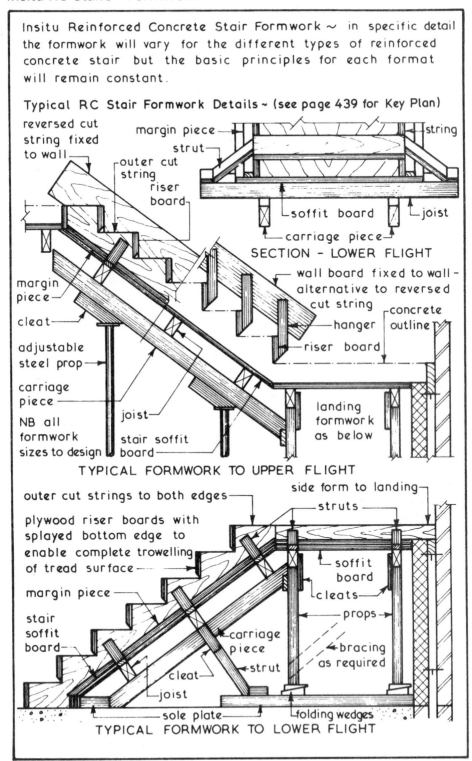

reversed cut string fixed to wall

margin piece

strut

string

outer cut string

riser board

SECTION - LOWER FLIGHT

margin piece

cleat

adjustable steel prop

carriage piece

NB all formwork sizes to design

joint

stair soffit board

soffit board

joist

carriage piece

wall board fixed to wall - alternative to reversed cut string

hanger

riser board

concrete outline

landing formwork as below

TYPICAL FORMWORK TO UPPER FLIGHT

outer cut strings to both edges

plywood riser boards with splayed bottom edge to enable complete trowelling of tread surface

margin piece

stair soffit board

cleat

joist

sole plate

side form to landing

struts

soffit board

cleats

props

bracing as required

carriage piece

strut

folding wedges

TYPICAL FORMWORK TO LOWER FLIGHT

440

Precast Concrete Stairs ~ these can be produced to most of the formats used for insitu concrete stairs and like those must comply with the appropriate requirements set out in Part K of the Building Regulations. To be economic the total production run must be sufficient to justify the costs of the moulds and therefore the designers choice may be limited to the stair types which are produced as a manufacturer's standard item.

Precast concrete stairs can have the following advantages :-

1. Good quality control of finished product.

2. Saving in site space since formwork fabrication and storage will not be required.

3. The stairs can be installed at any time after the floors have been completed thus giving full utilisation to the stair shaft as a lifting or hoisting space if required.

4. Hoisting, positioning and fixing can usually be carried out by semi - skilled labour.

Typical Example ~ Straight Flight Stairs

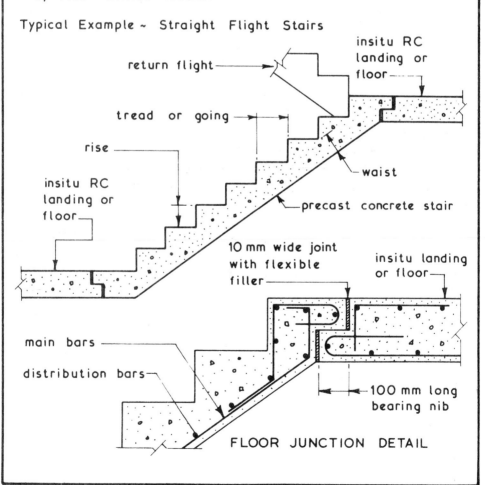

## Precast Concrete Stairs

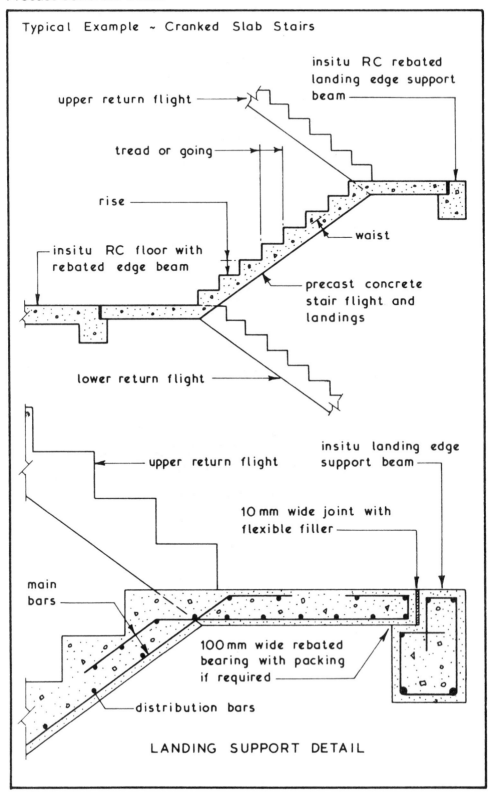

Typical Example ~ Cranked Slab Stairs

upper return flight

insitu RC rebated landing edge support beam

tread or going

rise

insitu RC floor with rebated edge beam

waist

precast concrete stair flight and landings

lower return flight

upper return flight

insitu landing edge support beam

10 mm wide joint with flexible filler

main bars

100 mm wide rebated bearing with packing if required

distribution bars

LANDING SUPPORT DETAIL

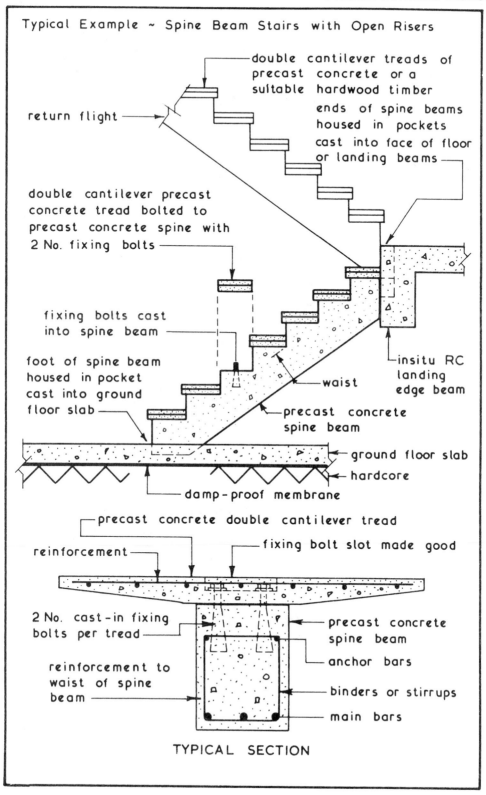

Typical Example ~ Spine Beam Stairs with Open Risers

double cantilever treads of precast concrete or a suitable hardwood timber

return flight

ends of spine beams housed in pockets cast into face of floor or landing beams

double cantilever precast concrete tread bolted to precast concrete spine with 2 No. fixing bolts

fixing bolts cast into spine beam

foot of spine beam housed in pocket cast into ground floor slab

insitu RC landing edge beam

waist

precast concrete spine beam

ground floor slab

hardcore

damp-proof membrane

precast concrete double cantilever tread

reinforcement

fixing bolt slot made good

2 No. cast-in fixing bolts per tread

precast concrete spine beam

anchor bars

reinforcement to waist of spine beam

binders or stirrups

main bars

TYPICAL SECTION

443

Precast Concrete Spiral Stairs ~ this form of stair is usually constructed with an open riser format using tapered treads which have a keyhole plan shape. Each tread has a hollow cylinder at the narrow end equal to the rise which is fitted over a central steel column usually filled with insitu concrete. The outer end of the tread has holes through which the balusters pass to be fixed on the underside of the tread below, a hollow spacer being used to maintain the distance between consecutive treads.

Typical Example ~

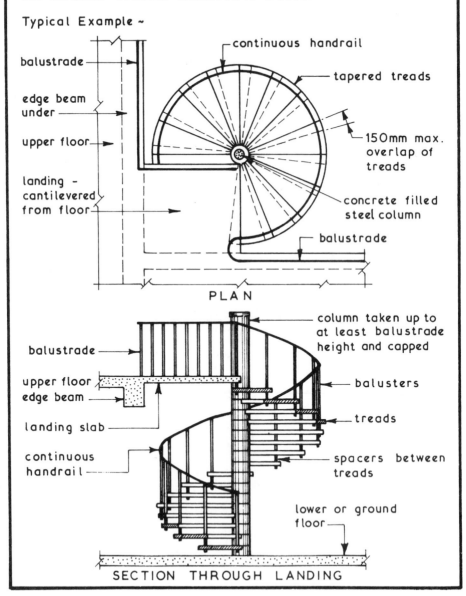

PLAN

SECTION THROUGH LANDING

Metal Stairs ~ these can be produced in cast iron, mild steel or aluminium alloy for use as escape stairs or for internal accommodation stairs. Most escape stairs are fabricated from cast iron or mild steel and must comply with the Building Regulation requirements for stairs in general and fire escape stairs in particular. Most metal stairs are purpose made and therefore tend to cost more than comparable concrete stairs. Their main advantage is the elimination of the need for formwork whilst the main disadvantage is the regular maintenance in the form of painting required for cast iron and mild steel stairs.

Typical Example ~ Straight Flight Steel External Escape Stair

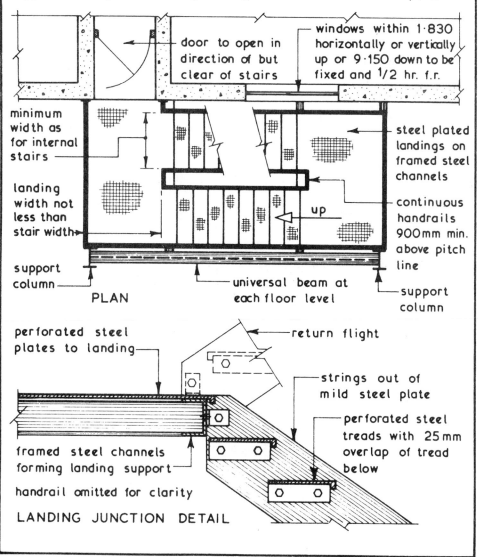

door to open in direction of but clear of stairs

windows within 1·830 horizontally or vertically up or 9·150 down to be fixed and 1/2 hr. f.r.

minimum width as for internal stairs

steel plated landings on framed steel channels

landing width not less than stair width

continuous handrails 900mm min. above pitch line

up

support column

universal beam at each floor level

support column

PLAN

perforated steel plates to landing

return flight

strings out of mild steel plate

framed steel channels forming landing support

perforated steel treads with 25mm overlap of tread below

handrail omitted for clarity

LANDING JUNCTION DETAIL

445

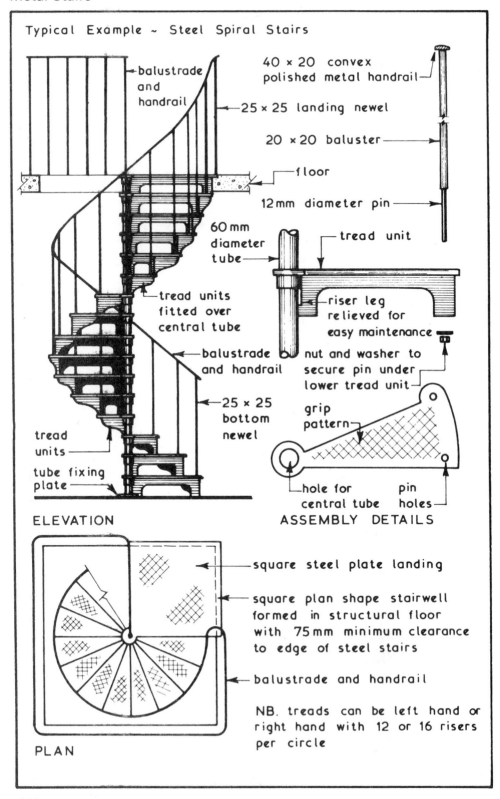

Typical Example ~ Steel Spiral Stairs

balustrade and handrail

25 × 25 landing newel

40 × 20 convex polished metal handrail

20 × 20 baluster

floor

12mm diameter pin

60 mm diameter tube

tread unit

tread units fitted over central tube

riser leg relieved for easy maintenance

balustrade and handrail

nut and washer to secure pin under lower tread unit

25 × 25 bottom newel

grip pattern

tread units

tube fixing plate

hole for central tube

pin holes

ELEVATION

ASSEMBLY DETAILS

square steel plate landing

square plan shape stairwell formed in structural floor with 75mm minimum clearance to edge of steel stairs

balustrade and handrail

NB. treads can be left hand or right hand with 12 or 16 risers per circle

PLAN

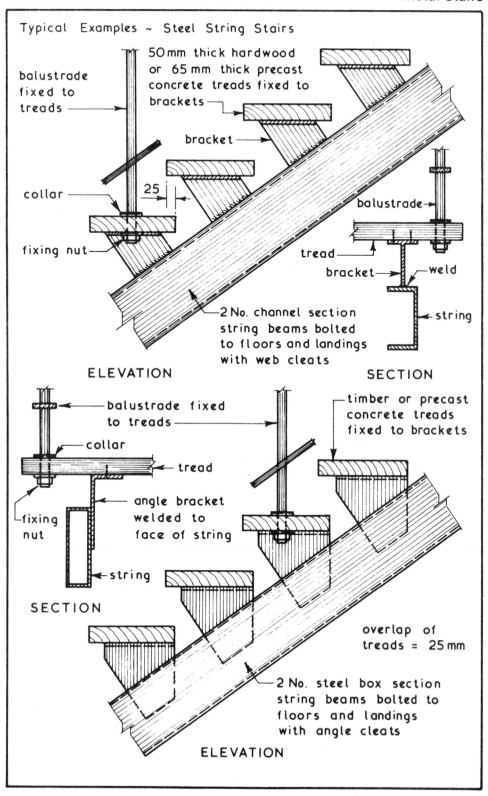

Typical Examples ~ Steel String Stairs

balustrade fixed to treads

50 mm thick hardwood or 65 mm thick precast concrete treads fixed to brackets

bracket

25

collar

fixing nut

balustrade

tread

bracket

weld

string

2 No. channel section string beams bolted to floors and landings with web cleats

ELEVATION

SECTION

balustrade fixed to treads

collar

tread

fixing nut

angle bracket welded to face of string

string

timber or precast concrete treads fixed to brackets

SECTION

overlap of treads = 25 mm

2 No. steel box section string beams bolted to floors and landings with angle cleats

ELEVATION

447

Balustrades and Handrails ~ these must comply in all respects with the requirements given in Part K of the Building Regulations and in the context of escape stairs are constructed of a non-combustible material with a handrail shaped to give a comfortable hand grip. The handrail may be covered or capped with a combustible material such as timber or plastic. Most balustrades are designed to be fixed after the stairs have been cast or installed by housing the balusters in a preformed pocket or by direct surface fixing.

Typical Details ~

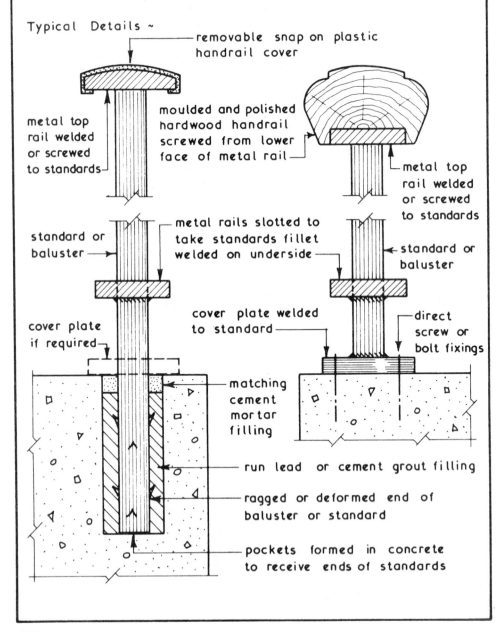

removable snap on plastic handrail cover

metal top rail welded or screwed to standards

moulded and polished hardwood handrail screwed from lower face of metal rail

metal top rail welded or screwed to standards

standard or baluster

metal rails slotted to take standards fillet welded on underside

standard or baluster

cover plate if required

cover plate welded to standard

direct screw or bolt fixings

matching cement mortar filling

run lead or cement grout filling

ragged or deformed end of baluster or standard

pockets formed in concrete to receive ends of standards

Functions ~ the main functions of any door are:-

1. Provide a means of access and egress.

2. Maintain continuity of wall function when closed.

3. Provide a degree of privacy and security.

Choice of door type can be determined by :-

1. Position - whether internal or external

2. Properties required - fire resistant, glazed to provide for borrowed light or vision through, etc.

3. Appearance - flush or panelled, painted or polished,etc.

Door Schedules ~ these can be prepared in the same manner and for the same purpose as that given for windows on page 275

Internal Doors ~ these are usually lightweight and can be fixed to a lining, if heavy doors are specified these can be hung to frames in a similar manner to external doors. An alternative method is to use door sets which are usually storey height and supplied with prehung doors.

Typical Door Lining Details ~

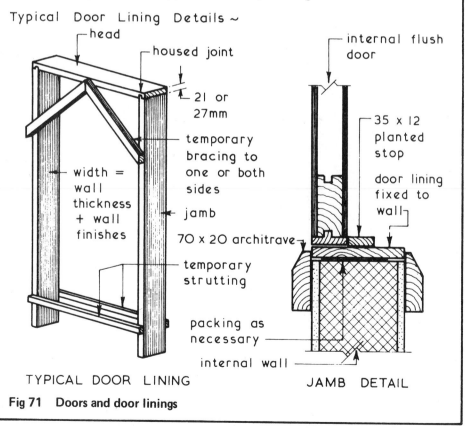

TYPICAL DOOR LINING          JAMB DETAIL

**Fig 71   Doors and door linings**

Internal Doors ~ these are similar in construction to the external doors but are usually thinner and therefore lighter in weight

Typical Examples ~

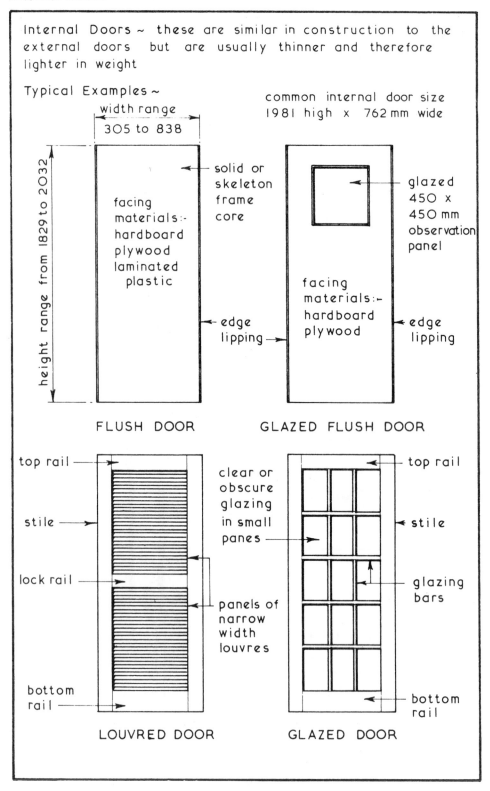

common internal door size 1981 high x 762 mm wide

width range
305 to 838

height range from 1829 to 2032

solid or skeleton frame core

facing materials:- hardboard plywood laminated plastic

edge lipping

glazed 450 x 450 mm observation panel

facing materials:- hardboard plywood

edge lipping

**FLUSH DOOR**

**GLAZED FLUSH DOOR**

top rail

stile

lock rail

bottom rail

clear or obscure glazing in small panes

panels of narrow width louvres

top rail

stile

glazing bars

bottom rail

**LOUVRED DOOR**

**GLAZED DOOR**

Internal Door Frames and Linings~ these are similar in construction to external door frames but usually have planted door stops and do not have a sill. The frames sized to be built in conjunction with various partition thicknesses and surface finishes. Linings with planted stops are usually employed for lightweight domestic doors.

Typical Examples~

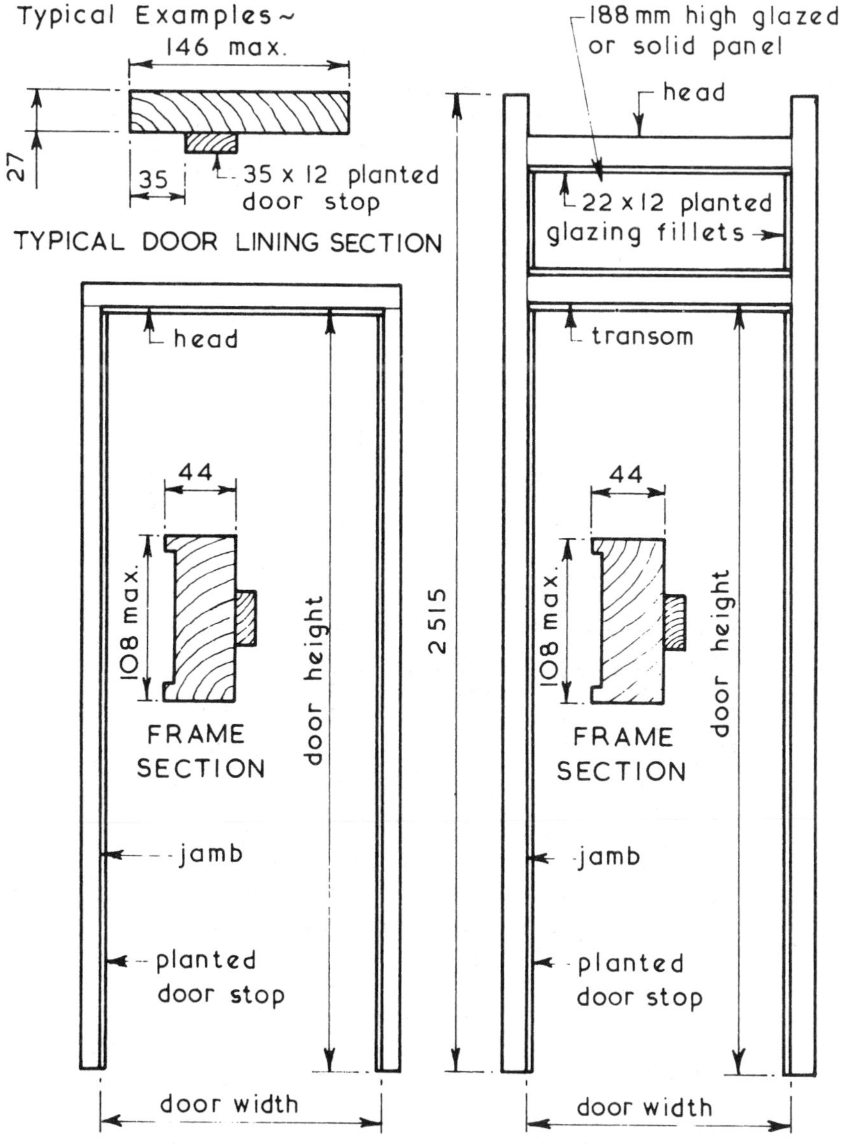

146 max.

27

35    35 x 12 planted door stop

TYPICAL DOOR LINING SECTION

188 mm high glazed or solid panel

head

22 x 12 planted glazing fillets

head

transom

44

108 max.

door height

2515

FRAME SECTION

jamb

planted door stop

door width

STANDARD FRAME

44

108 max.

door height

FRAME SECTION

jamb

planted door stop

door width

STOREY HEIGHT FRAME

Door Sets~ these are factory produced fully assembled prehung doors which are supplied complete with frame, architraves and ironmongery except for door furniture. The doors hung to the frames using pin butts for easy door removal. Prehung door sets are available in standard and storey height versions and are suitable for all internal door applications with normal wall and partition thicknesses.

Typical Examples~

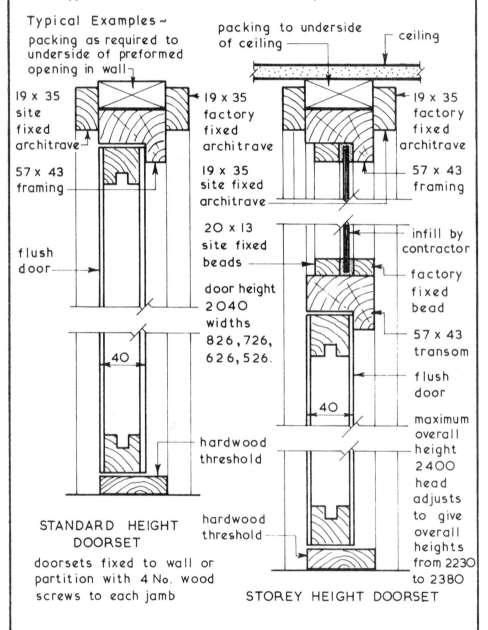

packing as required to underside of preformed opening in wall

19 x 35 site fixed architrave

57 x 43 framing

flush door

40

STANDARD HEIGHT DOORSET

doorsets fixed to wall or partition with 4 No. wood screws to each jamb

packing to underside of ceiling — ceiling

19 x 35 factory fixed architrave

19 x 35 site fixed architrave

20 x 13 site fixed beads

door height 2040 widths 826, 726, 626, 526.

hardwood threshold

STOREY HEIGHT DOORSET

19 x 35 factory fixed architrave

57 x 43 framing

infill by contractor

factory fixed bead

57 x 43 transom

flush door

40

maximum overall height 2400 head adjusts to give overall heights from 2230 to 2380

hardwood threshold

Half Hour Fire Check Doors ~ these are usually based on the recommendations given in PD 6512 : Part 1, a wide variety of door constructions are available from various manufacturers but generally they all have to be fitted to a similar frame.

A door's resistance to fire is measured by:—

1. Stability — resistance in minutes to collapse under simulated fire conditions

2. Integrity — resistance in minutes to the penetration of flame and hot gases under simulated fire conditions.

Half hour fire check doors have a stability / integrity rating of 30 / 20 whereas half hour fire resistant doors have a 30 / 30 rating.

Typical Details ~

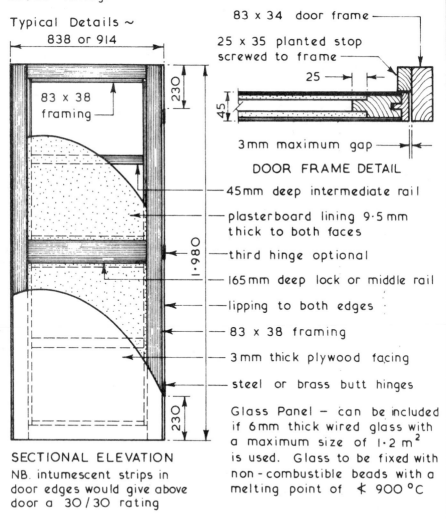

**DOOR FRAME DETAIL**

83 x 34 door frame

25 x 35 planted stop screwed to frame

25

3mm maximum gap

45

838 or 914

230

83 x 38 framing

1·980

230

—45mm deep intermediate rail

—plasterboard lining 9·5mm thick to both faces

—third hinge optional

—165mm deep lock or middle rail

—lipping to both edges

— 83 x 38 framing

— 3mm thick plywood facing

— steel or brass butt hinges

**SECTIONAL ELEVATION**

NB. intumescent strips in door edges would give above door a 30 / 30 rating

Glass Panel — can be included if 6mm thick wired glass with a maximum size of 1·2 m$^2$ is used. Glass to be fixed with non-combustible beads with a melting point of $\ngtr$ 900 °C

One Hour Fire Check Door ~ like the half hour fire check door shown on page 453 these doors are based on the recommendations given in PD 6512 : Part 1 which covers both door and frame. A wide variety of door constructions are available from various manufacturers but most of these are classified as a one hour fire resistant door with a stability / integrity rating of 60 / 60 as opposed to the fire check door rating of 60 /45.

Typical Details ~

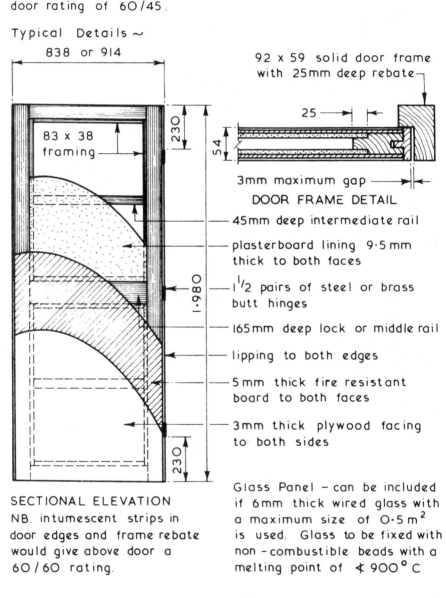

838 or 914

92 x 59 solid door frame with 25mm deep rebate

83 x 38 framing

230

25

54

3mm maximum gap

**DOOR FRAME DETAIL**

— 45mm deep intermediate rail

— plasterboard lining 9·5 mm thick to both faces

— $1\frac{1}{2}$ pairs of steel or brass butt hinges

1·980

— 165mm deep lock or middle rail

— lipping to both edges

— 5 mm thick fire resistant board to both faces

— 3mm thick plywood facing to both sides

230

**SECTIONAL ELEVATION**
NB. intumescent strips in door edges and frame rebate would give above door a 60 / 60 rating.

Glass Panel – can be included if 6mm thick wired glass with a maximum size of O·5 m$^2$ is used. Glass to be fixed with non - combustible beads with a melting point of ≮ 900°C

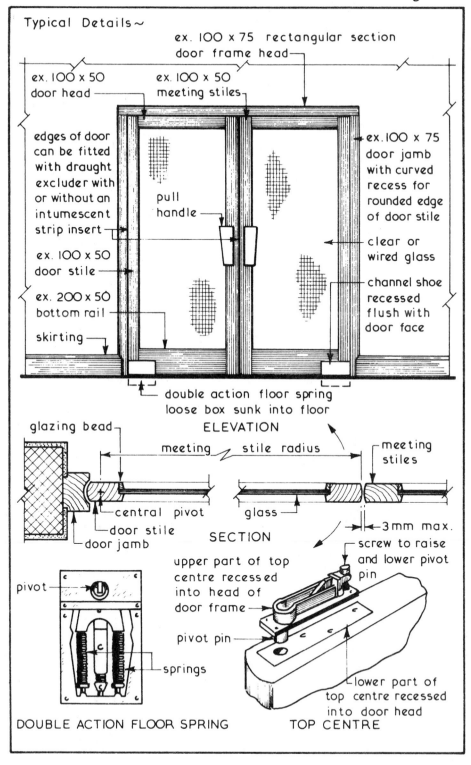

Typical Details~

ex. 100 x 75 rectangular section door frame head

ex. 100 x 50 door head

ex. 100 x 50 meeting stiles

edges of door can be fitted with draught excluder with or without an intumescent strip insert

pull handle

ex. 100 x 75 door jamb with curved recess for rounded edge of door stile

ex. 100 x 50 door stile

clear or wired glass

ex. 200 x 50 bottom rail

channel shoe recessed flush with door face

skirting

double action floor spring loose box sunk into floor

**ELEVATION**

glazing bead

meeting stile radius

meeting stiles

central pivot

door stile

door jamb

glass

3 mm max.

**SECTION**

screw to raise and lower pivot pin

upper part of top centre recessed into head of door frame

pivot pin

pivot

springs

lower part of top centre recessed into door head

**DOUBLE ACTION FLOOR SPRING**

**TOP CENTRE**

# Plasterboard Ceilings

Plasterboard~ this is a rigid board made with a core of gypsum sandwiched between face sheets of strong durable paper. In the context of ceilings two sizes can be considered –

1. Baseboard – 2·400 x 1·200 x 9·5 mm thick for supports at centres not exceeding 400 mm ; 2·400 x 1·200 x 12·7 mm for supports at centres not exceeding 600 mm.

   Baseboard has square edges and therefore the joints will need reinforcing with jute scrim at least 90 mm wide or alternatively a special tape to prevent cracking.

2. Gypsum Lath – 1·200 x 406 x 9·5 or 12·7 mm thick. Lath has rounded edges which eliminates the need to reinforce the joints.

Both types of board are available with an aluminium foil face which acts as a vapour barrier and /or a reflective surface to give an insulating plasterboard, for both uses the joints must be sealed with a self adhesive aluminium strip.

The boards are fixed to the underside of the floor or ceiling joists with galvanised or sheradised plasterboard nails at not more than 150 mm centres and are laid breaking the joint. Edge treatments consist of jute scrim reinforcement or a preformed plaster cove moulding.

Typical Details ~

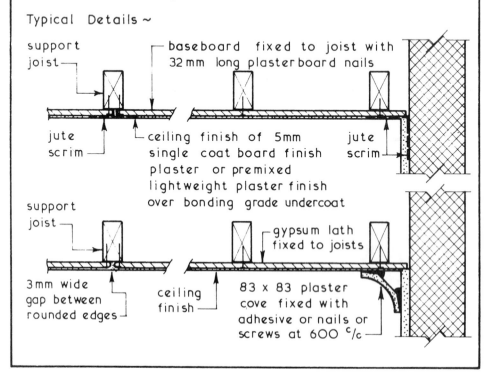

support joist

baseboard fixed to joist with 32 mm long plasterboard nails

jute scrim

ceiling finish of 5mm single coat board finish plaster or premixed lightweight plaster finish over bonding grade undercoat

jute scrim

support joist

gypsum lath fixed to joists

3 mm wide gap between rounded edges

ceiling finish

83 x 83 plaster cove fixed with adhesive or nails or screws at 600 $^c/_c$

Suspended Ceilings ~ these can be defined as ceilings which are fixed to a framework suspended from the main structure thus forming a void between the two components. The basic functional requirements of suspended ceilings are :-

1. They should be easy to construct, repair, maintain and clean.

2. So designed that an adequate means of access is provided to the void space for the maintenance of the suspension system, concealed services and/or light fittings.

3. Provide any required sound and/or thermal insulation.

4. Provide any required acoustic control in terms of absorption and reverberation.

5. Provide if required structural fire protection to structural steel beams supporting a concrete floor.

6. Conform with the minimum requirements set out in the Building Regulations and in particular the regulations governing the restriction of spread of flame over surfaces of ceilings and the exceptions permitting the use of certain plastic materials.

7. Design to be based on a planning module preferably a dimensional coordinated system with a first preference module of 300 mm.

Typical Suspended Ceiling Grid Framework Layout ~

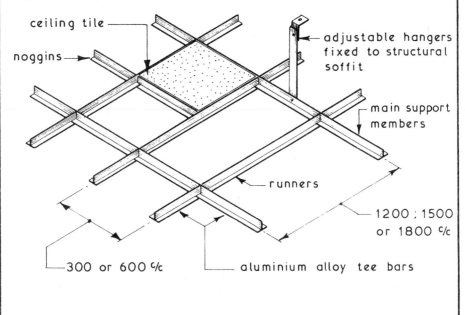

ceiling tile

noggins

adjustable hangers fixed to structural soffit

main support members

runners

1200 ; 1500 or 1800 c/c

300 or 600 c/c

aluminium alloy tee bars

# Suspended Ceilings

Classification of Suspended Ceiling ~ there is no standard method of classification since some are classified by their function such as illuminated and acoustic suspended ceilings others are classified by the materials used and classification by method of construction is also very popular. The latter method is simple since most suspended ceiling types can be placed in one of three groups :-

1. Jointless suspended ceilings.

2. Panelled suspended ceilings - see page 459

3. Decorative and open suspended ceilings - see page 460

Jointless Suspended Ceilings ~ these forms of suspended ceilings provide a continuous and jointless surface with the internal appearance of a conventional ceiling. They may be selected to fulfil fire resistance requirements or to provide a robust form of suspended ceiling. The two common ways of construction are a plasterboard or expanded metal lathing soffit with hand applied plaster finish or a sprayed applied rendering with a cement base.

Typical Details ~

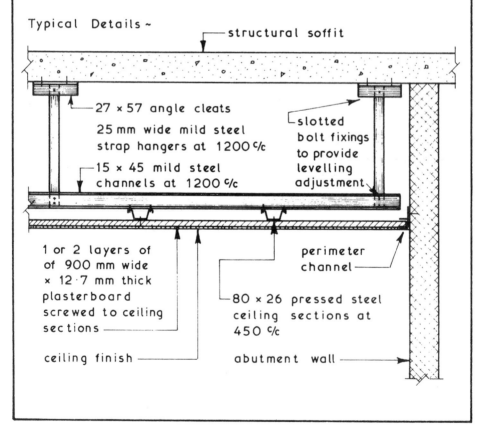

structural soffit

27 × 57 angle cleats

25 mm wide mild steel strap hangers at 1200 c/c

15 × 45 mild steel channels at 1200 c/c

slotted bolt fixings to provide levelling adjustment

1 or 2 layers of of 900 mm wide × 12·7 mm thick plasterboard screwed to ceiling sections

perimeter channel

80 × 26 pressed steel ceiling sections at 450 c/c

ceiling finish

abutment wall

Panelled Suspended Ceilings ~ these are the most popular form of suspended ceiling consisting of a suspended grid framework to which the ceiling covering is attached. The covering can be of a tile, tray, board or strip format in a wide variety of materials with an exposed or concealed supporting framework. Services such as luminares can usually be incorporated within the system. Generally panelled systems are easy to assemble and install using a water level or laser beam for initial and final levelling. Provision for maintenance access can be easily incorporated into most systems and layouts.

Typical Support Details ~

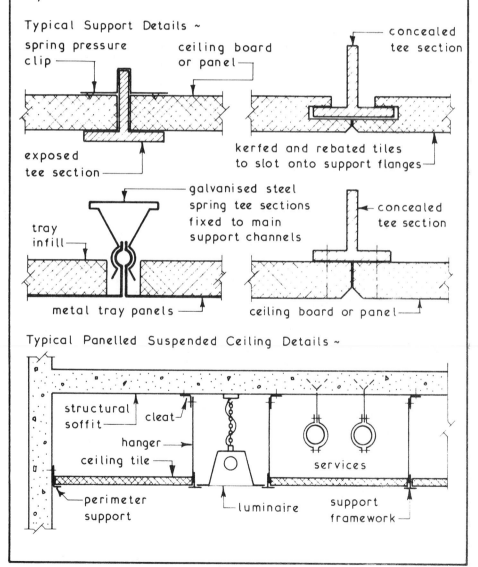

Typical Panelled Suspended Ceiling Details ~

# Suspended Ceilings

Decorative and Open Suspended Ceilings ~ these ceilings usually consist of an openwork grid or suspended shapes onto which the lights fixed at, above or below ceiling level can be trained thus creating a decorative and illuminated effect. Many of these ceilings are purpose designed and built as opposed to the proprietory systems associated with jointless and panelled suspended ceilings.

Typical Examples ~

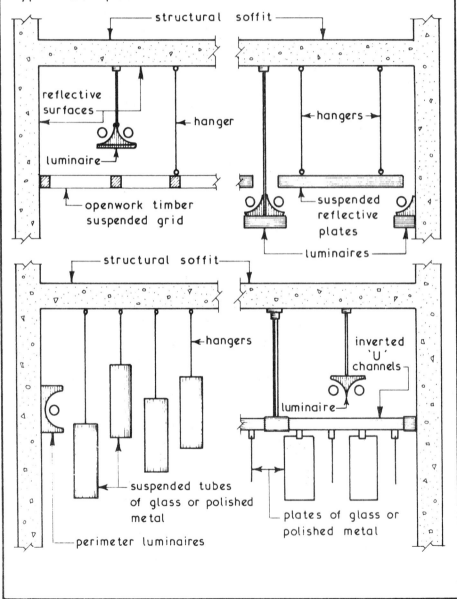

Functions ~ the main functions of paint are to provide -

1. An economic method of surface protection to building materials and components.

2. An economic method of surface decoration to building materials and components.

Composition ~ the actual composition of any paint can be complex but the basic components are -

1. Binder - this is the liquid vehicle or medium which dries to form the surface film and can be composed of linseed oil, drying oils, synthetic resins and water. The first function of a paint medium is to provide a means of spreading the paint over the surface and at the same time acting as a binder to the pigment.

2. Pigment - this provides the body, colour, durability and and corrosion protection properties of the paint. White lead pigments are very durable and moisture resistant but are poisonous and their use is generally restricted to priming and undercoating paints. If a paint contains a lead pigment the fact must be stated on the container. The general pigment used in paint is titanium dioxide which is not poisonous and gives good obliteration of the undercoats.

3. Solvents and Thinners - these are materials which can be added to a paint to alter its viscosity.

Paint Types - there is a wide range available but for most general uses the following can be considered -

1. Oil Based Paints - these are available in priming, undercoat and finishing grades. The latter can be obtained in a wide range of colours and finishes such as matt, semi-matt, eggshell, satin, gloss and enamel. Polyurethane paints have a good hardness and resistance to water and cleaning. Oil based paints are suitable for most applications if used in conjunction with correct primer and undercoat.

2. Water Based Paints - most of these are called emulsion paints the various finishes available being obtained by adding to the water medium additives such as alkyd resin & polyvinyl acetate (PVA). Finishes include matt, eggshell, semi-gloss and gloss. Emulsion paints are easily applied, quick drying and can be obtained with a washable finish and are suitable for most applications.

Supply ~ paint is usually supplied in metal containers ranging from 250 millilitres to 5 litres capacity to the colour ranges recommended in BS 381C (colours for specific purposes) and, BS 4800 (paint colours for building purposes).

Application ~ paint can be applied to almost any surface providing the surface preparation and sequence of paint coats are suitable. The manufacturers specification and /or the the recommendations of BS 6150 (painting of buildings) should be followed. Preparation of the surface to receive the paint is of the utmost importance since poor preparation is one of the chief causes of paint failure. The preparation consists basically of removing all dirt, grease, dust and ensuring that the surface will provide an adequate key for the paint which is to be applied. In new work the basic build-up of paint coats consists of –

1. Priming Coats – these are used on unpainted surfaces to obtain the necessary adhesion and to inhibit corrosion of ferrous metals. New timber should have the knots treated with a solution of shellac or other alcohol based resin called knotting prior to the application of the primer.

2. Undercoats – these are used on top of the primer after any defects have been made good with a suitable stopper or filler. The primary function of an undercoat is to give the opacity and build-up necessary for the application of the finishing coat(s).

3. Finish – applied directly over the undercoating in one or more coats to impart the required colour and finish.

Paint can applied by :-

1. Brush – the correct type, size and quality of brush such as those recommended in BS 2992 needs to be selected and used. To achieve a first class finish by means of brush application requires a high degree of skill.

2. Spray – as with brush application a high degree of skill is required to achieve a good finish. Generally compressed air sprays or airless sprays are used for building works.

3. Roller – simple and inexpensive method of quickly and cleanly applying a wide range of paints to flat and textured surfaces. Roller heads vary in size from 50 to 450 mm wide with various covers such as sheepskin, synthetic pile fibres, mohair and foamed polystyrene. All paint applicators must be thoroughly cleaned after use.

Painting ~ the main objectives of applying coats of paint to a surface are preservation, protection and decoration to give a finish which is easy to clean and maintain. To achieve these objectives the surface preparation and paint application must be adequate. The preparation of new and previously painted surfaces should ensure that prior to painting the surface is smooth, clean, dry and stable.

Basic Surface Preparation Techniques ~

Timber — to ensure a good adhesion of the paint film all timber should have a moisture content of less than 18 %. The timber surface should be prepared using an abrasive paper to produce a smooth surface brushed and wiped free of dust and any grease removed with a suitable spirit. Careful treatment of knots is essential either by sealing with two coats of knotting or in extreme cases cutting out the knot and replacing with sound timber. The stopping and filling of cracks and fixing holes with putty or an appropriate filler should be carried out after the application of the priming coat. Each coat of paint must be allowed to dry hard and be rubbed down with a fine abrasive paper before applying the next coat. On previously painted surfaces if the paint is in a reasonable condition the surface will only require cleaning and rubbing down before repainting, when the paint is in a poor condition it will be necessary to remove completely the layers of paint and then prepare the surface as described above for new timber.

Building Boards — most of these boards require no special preparation except for the application of a sealer as specified by the manufacturer.

Iron and Steel — good preparation is the key to painting iron and steel successfully and this will include removing all rust, mill scale, oil, grease and wax. This can be achieved by wire brushing, using mechanical means such as shot blasting, flame cleaning and chemical processes and many of these processes are often carried out in the steel fabrication works prior to shop applied priming.

Plaster – the essential requirement of the preparation is to ensure that the plaster surface is perfectly dry, smooth and free of defects before applying any coats of paint especially when using gloss paints. Plaster which contains lime can be alkaline and such surfaces should be treated with an alkali resistant primer when the surface is dry before applying the final coats of paint.

Paint Defects ~ these may be due to poor or incorrect preparation of the surface, poor application of the paint and/or chemical reactions. The general remedy is to remove all the affected paint and carry out the correct preparation of the surface before applying in the correct manner new coats of paint. Most paint defects are visual and therefore an accurate diagnosis of the cause must be established before any remedial treatment is undertaken.

Typical Paint Defects ~

1. Bleeding – staining and disruption of the paint surface by chemical action, usually caused by applying an incorrect paint over another. Remedy is to remove affected paint surface and repaint with correct type of overcoat paint.

2. Blistering – usually caused by poor presentation allowing resin or moisture to be entrapped, the subsequent expansion causing the defect. Remedy is to remove all the coats of paint and ensure that the surface is dry before repainting.

3. Blooming – mistiness usually on high gloss or varnished surfaces due to the presence of moisture during application and can be avoided by not painting under these conditions. Remedy is to remove affected paint and repaint.

4. Chalking – powdering of the paint surface due to natural aging or the use of poor quality paint. Remedy is to remove paint if necessary; prepare surface and repaint.

5. Cracking and Crazing – usually due to unequal elasticity of successive coats of paint. Remedy is to remove affected paint and repaint with compatible coats of paint.

6. Flaking and Peeling – can be due to poor adhesion, presence of moisture, painting over unclean areas or poor preparation. Remedy is to remove defective paint, prepare surface and repaint.

7. Grinning - due to poor opacity of paint film allowing paint coat below or background to show through, could be the result of poor application incorrect thinning or the use of the wrong colour. Remedy is to apply further coats of paint to obtain a satisfactory surface.

8. Saponification - formation of soap from alkali present in or on surface painted. The paint is ultimately destroyed and a brown liquid appears on the surface. Remedy is to remove the paint films and seal the alkaline surface before repainting.

Joinery Production ~ this can vary from the flow production where one product such as flush doors is being made usually with the aid of purpose designed and built machines to batch production where a limited number of similar items are being made with the aid of conventional woodworking machines. Purpose made joinery is very often largely hand made with a limited use of machines and is considered when special and/or high class joinery components are required.

Woodworking Machines ~ except for the portable electric tools such as drills, routers, jigsaws and sanders most woodworking machines need to be fixed to a solid base and connected to an extractor system to extract and collect the sawdust and chippings produced by the machines.

Saws - basically three formats are available namely the circular cross cut and band saws. Circular are general purpose saws and usually have tungsten carbide tipped teeth with feed rates of up to 60·000 per minute. Cross cut saws usually have a long bench to support the timber, the saw being mounted on a radial arm enabling the circular saw to be drawn across the timber to be cut. Band saws consist of an endless thin band or blade with saw teeth and a table on which to support the timber and are generally used for curved work.

Planers - most of these machines are combined planers and thicknessers, the timber being passed over the table surface for planing and the table or bed for thicknessing. The planer has a guide fence which can be tilted for angle planing and usually the rear bed can be lowered for rebating operations. The same rotating cutter block is used for all operations. Planing speeds are dependent upon the operator since it is a hand fed operation whereas thicknessing is mechanically fed with a feed speed range of 6·000 to 20·000 per minute. Maximum planing depth is usually 10 mm per passing.

Morticing Machines - these are used to cut mortices up to 25mm wide and can be either a chisel or chain morticer. The former consists of a hollow chisel containing a bit or auger whereas the latter has an endless chain cutter.

Tenoning Machines - these machines with their rotary cutter blocks can be set to form tenon and scribe. In most cases they can also be set for trenching, grooving and cross cutting.

Spindle Moulder - this machine has a horizontally rotating cutter block into which standard or purpose made cutters are fixed to reproduce a moulding on timber passed across the cutter.

Purpose Made Joinery ~ joinery items in the form of doors, windows, stairs and cupboard fitments can be purchased as stock items from manufacturers but there is also a need for purpose made joinery to fulfil client / designer / user requirement, to suit a specific need, to fit into a non-standard space, as a specific decor requirement or to create a particular internal environment. These purpose made joinery items can range from the simple to the complex which require high degrees of shop and site skills.

Typical Purpose Made Counter Details ~

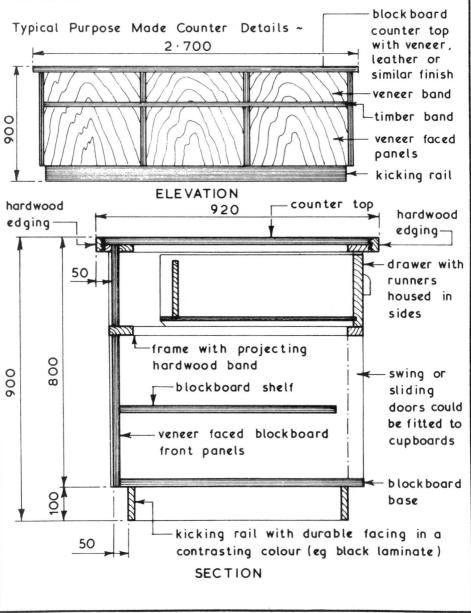

block board
counter top
with veneer,
leather or
similar finish

veneer band

timber band

veneer faced
panels

kicking rail

2·700

900

ELEVATION

920

counter top

hardwood
edging

hardwood
edging

50

drawer with
runners
housed in
sides

frame with projecting
hardwood band

blockboard shelf

swing or
sliding
doors could
be fitted to
cupboards

veneer faced blockboard
front panels

blockboard
base

kicking rail with durable facing in a
contrasting colour (eg black laminate)

SECTION

900

800

100

50

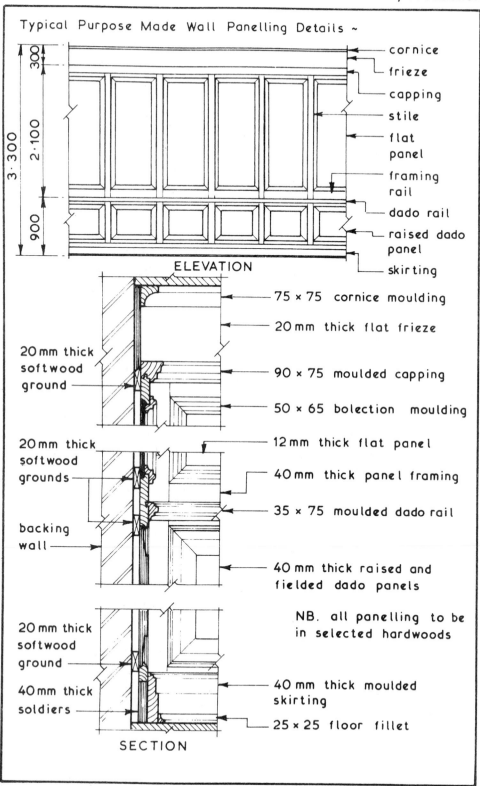

Typical Purpose Made Wall Panelling Details ~

ELEVATION

- cornice
- frieze
- capping
- stile
- flat panel
- framing rail
- dado rail
- raised dado panel
- skirting

3·300
300
2·100
900

20 mm thick softwood ground

20 mm thick softwood grounds

backing wall

20 mm thick softwood ground

40 mm thick soldiers

SECTION

- 75 × 75 cornice moulding
- 20 mm thick flat frieze
- 90 × 75 moulded capping
- 50 × 65 bolection moulding
- 12 mm thick flat panel
- 40 mm thick panel framing
- 35 × 75 moulded dado rail
- 40 mm thick raised and fielded dado panels

NB. all panelling to be in selected hardwoods

- 40 mm thick moulded skirting
- 25 × 25 floor fillet

467

Joinery Timbers ~ both hardwoods and softwoods can be used for joinery works. Softwoods can be selected for their stability, durability and/or workability if the finish is to be paint but if it is left in its natural colour with a sealing coat the grain texture and appearance should be taken into consideration. Hardwoods are usually left in their natural colour and treated with a protective clear sealer or polish therefore texture, colour and grain pattern are important when selecting hardwoods for high class joinery work.

Typical Softwoods Suitable for Joinery Work ~

1. Douglas Fir - sometimes referred to as Columbian Pine or Oregon Pine. It is available in long lengths and has a straight grain. Colour is reddish brown to pink. Suitable for general and high class joinery. Approximate density 530 kg/m³.

2. Redwood - also known as Scots Pine, Red Pine, Red Deal and Yellow Deal. It is a widely used softwood for general joinery work having good durability a straight grain and is reddish brown to straw in colour. Approximate density 430 kg/m³.

3. European Spruce - similar to redwood but with a lower durability. It is pale yellow to pinkish white in colour and is used mainly for basic framing work and simple internal joinery. Approximate density 460 kg/m³.

4. Pitch Pine - durable softwood suitable for general joinery work. It is light red to reddish yellow in colour and tends to have large knots which in some cases can be used as a decorative effect. Approximate density 650 kg/m³.

5. Parana Pine - moderately durable straight grained timber available in a good range of sizes. Suitable for general joinery work especially timber stairs. Light to dark brown in colour with, the occasional pink stripe. Approximate density 560 kg/m³.

6. **Western Hemlock** - durable softwood suitable for interior joinery work such as panelling   Light yellow to reddish brown in colour. Approximate density 500 kg/m³.

For typical hardwoods see page 469

**Fig 98   Joinery production—4**

Typical Hardwoods Suitable for Joinery Works ~

1. Beech – hard close grained timber with some silver grain in the predominately reddish yellow to light brown colour. Suitable for all internal joinery. Approximate density 700 kg/m³.

2. Iroko – hard durable hardwood with a figured grain and is usually golden brown in colour. Suitable for all forms good class joinery. Approximate density 660 kg/m³.

3. Mahogany ( African ) – interlocking grained hardwood with good durability. It has an attractive light brown to deep red colour and is suitable for panelling and all high class joinery work. Approximate density 560 kg/m³.

4. Mahogany ( Honduras ) – durable hardwood usually straight grained but can have a mottled or swirl pattern. It is light red to pale reddish brown in colour and is suitable for all good class joinery work. Approximate density 530 kg/m³.

5. Mahogany ( South American) – a well figured, stable and durable hardwood with a deep red or brown colour which is suitable for all high class joinery particularly where a high polish is required. Approximate density 550 kg/m³.

6. Oak ( English) – very durable hardwood with a wide variety of grain patterns. It is usually a light yellow brown to a warm brown in colour and is suitable for all forms of joinery but should not be used in conjunction with ferrous metals due to the risk of staining caused by an interaction of the two materials. ( The gallic acid in oak causes corrosion in ferrous metals.) Approximate density 720 kg/m³.

7. Sapele – close texture timber of good durability, dark reddish brown in colour with a varied grain pattern. It is suitable for most internal joinery work especially where a polished finish is required. Approximate density 640 kg/m³.

8. Teak – very strong and durable timber but hard to work. It is light golden brown to dark golden yellow in colour which darkens with age and is suitable for high class joinery work and laboratory fittings. Approximate density 650 kg/m³.

# Plastics in Building

Plastics ~ the term plastic can be applied to any group of substances based on synthetic or modified natural polymers which during manufacture are moulded by heat and/or pressure into the required form. Plastics can be classified by their overall grouping such as polyvinyl chloride (PVC) or they can be classified as thermoplastic or thermosetting. The former soften on heating whereas the latter are formed into permanent non-softening materials. The range of plastics available give the designer and builder a group of materials which are strong reasonably durable, easy to fit and maintain and since most are mass produced of relative low cost.

Typical Applications of Plastics in Buildings ~

| Application | Plastics Used |
|---|---|
| Rainwater goods | rigid PVC. |
| Soil, waste, water and gas pipes and fittings | rigid PVC; polyethylene; acrylonitrile; butadiene styrene (ABS). |
| Hot and cold water pipes | chlorinated PVC; ABS; polypropylene; polyethylene; PVC. |
| Bathroom and kitchen fittings | glass fibre reinforced polyester (GRP); acrylic resins. |
| Cold water cisterns | rigid PVC; polystyrene; polyethylene. |
| Rooflights and sheets | rigid PVC; GRP; acrylic resins. |
| DPC's and membranes, vapour barriers | low density polythene (polyethylene) film; PVC film. |
| Doors and windows | rigid PVC; GRP. |
| Electrical conduit and fittings | plasticized PVC; rigid PVC; phenolic resins |
| Thermal insulation | generally cellular plastics such as expanded polystyrene bead and boards; expanded PVC; foamed polyurethane; foamed phenol formaldehyde; foamed urea formaldehyde. |
| Floor finishes | rigid and plasticized PVC tiles and sheets; resin based floor paints. |
| Wall claddings and internal linings | unplasticized PVC; polyvinyl flouride film laminate; melamine resins; rigid PVC; expanded polystyrene tiles & sheets. |

# 7 DOMESTIC SERVICES

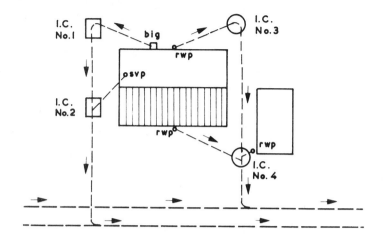

SUBSOIL DRAINAGE

SURFACE WATER REMOVAL

ROAD DRAINAGE

RAINWATER INSTALLATIONS

DRAINAGE SYSTEMS

WATER SUPPLY

COLD WATER INSTALLATIONS

HOT WATER INSTALLATIONS

CISTERNS AND CYLINDERS

SANITARY FITTINGS

SINGLE AND VENTILATED STACK SYSTEMS

DOMESTIC HOT WATER HEATING SYSTEMS

ELECTRICAL SUPPLY AND INSTALLATION

GAS SUPPLY AND GAS FIRES

OPEN FIREPLACES AND FLUES

Effluent ~ can be defined as that which flows out – in building drainage terms there are three main forms of effluent :-

1. Subsoil Water ~ water collected by means of special drains from the earth primarily to lower the water table level in the subsoil. It is considered to be clean and therefore requires no treatment and can be discharged direct into an approved water course.

2. Surface Water ~ effluent collected from surfaces such as roofs and paved areas and like subsoil water is considered to be clean and can be discharged direct into an approved water course or soakaway.

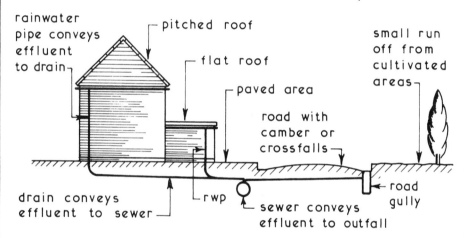

rainwater pipe conveys effluent to drain

pitched roof

flat roof

paved area

small run off from cultivated areas

road with camber or crossfalls

road gully

drain conveys effluent to sewer

rwp

sewer conveys effluent to outfall

3. Foul or Soil Water ~ effluent contaminated by domestic or trade waste and will require treatment to render it clean before it can be discharged into an approved water course.

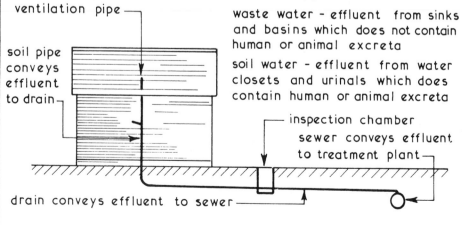

ventilation pipe

soil pipe conveys effluent to drain

waste water – effluent from sinks and basins which does not contain human or animal excreta

soil water – effluent from water closets and urinals which does contain human or animal excreta

inspection chamber sewer conveys effluent to treatment plant

drain conveys effluent to sewer

473

Subsoil Drainage ~ Building Regulation C3 requires that subsoil drainage shall be provided if it is needed to avoid :-
a) the passage of ground moisture into the interior of the building or
b) damage to the fabric of the building.

Subsoil drainage can also be used to improve the stability of the ground, lower the humidity of the site and enhance its horticultural properties. Subsoil drains consist of porous or perforated pipes laid dry jointed in a rubble filled trench. Porous pipes allow the subsoil water to pass through the body of the pipe whereas perforated pipes which have a series of holes in the lower half allow the subsoil water to rise into the pipe. This form of ground water control is only economic up to a depth of 1·500, if the water table needs to be lowered to a greater depth other methods of ground water control should be considered (see pages 233 to 237).

The water collected by a subsoil drainage system has to be conveyed to a suitable outfall such as a river, lake or surface water drain or sewer. In all cases permission to discharge the subsoil water will be required from the authority or owner and in the case of streams, rivers and lakes, bank protection at the outfall may be required to prevent erosion. (see page 475)

Typical Subsoil Drain Details ~

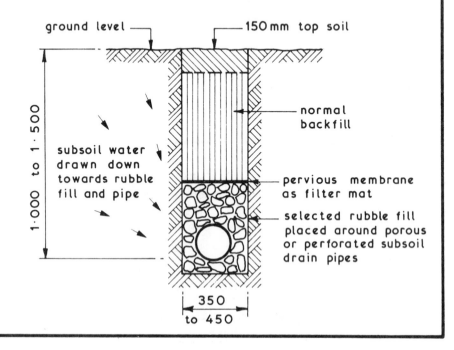

ground level ——                  —— 150 mm top soil

1·000 to 1·500

subsoil water drawn down towards rubble fill and pipe

normal backfill

pervious membrane as filter mat

selected rubble fill placed around porous or perforated subsoil drain pipes

350 to 450

Subsoil Drainage Systems ~ the layout of subsoil drains will depend on whether it is necessary to drain the whole site or if it is only the substructure of the building which needs to be protected. The latter is carried out by installing a cut off drain around the substructure to intercept the flow of water and divert it away from the site of the building. Junctions in a subsoil drainage system can be made using standard fittings or by placing the end of the branch drain onto the crown of the main drain.

Typical Examples ~

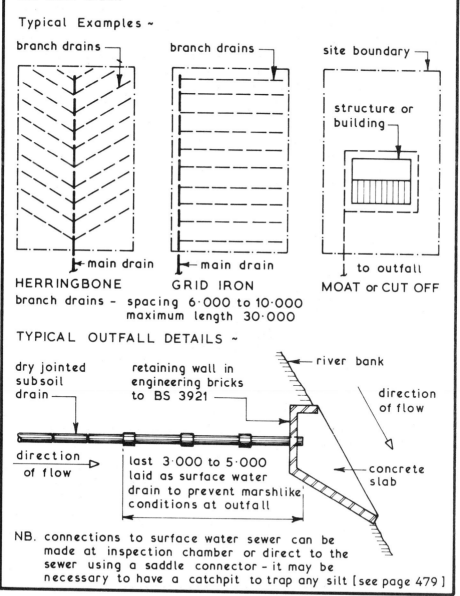

branch drains —

**HERRINGBONE**

branch drains —

**GRID IRON**

site boundary —

structure or building —

↓ to outfall

**MOAT or CUT OFF**

branch drains - spacing 6·000 to 10·000
maximum length 30·000

TYPICAL OUTFALL DETAILS ~

dry jointed subsoil drain —

retaining wall in engineering bricks to BS 3921 —

river bank

direction of flow

direction of flow

last 3·000 to 5·000 laid as surface water drain to prevent marshlike conditions at outfall

concrete slab

NB. connections to surface water sewer can be made at inspection chamber or direct to the sewer using a saddle connector - it may be necessary to have a catchpit to trap any silt [see page 479]

475

General Principles ~ a roof must be designed with a suitable fall towards the surface water collection channel or gutter which in turn is connected to vertical rainwater pipes which conveys the collected discharge to the drainage system. The fall of the roof will be determined by the chosen roof covering or the chosen pitch will limit the range of coverings which can be selected.

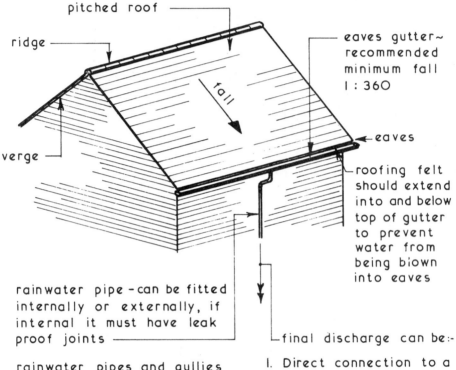

pitched roof

ridge

eaves gutter ~ recommended minimum fall 1 : 360

fall

eaves

verge

roofing felt should extend into and below top of gutter to prevent water from being blown into eaves

rainwater pipe - can be fitted internally or externally, if internal it must have leak proof joints

final discharge can be :-

rainwater pipes and gullies must be arranged so as not to cause dampness or damage to any part of the building

1. Direct connection to a drain discharging into a soakaway

2 Direct connection to a drain discharging into a surface water sewer

Minimum Roof Pitches ~

Slates – depends on width from 25°
Hand made plain tiles – 45°
Machine made plain tiles – 35°
Single lap and interlocking tiles - depends on type from 12½°
Thatch – 52°
Timber shingles – 14°

3. Indirect connection to a drain by means of a trapped gully if drain discharges into a combined sewer

**See page 482 for details**

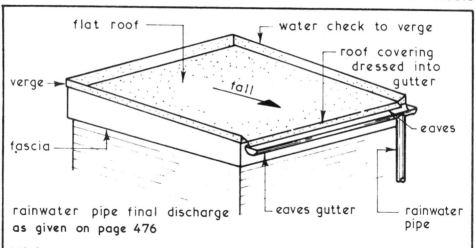

verge →

fall

flat roof

water check to verge

roof covering
dressed into
gutter

eaves

fascia

rainwater pipe final discharge
as given on page 476

eaves gutter

rainwater
pipe

Minimum Recommended Falls for Various Finishes ~
Aluminium — 1 : 60      Lead — 1 : 120      Copper — 1 : 60
Built-up roofing felts — 1 : 60      Mastic asphalt — 1 : 80

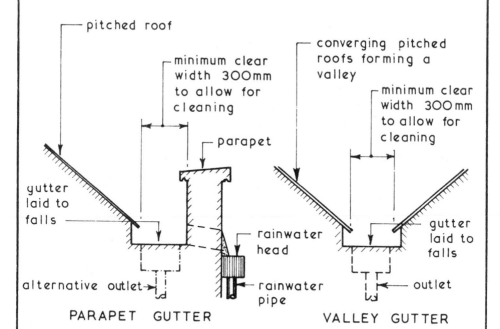

pitched roof

minimum clear
width 300mm
to allow for
cleaning

parapet

converging pitched
roofs forming a
valley

minimum clear
width 300mm
to allow for
cleaning

gutter
laid to
falls

rainwater
head

gutter
laid to
falls

alternative outlet →

rainwater
pipe

outlet

## PARAPET GUTTER

gutter formed to discharge
into internal rainwater pipes
or to external rainwater
pipes via outlets through
the parapet

## VALLEY GUTTER

gutter formed to discharge
into internal rainwater pipes
or to external rainwater
pipes sited at the gable
ends

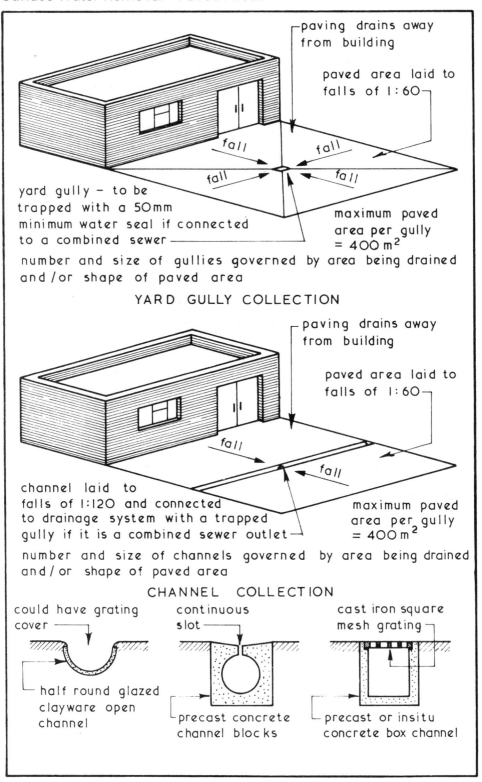

**paving drains away from building**

**paved area laid to falls of 1:60**

fall     fall

fall     fall

yard gully – to be trapped with a 50mm minimum water seal if connected to a combined sewer

maximum paved area per gully = 400 m²

number and size of gullies governed by area being drained and/or shape of paved area

YARD GULLY COLLECTION

**paving drains away from building**

**paved area laid to falls of 1:60**

fall

fall

channel laid to falls of 1:120 and connected to drainage system with a trapped gully if it is a combined sewer outlet

maximum paved area per gully = 400 m²

number and size of channels governed by area being drained and/or shape of paved area

CHANNEL COLLECTION

could have grating cover

half round glazed clayware open channel

continuous slot

precast concrete channel blocks

cast iron square mesh grating

precast or insitu concrete box channel

Highway Drainage ~ the stability of a highway or road relies on two factors –

1. Strength and durability of upper surface.
2. Strength and durability of subgrade which is the subsoil on which the highway construction is laid.

The above can be adversely affected by water therefore it may be necessary to install two drainage systems. One system (subsoil drainage) to reduce the flow of subsoil water through the subgrade under the highway construction and a system of surface water drainage.

Typical Highway Subsoil Drainage Methods ~

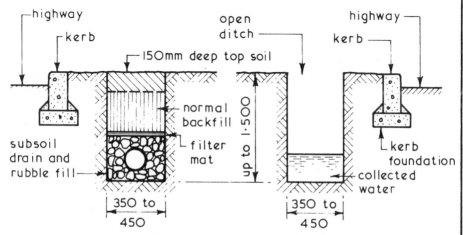

Subsoil Drain - acts as a cut off drain and can be formed using perforated or porous drain pipes. If filled with rubble only it is usually called a French or rubble drain.

Open Ditch - acts as a cut off drain and could also be used to collect surface water discharged from a rural road where there is no raised kerb or surface water drains.

Surface Water Drainage Systems ~

road gully — sw drain from building combined sewer to treatment outfall — I.C. — carrier drain — gully — catch pit — subsoil drain

road gully — sw drain from building separate sewer to river outfall — I.C. — carrier drain — gully — catch pit — subsoil drain

Road Drainage ~ this consists of laying the paved area or road to a suitable crossfall or gradient to direct the run-off of surface water towards the drainage channel or gutter. This is usually bounded by a kerb which helps to convey the water to the road gullies which are connected to a surface water sewer. For drains or sewers under 900 mm internal diameter inspection chambers will be required as set out in the Building Regulations. The actual spacing of road gullies is usually determined by the local highway authority based upon the carriageway gradient and the area to be drained into one road gully. Alternatively the following formula could be used :-

$$D = \frac{280 \sqrt{S}}{W}$$

where
D = gully spacing in metres
S = carriageway gradient (per cent)
W = width of carriageway in metres

∴ if S = 1 : 60 = 1·66% and D = 4·500

$$D = \frac{280 \sqrt{1·66}}{4·500} = \text{say } 80·000$$

Typical Road Gully Detail ~

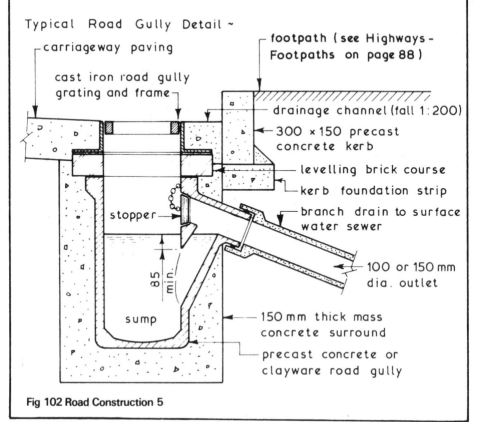

- carriageway paving
- cast iron road gully grating and frame
- footpath (see Highways - Footpaths on page 88)
- drainage channel (fall 1 : 200)
- 300 × 150 precast concrete kerb
- levelling brick course
- kerb foundation strip
- branch drain to surface water sewer
- stopper
- 85 min.
- 100 or 150 mm dia. outlet
- sump
- 150 mm thick mass concrete surround
- precast concrete or clayware road gully

**Fig 102 Road Construction 5**

Materials ~ the traditional material for domestic eaves gutters and rainwater pipes is cast iron but UPVC systems are very often specified today because of their simple installation and low maintenance costs. Other materials which could be considered are asbestos cement, aluminium alloy, galvanised steel and stainless steel but whatever material is chosen it must be of adequate size, strength and durability.

Typical Eaves Details ~

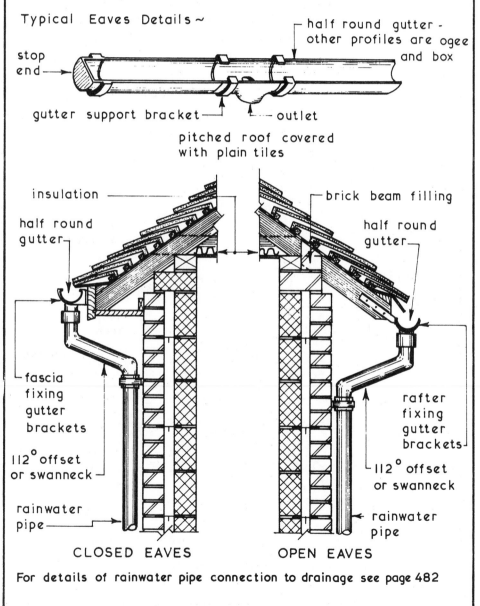

half round gutter - other profiles are ogee and box

stop end

gutter support bracket — outlet

pitched roof covered with plain tiles

insulation

brick beam filling

half round gutter

half round gutter

fascia fixing gutter brackets

rafter fixing gutter brackets

112° offset or swanneck

112° offset or swanneck

rainwater pipe

rainwater pipe

CLOSED EAVES

OPEN EAVES

For details of rainwater pipe connection to drainage see page 482

481

# Rainwater Installation Details

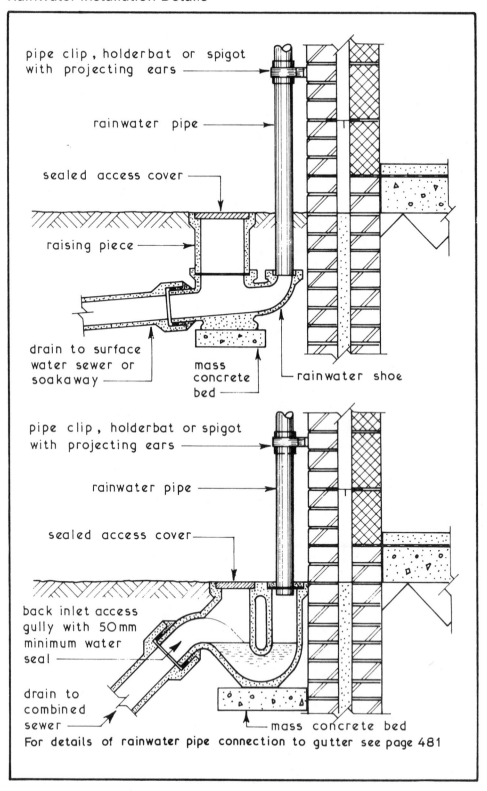

pipe clip, holderbat or spigot
with projecting ears

rainwater pipe

sealed access cover

raising piece

drain to surface
water sewer or
soakaway

mass
concrete
bed

rainwater shoe

pipe clip, holderbat or spigot
with projecting ears

rainwater pipe

sealed access cover

back inlet access
gully with 50mm
minimum water
seal

drain to
combined
sewer

mass concrete bed

For details of rainwater pipe connection to gutter see page 481

Drains ~ these can be defined as a means of conveying surface water or foul water below ground level.

Sewers ~ these have the same functions as drains but collect the discharge from a number of drains and convey it to the final outfall. They can be a private or public sewer depending on who is responsible for the maintenance.

Basic Principles ~ to provide a drainage system which is simple efficient and economic by laying the drains to a gradient which will render them self cleansing and will convey the effluent to a sewer without danger to health or giving nuisance. To provide a drainage system which will comply with the minimum requirements given in **Part H of the Building Regulations**

Typical Basic Requirements ~

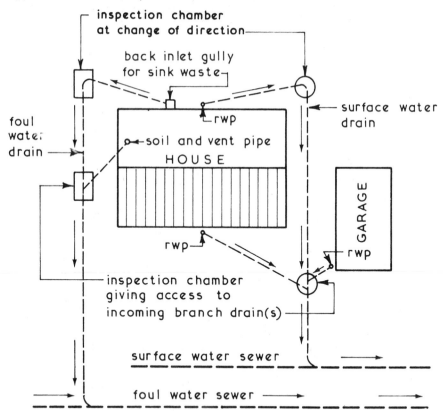

all junctions should be oblique and in direction of flow

there must be an access point at a junction unless each run can be cleared from another access point

Separate System ~ the most common drainage system in use where the surface water discharge is conveyed in separate drains and sewers to that of foul water discharges and therefore receives no treatment before the final outfall.

Typical Example ~

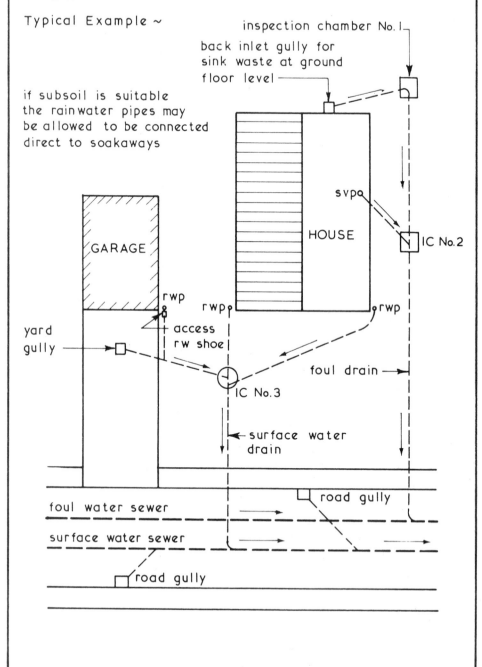

Combined System ~ this is the simplest and cheapest system to design and install but since all forms of discharge are conveyed in the same sewer the whole effluent must be treated unless a sea outfall is used to discharge the untreated effluent.

Typical Example ~

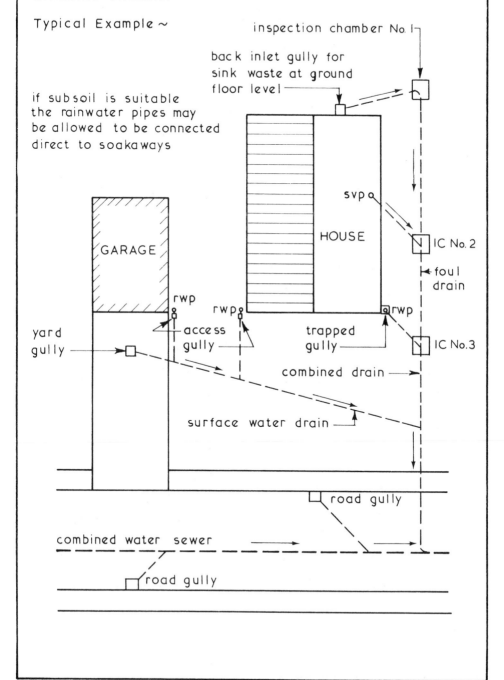

inspection chamber No. 1

back inlet gully for
sink waste at ground
floor level

if subsoil is suitable
the rainwater pipes may
be allowed to be connected
direct to soakaways

svp

GARAGE

HOUSE

IC No. 2

foul
drain

rwp

yard
gully

rwp

access
gully

trapped
gully

rwp

IC No. 3

combined drain

surface water drain

road gully

combined water sewer

road gully

485

Partially Separate System ~ a compromise system - there are two drains, one to convey only surface water and a combined drain to convey the total foul discharge and a proportion of the surface water.

Typical Example ~

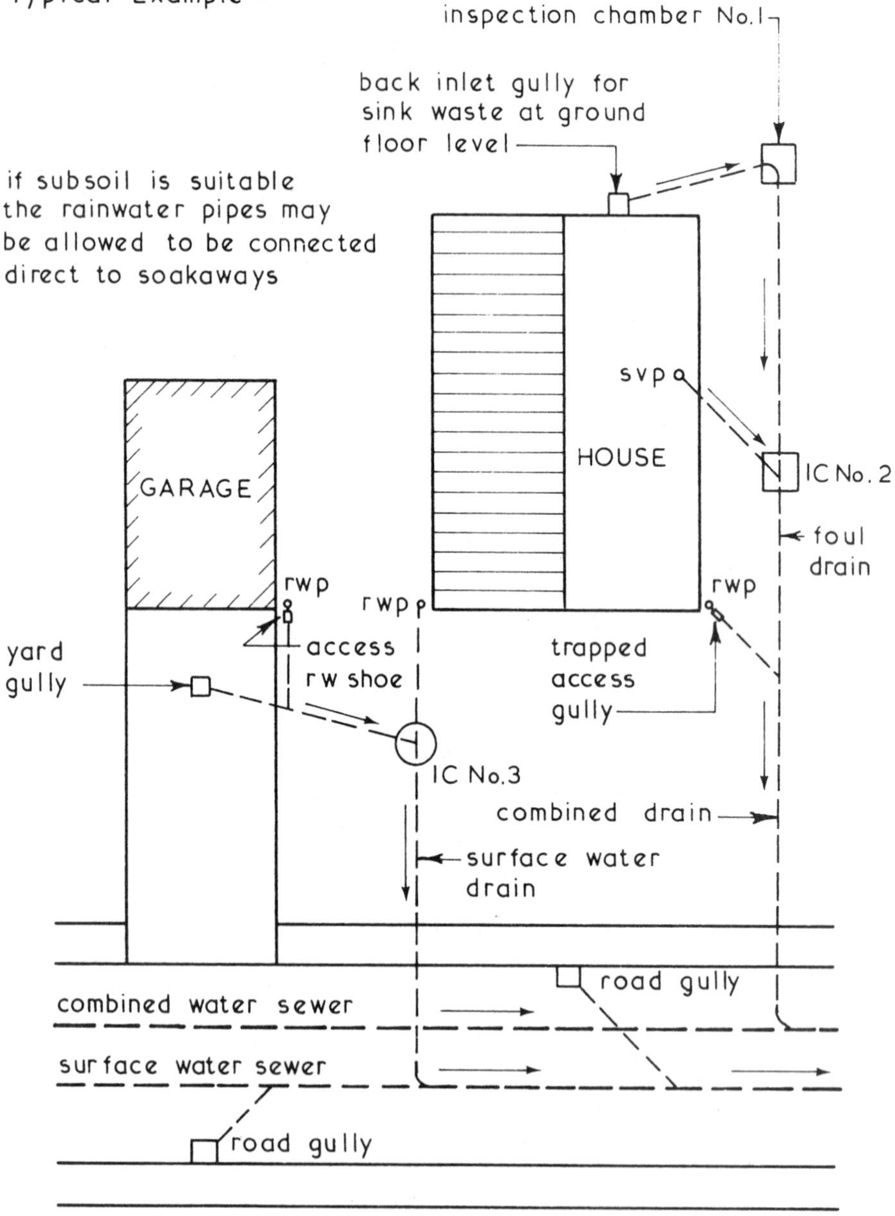

inspection chamber No.1

back inlet gully for sink waste at ground floor level

if subsoil is suitable the rainwater pipes may be allowed to be connected direct to soakaways

svp

HOUSE

IC No. 2

foul drain

GARAGE

rwp

rwp

yard gully

access rw shoe

rwp

trapped access gully

IC No.3

combined drain

surface water drain

road gully

combined water sewer

surface water sewer

road gully

Inspection Chambers ~ these are sometimes called manholes and provide a means of access to a drainage system for the purpose of inspection and maintenance. They should be **positioned in accordance with the requirements of Part H** of the Building Regulations. In domestic work inspection chambers can be of brick, precast concrete or preformed in plastic for use with patent drainage systems. The size of an inspection chamber depends on depth to invert level, size and number of branch drains to be accommodated **within the chamber - guidance to sizing is given in BS 8301**

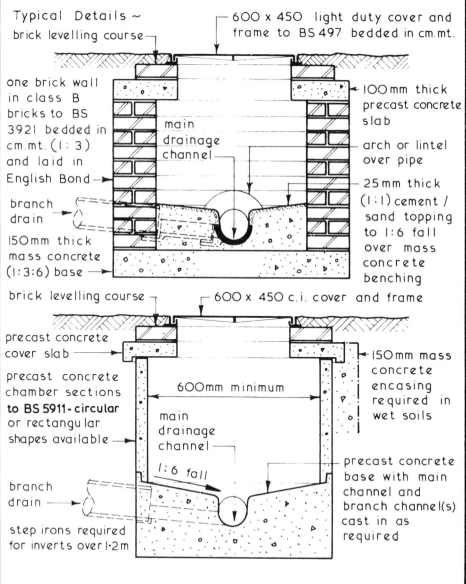

Typical Details ~
brick levelling course

— 600 x 450 light duty cover and frame to BS 497 bedded in cm.mt.

one brick wall in class B bricks to BS 3921 bedded in cm.mt.(1: 3) and laid in English Bond →

main drainage channel

100 mm thick precast concrete slab

arch or lintel over pipe

branch drain →

25 mm thick (1:1) cement / sand topping to 1:6 fall over mass concrete benching

150 mm thick mass concrete (1:3:6) base →

brick levelling course

— 600 x 450 c.i. cover and frame

precast concrete cover slab →

precast concrete chamber sections **to BS 5911- circular** or rectangular shapes available →

main drainage channel

600 mm minimum

150 mm mass concrete encasing required in wet soils

1:6 fall

branch drain →

step irons required for inverts over 1·2 m

precast concrete base with main channel and branch channel(s) cast in as required

## Simple Drainage—Jointing and Bedding Pipes

Excavations ~ drains are laid in trenches which are set out, excavated and supported in a similar manner to foundation trenches except for the base of the trench which is cut to the required gradient or fall.

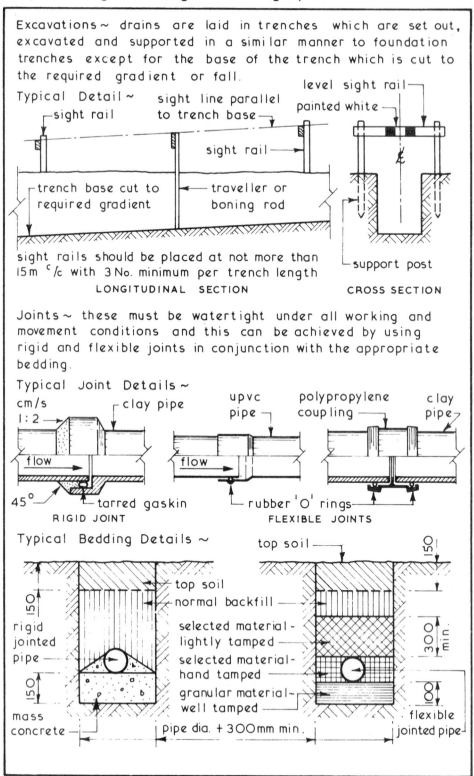

Typical Detail ~

sight rail

sight line parallel to trench base

level sight rail

painted white

sight rail

trench base cut to required gradient

traveller or boning rod

support post

sight rails should be placed at not more than 15m ᶜ/c with 3No. minimum per trench length

LONGITUDINAL SECTION

CROSS SECTION

Joints ~ these must be watertight under all working and movement conditions and this can be achieved by using rigid and flexible joints in conjunction with the appropriate bedding.

Typical Joint Details ~

cm/s 1:2

clay pipe

flow

45°  tarred gaskin

RIGID JOINT

upvc pipe

polypropylene coupling

clay pipe

flow

rubber 'O' rings

FLEXIBLE JOINTS

Typical Bedding Details ~

top soil

150

rigid jointed pipe

top soil

normal backfill

selected material - lightly tamped

selected material - hand tamped

granular material ~ well tamped

150

300 min.

100

mass concrete

pipe dia. + 300mm min.

flexible jointed pipe

488

Water Supply ~ an adequate supply of cold water of drinking quality should be provided to every residential building and a drinking water tap installed within the building. The installation should be designed to prevent waste, undue consumption, misuse, contamination of general supply, be protected against corrosion and frost damage and be accessible for maintenance activities. The intake of a cold water supply to a building is owned jointly by the water authority and the consumer who therefore have joint maintenance responsibilities.

Typical Water Supply Arrangement ~

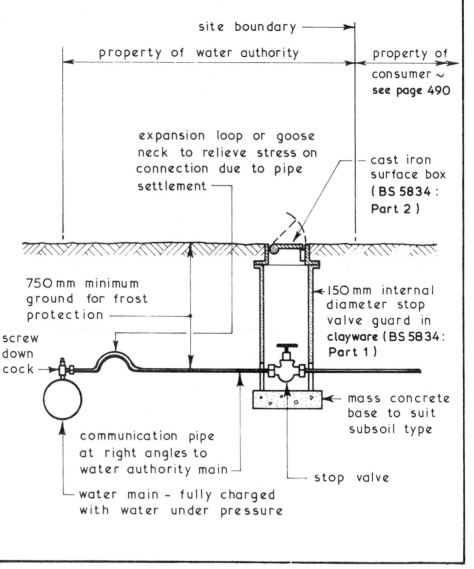

site boundary

property of water authority

property of consumer ~ **see page 490**

expansion loop or goose neck to relieve stress on connection due to pipe settlement

cast iron surface box (BS 5834 : Part 2)

750 mm minimum ground for frost protection

150 mm internal diameter stop valve guard in **clayware (BS 5834 : Part 1)**

screw down cock

mass concrete base to suit subsoil type

communication pipe at right angles to water authority main

stop valve

water main - fully charged with water under pressure

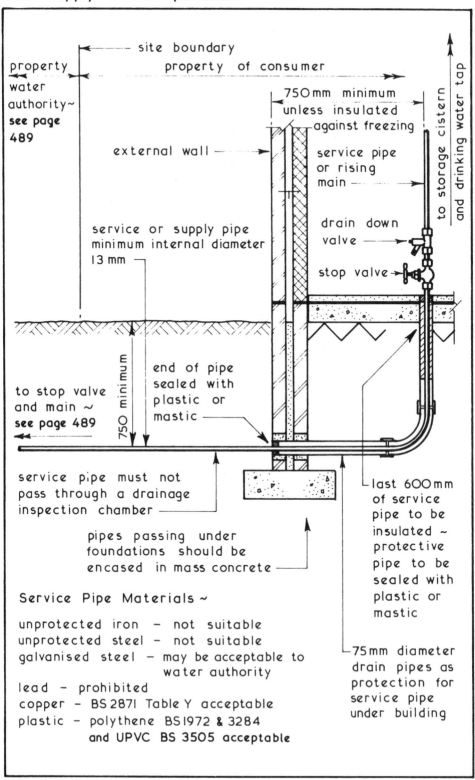

site boundary

property of consumer

property

water authority~ **see page 489**

750 mm minimum
unless insulated
against freezing

to storage cistern
and drinking water tap

external wall —

service pipe
or rising
main —

service or supply pipe
minimum internal diameter
13 mm —

drain down
valve —

stop valve —

750 minimum

end of pipe
sealed with
plastic or
mastic —

to stop valve
and main ~
**see page 489**

service pipe must not
pass through a drainage
inspection chamber —

pipes passing under
foundations should be
encased in mass concrete —

last 600 mm
of service
pipe to be
insulated ~
protective
pipe to be
sealed with
plastic or
mastic

75 mm diameter
drain pipes as
protection for
service pipe
under building

Service Pipe Materials ~

unprotected iron — not suitable
unprotected steel — not suitable
galvanised steel — may be acceptable to
water authority
lead — prohibited
copper — BS 2871 Table Y acceptable
plastic — polythene BS 1972 & 3284
and UPVC BS 3505 acceptable

General ~ when planning or designing any water installation the basic physical laws must be considered :-

1. Water is subject to the force of gravity and will find its own level.

2. To overcome friction within the conveying pipes water which is stored prior to distribution will require to be under pressure and this is normally achieved by storing the water at a level above the level of the outlets. The vertical distance between the levels is usually called the head.

3. Water becomes less dense as its temperature is raised therefore warm water will always displace colder water whether in a closed or open circuit.

Direct Cold Water Systems ~ the cold water is supplied to the outlets at mains pressure the only storage requirement is a small capacity cistern to feed the hot water storage tank. These systems are suitable for districts which have high level reservoirs with a good supply and pressure. Main advantage is that drinking water is available from all cold water outlets, disadvantages include lack of reserve in case of supply cut off, risk of back syphonage due to negative mains pressure and a risk of reduced pressure during peak demand periods.

Typical Direct Cold Water System ~

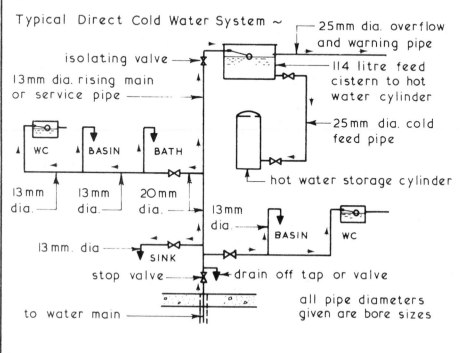

## Cold Water Installations

Indirect Systems ~ cold water is supplied to all outlets from a cold water storage cistern except for the cold water supply to the sink(s) where the drinking water tap is connected directly to the incoming supply from the main. This system requires more pipework than the indirect system but reduces the risk of back syphonage and provides a reserve of water should the mains supply fail or be cut off. The local water authority will stipulate the system to be used in their area.

Typical Indirect Cold Water System ~

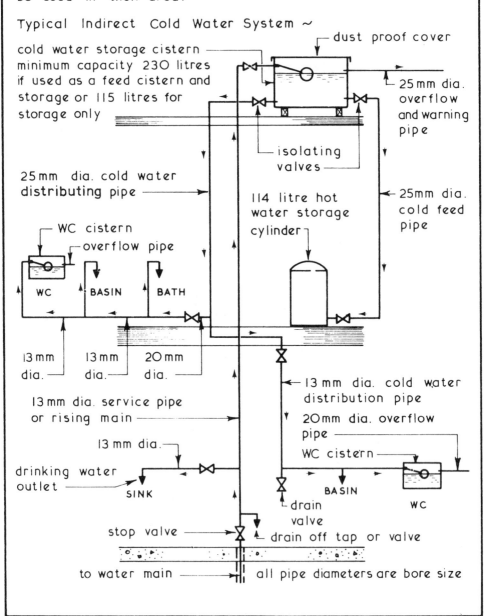

dust proof cover

cold water storage cistern minimum capacity 230 litres if used as a feed cistern and storage or 115 litres for storage only

25mm dia. overflow and warning pipe

isolating valves

25mm dia. cold water distributing pipe

114 litre hot water storage cylinder

25mm dia. cold feed pipe

WC cistern

overflow pipe

WC    BASIN    BATH

13 mm dia.    13 mm dia.    20 mm dia.

13 mm dia. service pipe or rising main

13 mm dia.

drinking water outlet

SINK

13 mm dia. cold water distribution pipe

20mm dia. overflow pipe

WC cistern

BASIN

WC

drain valve

stop valve

drain off tap or valve

to water main

all pipe diameters are bore size

Direct System~ this is the simplest and cheapest system of hot water installation. The water is heated in the boiler and the hot water rises by convection to the hot water storage tank or cylinder to be replaced by the cooler water from the bottom of the storage vessel. Hot water drawn from storage is replaced with cold water from the cold water storage cistern. Direct systems are suitable for soft water areas and for installations which are not supplying a central heating circuit.

Typical Direct Hot Water System ~

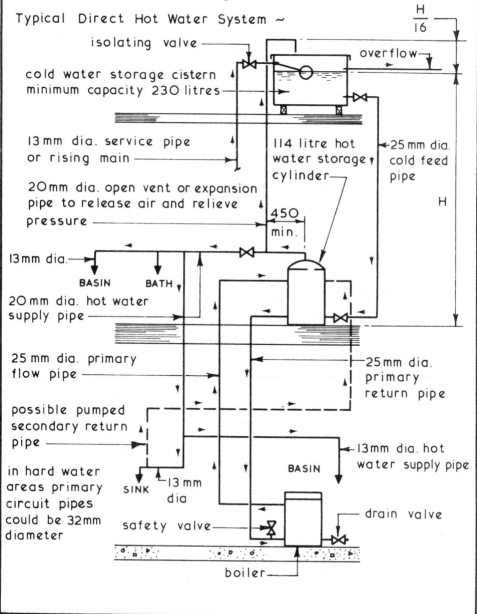

Indirect System ~ this is a more complex system than the direct system but does overcome the problem of furring which can occur in direct hot water systems. This method is therefore suitable for hard water areas and in all systems where a central heating circuit is to be part of the hot water installation. Basically the pipe layouts of the two systems are similar but in the indirect system a separate small capacity feed cistern is required to charge and top up the primary circuit. In this system the hot water storage tank or cylinder is in fact a heat exchanger - see page 497

Typical Indirect Hot Water System ~

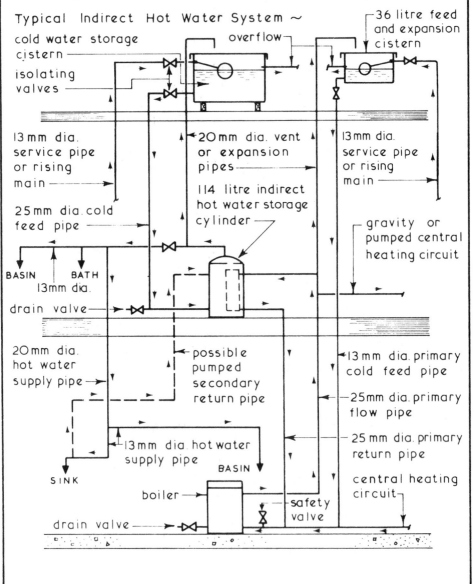

cold water storage cistern

overflow

36 litre feed and expansion cistern

isolating valves

13 mm dia. service pipe or rising main

20 mm dia. vent or expansion pipes

13 mm dia. service pipe or rising main

25 mm dia. cold feed pipe

114 litre indirect hot water storage cylinder

gravity or pumped central heating circuit

BASIN          BATH
13mm dia.

drain valve

20 mm dia. hot water supply pipe

possible pumped secondary return pipe

13 mm dia. primary cold feed pipe

25mm dia. primary flow pipe

13 mm dia. hot water supply pipe

25 mm dia. primary return pipe

SINK

BASIN

central heating circuit

boiler

safety valve

drain valve

494

Flow Controls ~ these are valves inserted into a water installation to control the water flow along the pipes or to isolate a branch circuit or to control the draw-off of water from the system.

Typical Examples ~

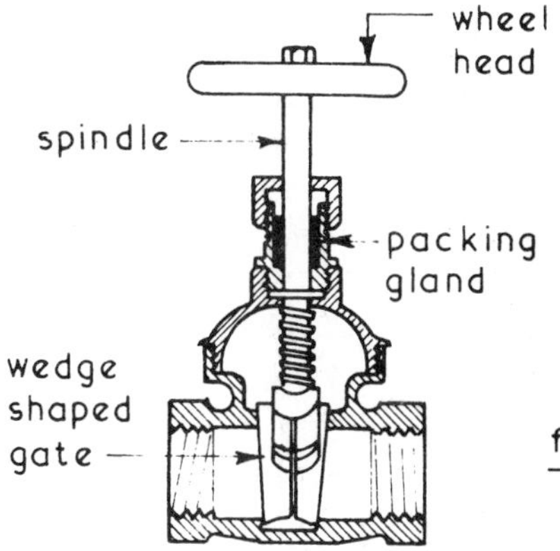

**GATE VALVE**
used to control flow of water

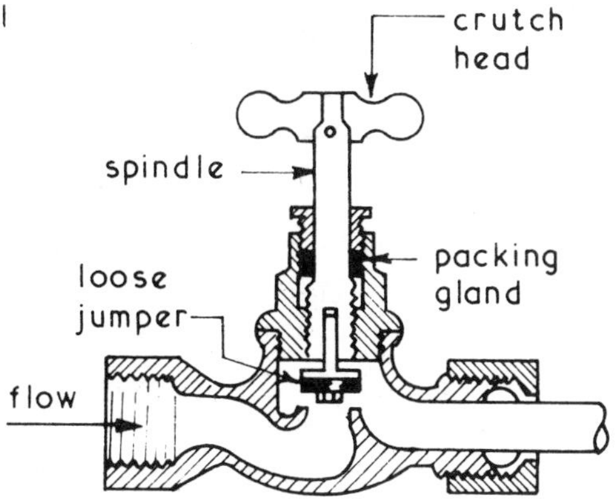

**STOP VALVE**
used to stop flow of water

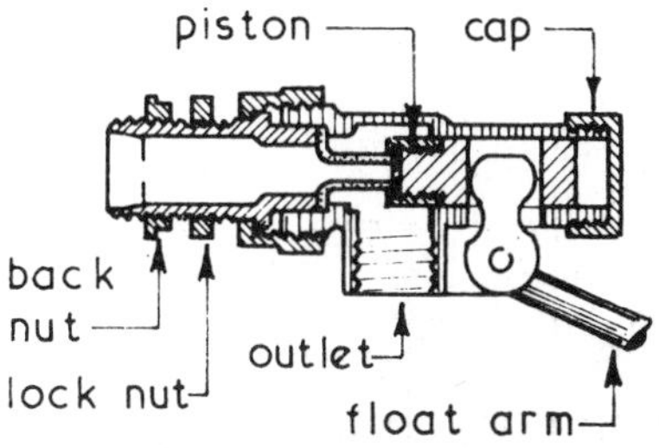

**PORTSMOUTH BALLVALVE**

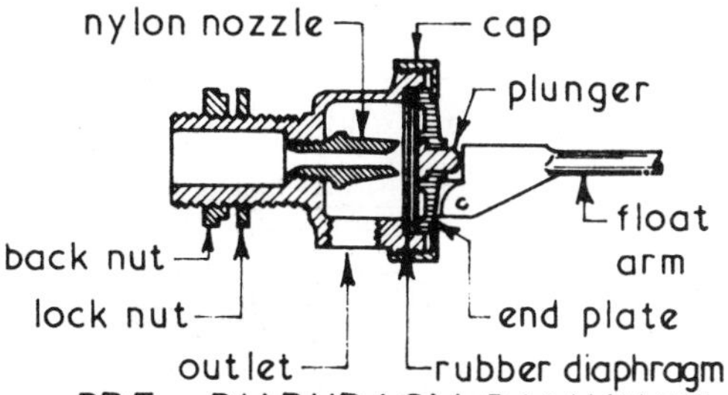

**BRE DIAPHRAGM BALLVALVE**

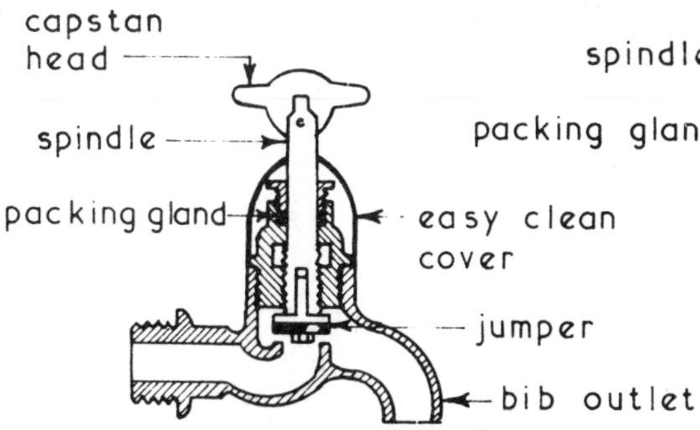

**BIB TAP**
horizontal inlet - used over sinks and for hose pipe outlets

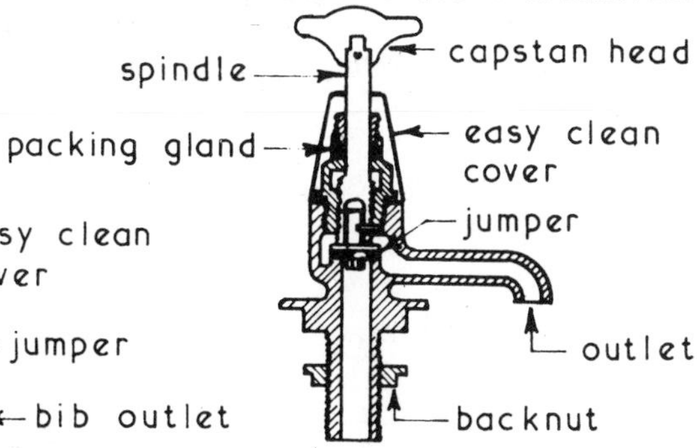

**PILLAR TAP**
vertical inlet - used in conjunction with fittings.

Cisterns ~ these are fixed containers used for storing water at atmospheric pressure. The inflow of water is controlled by a ballvalve which is adjusted to shut off the water supply when it has reached the designed level within the cistern. The capacity of the cistern depends on the draw off demand and whether the cistern feeds both hot and cold water systems. Domestic cold water cisterns should be placed at least 750mm away from an external wall or roof surface and in such a position that it can be inspected, cleaned and maintained. A minimum clear space of 300mm is required over the cistern for ballvalve maintenance. An overflow or warning pipe of not less than 19mm diameter must be fitted to fall away in level to discharge in a conspicuous position. All draw off pipes must be fitted with a stop valve positioned as near to the cistern as possible.

Cisterns are available in a variety of sizes and materials such as galvanised mild steel ( BS 417 ), asbestos cement ( BS 2777 ), plastic moulded ( BS 4213 ) and glass fibre. If the cistern and its associated pipework are to be housed in a cold area such as a roof they should be insulated against freezing.

Typical Details ~

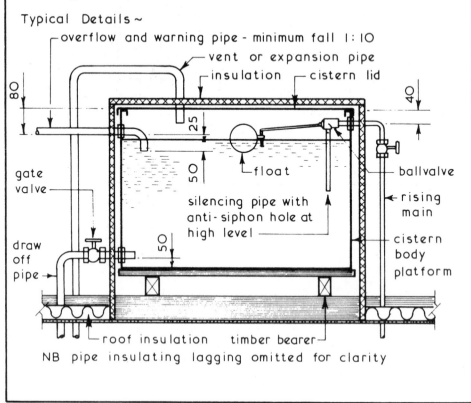

NB pipe insulating lagging omitted for clarity

Indirect Hot Water Cylinders ~ these cylinders are a form of heat exchanger where the primary circuit of hot water from the boiler flows through a coil or annulus within the storage vessel and transfers the heat to the water stored within. An alternative hot water cylinder for small installations is the single feed or 'Primatic' cylinder which is self venting and relies on two air locks to separate the primary water from the secondary water. This form of cylinder is connected to pipework in the same manner as **for a direct system (see page 493) and therefore gives** savings in both pipework and fittings. Indirect cylinders usually conform to the recommendations of BS1565 (galvanised mild steel) or BS1566 (copper).

Typical Examples ~

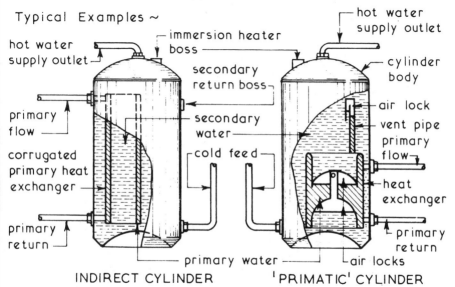

INDIRECT CYLINDER          'PRIMATIC' CYLINDER

'Primatic' Cylinders ~

1. Cylinder is filled in the normal way and the primary system is filled via the heat exchanger, as the initial filling continues air locks are formed in the upper and lower chambers of the heat exchanger and in the vent pipe.

2. The two air locks in the heat exchanger are permanently maintained and are self-recuperating in operation. These air locks isolate the primary water from the secondary water almost as effectively as a mechanical barrier.

3. The expansion volume of total primary water at a flow temperature of 82° C is approximately $1/25$ and is accommodated in the upper expansion chamber by displacing air into the lower chamber, upon contraction reverse occurs.

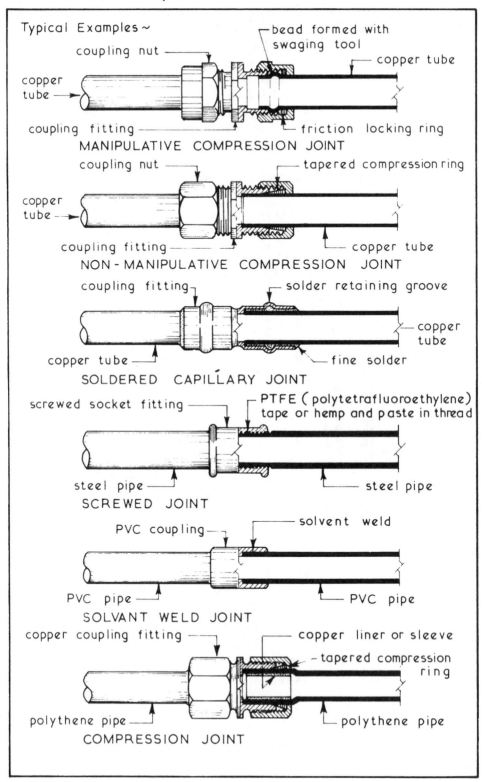

Typical Examples ~

MANIPULATIVE COMPRESSION JOINT

NON-MANIPULATIVE COMPRESSION JOINT

SOLDERED CAPILLARY JOINT

SCREWED JOINT

SOLVANT WELD JOINT

COMPRESSION JOINT

Typical Examples ~

— weir overflow

└ outlet for 38mm
diameter trap and
pipe

**BELFAST PATTERN SINK**

tap holes ┐  ┌ flutes

overflow ┘  └ 180 or 200mm
deep bowl

**SINGLE DRAINER STAINLESS STEEL SINK**

Fireclay Sinks ( BS 1206 ) -
these are white glazed sinks
and are available in a wide
range of sizes from 460 x
380 x 200 deep up to 1220 x
610 x 305 deep and can be
obtained with an integral
drainer. They should be fixed
at a height between 850 and
920mm and supported by legs,
cantilever brackets or dwarf
brick walls.

Metal Sinks ( BS 1244 ) -
these can be made of enamelled
cast iron, enamelled pressed
steel or stainless steel with
single or double drainers in
sizes ranging from 1070 x
460 to 1600 x 530 supported
on cantilever brackets or sink
cupboards.

Ceramic Wash Basins ( BS 1188 )

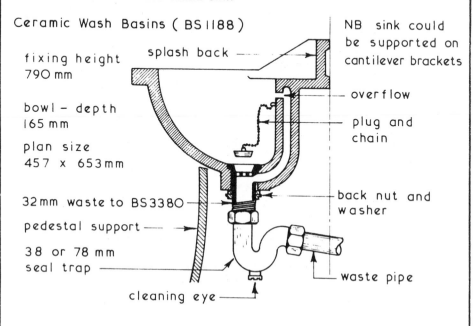

splash back

fixing height
790 mm

bowl - depth
165 mm

plan size
457 x 653mm

32mm waste to BS 3380

pedestal support

38 or 78 mm
seal trap

cleaning eye

NB sink could
be supported on
cantilever brackets

— overflow

plug and
chain

back nut and
washer

waste pipe

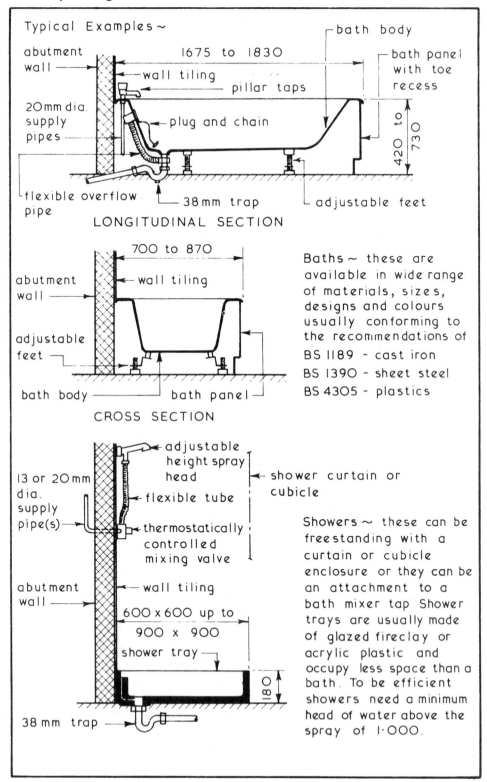

Typical Examples ~

abutment wall

1675 to 1830

bath body

wall tiling

pillar taps

bath panel with toe recess

20 mm dia. supply pipes

plug and chain

420 to 730

flexible overflow pipe

38 mm trap

adjustable feet

LONGITUDINAL SECTION

700 to 870

abutment wall

wall tiling

adjustable feet

bath body

bath panel

CROSS SECTION

Baths ~ these are available in wide range of materials, sizes, designs and colours usually conforming to the recommendations of

BS 1189 - cast iron
BS 1390 - sheet steel
BS 4305 - plastics

13 or 20 mm dia. supply pipe(s)

adjustable height spray head

shower curtain or cubicle

flexible tube

thermostatically controlled mixing valve

abutment wall

wall tiling

600 x 600 up to 900 x 900

shower tray

180

38 mm trap

Showers ~ these can be freestanding with a curtain or cubicle enclosure or they can be an attachment to a bath mixer tap Shower trays are usually made of glazed fireclay or acrylic plastic and occupy less space than a bath. To be efficient showers need a minimum head of water above the spray of 1·000.

Typical Examples ~

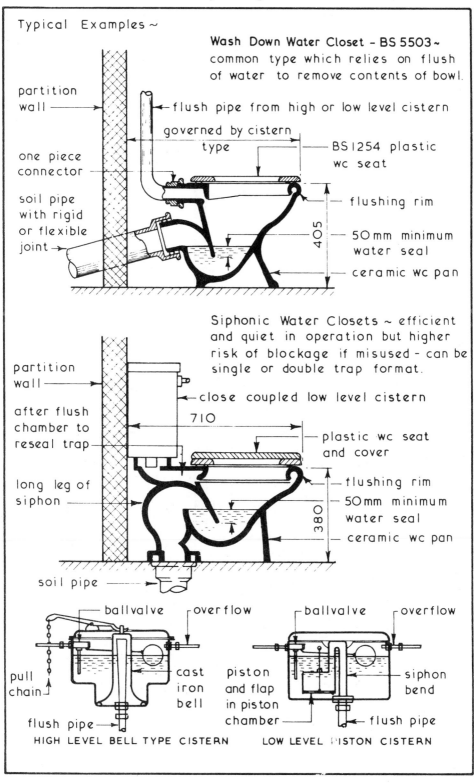

**Wash Down Water Closet - BS 5503 ~** common type which relies on flush of water to remove contents of bowl.

partition wall

one piece connector

soil pipe with rigid or flexible joint →

flush pipe from high or low level cistern

governed by cistern type

BS 1254 plastic wc seat

flushing rim

405

50 mm minimum water seal

ceramic wc pan

Siphonic Water Closets ~ efficient and quiet in operation but higher risk of blockage if misused - can be single or double trap format.

partition wall

after flush chamber to reseal trap

long leg of siphon

close coupled low level cistern

710

plastic wc seat and cover

flushing rim

50 mm minimum water seal

380

ceramic wc pan

soil pipe

ballvalve — overflow

pull chain

cast iron bell

flush pipe →

**HIGH LEVEL BELL TYPE CISTERN**

ballvalve — overflow

piston and flap in piston chamber

siphon bend

flush pipe

**LOW LEVEL PISTON CISTERN**

# Single Stack Discharge System

Single Stack System ~ method developed by the Building Research Establishment to eliminate the need for ventilating pipework to maintain the water seals in traps to sanitary fittings. The slope and distances of the branch connections must be kept within the design limitations given below. This system is only possible when the sanitary appliances are closely grouped around the discharge stack.

Typical Details ~

100mm diameter soil and vent pipe

balloon cage to top of pipe above roof level

32mm dia. waste pipe with 75mm min. seal trap

maximum length 1·700

no limit

slope determined by length
**slope 20 to 120mm/m**

BASIN

BATH

wc with 50mm minimum water seal

WATER CLOSET

3·000 maximum

**slope 18 to 90mm/m**

40mm dia. waste pipe with 75mm min. seal trap

50mm dia. parallel branch

104° branch with 50mm radius at junction

wc branch

SINK

3·000 maximum
**slope 18 to 90mm/m**

svp

200

50mm radius

40mm dia. waste pipe with 75mm min. seal trap

minimum vertical distance from lowest connection to drain invert **450mm**

large radius bends

* no connection to be made within shaded portion

Ventilated Stack Systems ~ where the layout of sanitary appliances is such that they do not conform to the requirements for the single stack system shown on page 502 ventilating pipes will be required to maintain the water seals in the traps. Three methods are available to overcome the problem namely a fully ventilated system, a ventilated stack system and a modified single stack system which can be applied over any number of stories.

Typical Examples ~

opening into building

900 min.

cage or cover

termination of ventilating part of stack

less than 3·000

cross connection above spillover level of highest appliance

ventilating pipes

300 max.

300 max.

BASIN

SINK

300 max.

branch discharge pipes

BATH

WC

ventilating pipes

branch discharge pipe

ventilating stack

cross connection as alternative to WC branch ventilating pipe

discharge stack

**FULLY VENTILATED SYSTEM**
used where there are a large number of appliances which are widely dispersed or grouped in ranges

branch discharge pipe

BASIN

branch discharge pipe

SINK

branch discharge pipe

BATH

WC

cross connection

ventilating stack

discharge stack

**VENTILATED STACK SYSTEM**
used where grouping of appliances makes individual venting unnecessary ~ seals retained by cross venting to a separate ventilating stack

ventilating pipe

300 max.

BASIN

branch discharge pipe

SINK

branch discharge pipes

BATH

WC

ventilating stack

discharge stack

**MODIFIED SINGLE STACK SYSTEM**
used to ventilate only those branch pipes exceeding length required for a single stack system

Minimum diameter for branch ventilating pipes = 25mm.

# Hot Water Heating Systems

One Pipe System ~ the hot water is circulated around the system by means of a centrifugal pump. The flow pipe temperature being about 80°C and the return pipe temperature being about 60 to 70°C. The one pipe system is simple in concept and easy to install but has the main disadvantage that the hot water passing through each heat emitter flows onto the next heat emitter or radiator therefore the average temperature of successive radiators is reduced unless the radiators are carefully balanced or the size of the radiators at the end of the circuit are increased to compensate for the temperature drop.

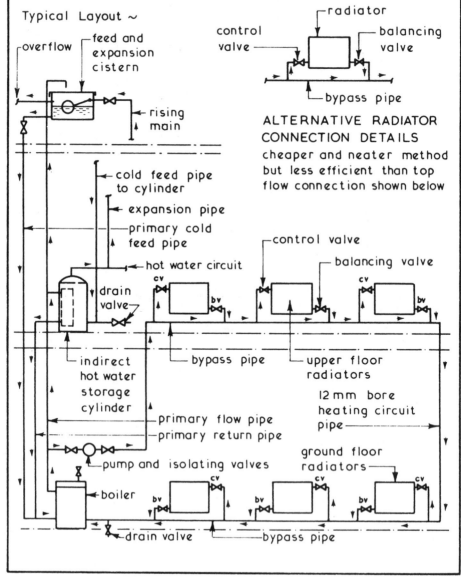

Typical Layout ~

ALTERNATIVE RADIATOR CONNECTION DETAILS
cheaper and neater method but less efficient than top flow connection shown below

Two Pipe System ~ this is a dearer but much more efficient system than the one pipe system shown on page 504. It is easier to balance since each radiator or heat emitter receives hot water at approximately the same temperature since the hot water leaving the radiator is returned to the boiler via the return pipe without passing through another radiator.

Typical Layout ~

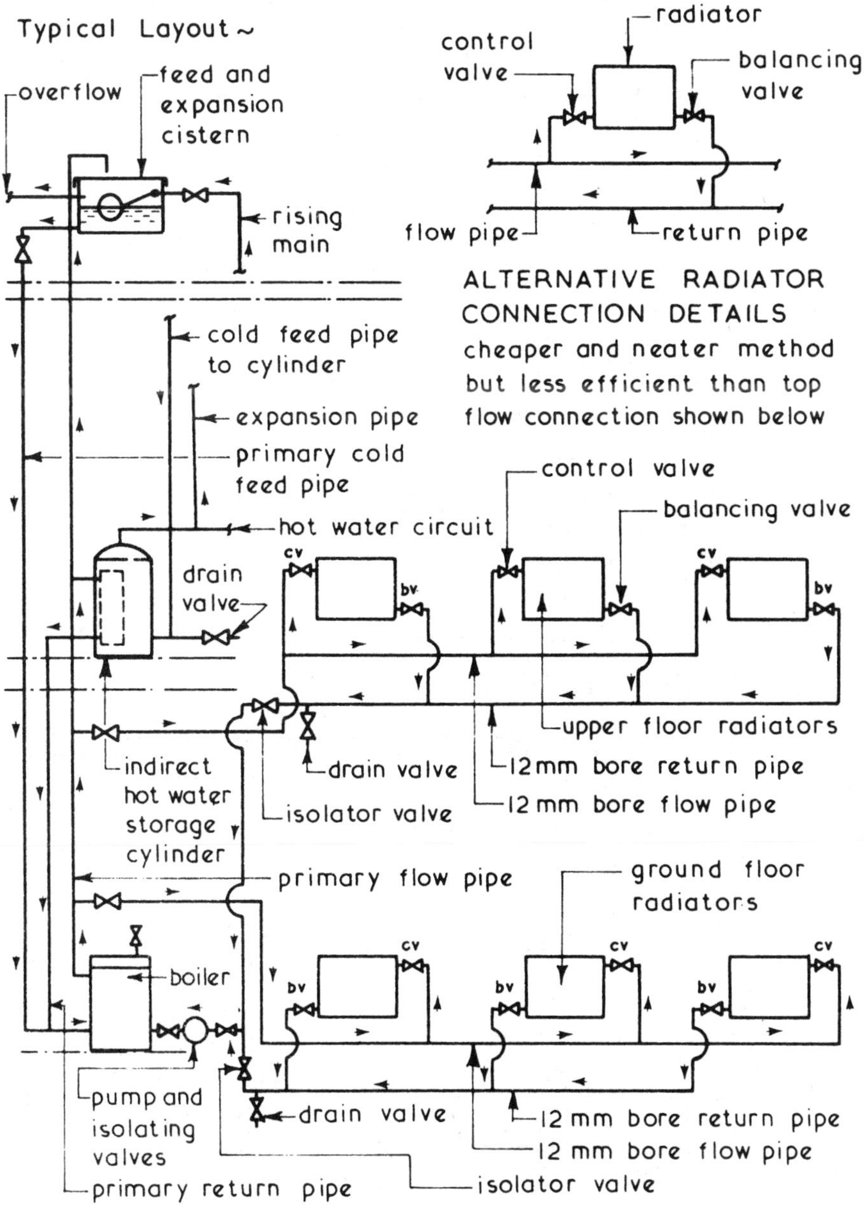

overflow

feed and expansion cistern

rising main

cold feed pipe to cylinder

expansion pipe

primary cold feed pipe

hot water circuit

drain valve

indirect hot water storage cylinder

primary flow pipe

boiler

pump and isolating valves

primary return pipe

radiator

control valve

balancing valve

flow pipe

return pipe

ALTERNATIVE RADIATOR CONNECTION DETAILS cheaper and neater method but less efficient than top flow connection shown below

control valve

balancing valve

cv

bv

upper floor radiators

12 mm bore return pipe

12 mm bore flow pipe

drain valve

isolator valve

ground floor radiators

cv

bv

12 mm bore return pipe

12 mm bore flow pipe

drain valve

isolator valve

# Hot Water Heating Systems

Micro Bore System ~ this system uses 6 to 12 mm diameter soft copper tubing with an individual flow and return pipe to each heat emitter or radiator from a 22 mm diameter manifold. The flexible and unobstrusive pipework makes this system easy to install in awkward situations but it requires a more powerful pump than that used in the traditional small bore systems. The heat emitter or radiator valves can be as used for the one or two pipe small bore systems alternatively a double entry valve can be used.

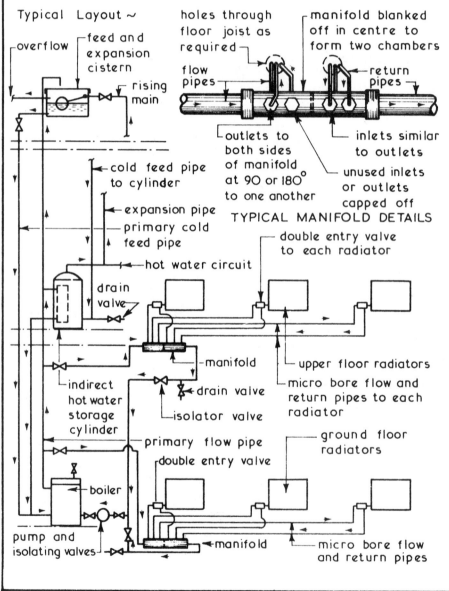

Typical Layout ~

overflow

feed and expansion cistern

rising main

holes through floor joist as required

manifold blanked off in centre to form two chambers

flow pipes

return pipes

outlets to both sides of manifold at 90 or 180° to one another

inlets similar to outlets

unused inlets or outlets capped off

cold feed pipe to cylinder

expansion pipe

primary cold feed pipe

hot water circuit

drain valve

manifold

indirect hot water storage cylinder

drain valve

isolator valve

primary flow pipe

double entry valve

boiler

pump and isolating valves

manifold

TYPICAL MANIFOLD DETAILS

double entry valve to each radiator

upper floor radiators

micro bore flow and return pipes to each radiator

ground floor radiators

micro bore flow and return pipes

Controls ~ the range of controls available to regulate the heat output and timing operations for a domestic hot water heating system is considerable ranging from thermostatic radiator control valves to programmers and controllers.

Typical Example ~

Boiler — fitted with a thermostat to control the temperature of the hot water leaving the boiler.

Heat Emitters or Radiators — fitted with thermostatically controlled radiator valves to control flow of hot water to the radiators to keep room at desired temperature.

Programmer / Controller — this is basically a time switch which can usually be set for 24 hours, once daily or twice daily time periods and will generally give separate programme control for the hot water supply and central heating systems. It receives signals from the hot water cylinder and room thermostats to control the pump and motorised valve action.

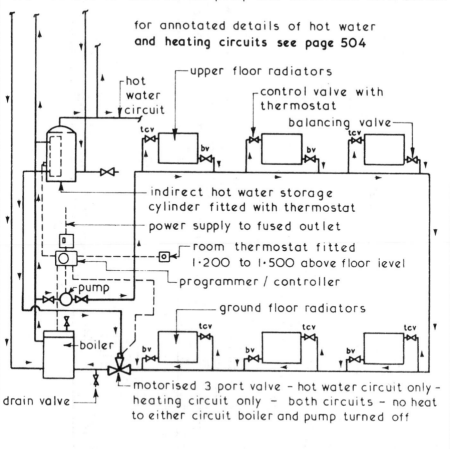

for annotated details of hot water **and heating circuits see page 504**

hot water circuit — upper floor radiators — control valve with thermostat — balancing valve

indirect hot water storage cylinder fitted with thermostat

power supply to fused outlet

room thermostat fitted 1·200 to 1·500 above floor level

programmer / controller

pump

ground floor radiators

boiler

drain valve — motorised 3 port valve – hot water circuit only – heating circuit only – both circuits – no heat to either circuit boiler and pump turned off

Electrical Supply ~ in England and Wales electricity is generated and supplied by the Central Electricity Generating Board and distributed through Area Electricity Boards, whereas in Scotland it is generated, supplied and distributed by the South of Scotland Electricity Board and the North of Scotland Hydro-Electric Board. The electrical supply to a domestic installation is usually 240 volt single phase and is designed with the following basic aims :-

1. Proper earthing to avoid shocks to occupants.
2. Prevention of current leakage.
3. Prevention of outbreak of fire.

Typical Electrical Supply Intake Details ~

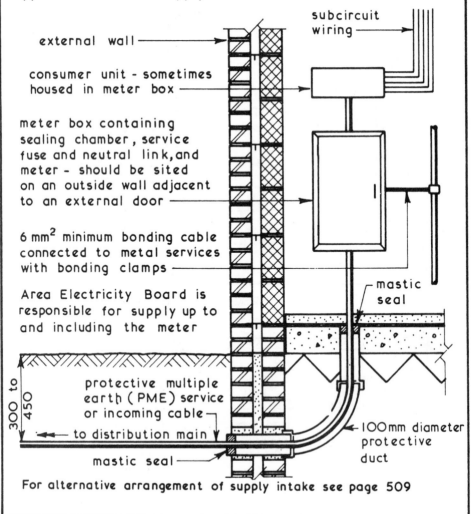

subcircuit wiring

external wall

consumer unit - sometimes housed in meter box

meter box containing sealing chamber, service fuse and neutral link, and meter - should be sited on an outside wall adjacent to an external door

6 mm² minimum bonding cable connected to metal services with bonding clamps

Area Electricity Board is responsible for supply up to and including the meter

mastic seal

300 to 450

protective multiple earth (PME) service or incoming cable

to distribution main

100mm diameter protective duct

mastic seal

For alternative arrangement of supply intake see page 509

508

Electrical Supply Intake ~ although the electrical supply intake can be terminated in a meter box situated within a dwelling most Area Electricity Boards prefer to use the external meter box to enable the meter to be read without the need to enter the premises.

Typical Electrical Supply Intake Details ~

subcircuit wiring

external wall

750mm minimum clear space in front of meter box

consumer unit

cavity tray

external meter box containing sealing chamber, service fuse and neutral link, and meter - requires brick opening 400 wide x 600 high

$6\,mm^2$ minimum bonding cable connected to metal services with bonding clamps

31

670 to 1070

PVC or similar sheet behind meter box

Area Electricity Board responsible for supply up to and including the meter

300 to 450

PME service cable in 38mm diameter PVC duct sealed at ends to prevent gas entry

duct with 350mm minimum bending radius built into wall as work proceeds

to distribution main

For alternative arrangement of supply intake see page 508

Entry and Intake of Electrical Service ~ the Area Electricity Board is responsible for providing the electrical supply up to and including the meter but the consumer is responsible for the safety and protection of the Boards equipment. The Area Electricity Board will install the service cable up to the meter position where their termination equipment is installed. This equipment may be located internally ·or fixed externally on a wall the latter being preferred since it gives easy access for reading **the meter - for intake details see Basic Requirements for Electrical Supply on page 508**

Meter Boxes - generally the Area Electricity Boards meters and termination equipment are housed in a meter box which are available in fibreglass and steel ranging in size from 450mm wide x 638mm high to 585mm wide x 815mm high with an overall depth of 177mm.

Consumer Control Unit - this provides a uniform, compact and effective means of efficiently controlling and distributing electrical energy within a dwelling. The control unit contains a main double pole isolating switch controlling the live phase and solid metal conductors called bus bars for the neutral phase which in turn is connected to the fuses or miniature circuit breakers protecting the final subcircuits.

Typical Layout ~

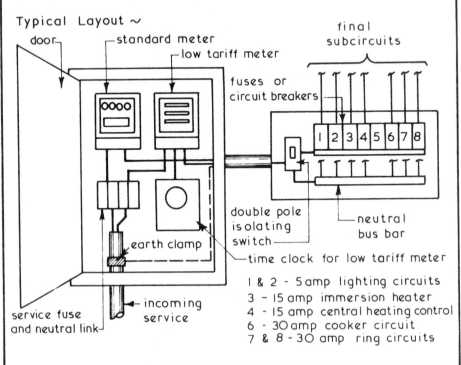

door
standard meter
low tariff meter
final subcircuits
fuses or circuit breakers
double pole isolating switch
time clock for low tariff meter
neutral bus bar
earth clamp
service fuse and neutral link
incoming service

1 & 2 - 5 amp lighting circuits
3 - 15 amp immersion heater
4 - 15 amp central heating control
6 - 30 amp cooker circuit
7 & 8 - 30 amp ring circuits

Electric Cables ~ these are made up of copper or aluminium wires called conductors surrounded by an insulating material such as PVC or rubber.

Typical Examples ~

rubber or PVC outer sheath

live – brown
earth – green and yellow
neutral – blue

rubber or PVC insulation

aluminium or copper conductors

### SHEATHED CABLES

ductile seamless solid drawn copper sheath which acts as an earth conductor

high conductivity copper conductors

densely packed pure magnesium oxide insulation

NB. magnesium oxide is hygroscopic therefore ends of cable must be fitted with special sealing pots

### MINERAL INSULATED CABLE

Conduits ~ these are steel or plastic tubes which protect the cables. Steel conduits act as an earth conductor whereas plastic conduits will require a separate earth conductor drawn in. Conduits enable a system to be rewired without damage or interference of the fabric of the building. The cables used within conduits are usually insulated only whereas in non-rewireable systems the cables have a protective outer sheath.

Typical Conduit Fittings ~

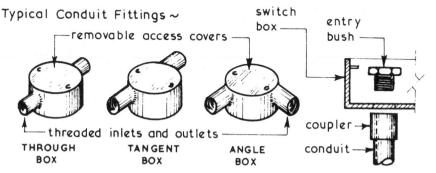

removable access covers

switch box

entry bush

coupler

threaded inlets and outlets

conduit

**THROUGH BOX**  **TANGENT BOX**  **ANGLE BOX**

Trunking – alternative to conduit and consists of a preformed cable carrier which is surface mounted and is fitted with a removable or 'snap on' cover which can have the dual function of protection and trim or surface finish.

Wiring systems~ rewireable systems housed in horizontal conduits can be cast into the structural floor slab or sited within the depth of the floor screed. To ensure that such a system is rewireable draw-in boxes must be incorporated at regular intervals and not more than two right angle boxes to be included between draw-in points. Vertical conduits can be surface mounted or housed in a chase cut into a wall provided the depth of the chase is not more than one third of the wall thickness. A horizontal non-rewireable system can be housed within the depth of the timber joists to a suspended floor whereas vertical cables can be surface mounted or housed in a length of conduit as described for rewireable systems.

Typical Examples~

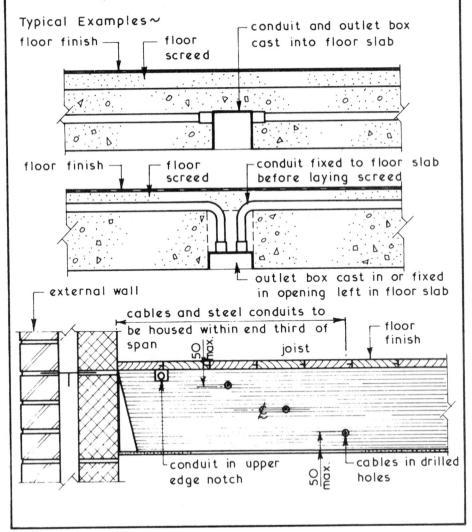

floor finish — floor screed — conduit and outlet box cast into floor slab

floor finish — floor screed — conduit fixed to floor slab before laying screed

— outlet box cast in or fixed in opening left in floor slab

— external wall

cables and steel conduits to be housed within end third of span

50 max.

joist

floor finish

— conduit in upper edge notch

50 max.

— cables in drilled holes

Cable Sizing ~ the size of a conductor wire can be calculated taking into account the maximum current the conductor will have to carry (which is limited by the heating effect caused by the resistance to the flow of electricity through the conductor) and the voltage drop which will occur when the current is carried. For domestic electrical installations the following cable ratings are usually suitable-

Lighting Circuits -

neutral - as for live conductor

live conductor - 1·13 mm diameter 1 mm² cross section area

Immersion Heater -

neutral - as for live conductor

live conductor - 1·38 mm diameter 1·5 mm² cross section area

Power Ring Circuits -

neutral - as for live conductor

live conductor - 1·78 mm diameter 2·5 mm² cross section area

30 amp Cooker Circuit -

neutral - as for live conductor

live conductor - 7 No. 1 04 mm diameter wires 6 mm² total cross section area

All the above ratings are for one twin cable with or without an earth conductor

Electrical Accessories ~ for power circuits these include cooker control units and fused connector units for fixed appliances such as immersion heaters, water heaters and refrigerators.

Socket Outlets ~ these may be single or double outlets, switched or unswitched surface or flush mounted and may be fitted with indicator lights. Recommended fixing heights are —

150 min.

floor

GENERAL

825 - 900

1200

if over work surface

door handle height 1040 with 300 min.

floor

FOR THE ELDERLY

floor

FOR THE DISABLED

Plugs ~

cap screw

BS 1363 plug

earth - green / yellow cable

cartridge fuse — up to 720 watt - 3 amp up to 3000 watt - 13 amp

neutral - blue cable

cable grip

live - brown cable

# Electrical Installations

Power Circuits ~ in new domestic electrical installations the ring main system is usually employed instead of the older system of having each socket outlet on its own individual fused circuit with unfused round pin plugs. Ring circuits consist of a fuse or miniature circuit breaker protected subcircuit with a 30 amp. rating of a live conductor, neutral conductor and an earth looped from socket outlet to socket outlet. Metal conduit systems do not require an earth wire providing the conduit is electrically sound and earthed. The number of socket outlets per ring main is unlimited but a separate circuit must be provided for every 100 m² of floor area. To conserve wiring spur outlets can be used as long as the total number of spur outlets does not exceed the total number of outlets connected to the ring and that there is not more than two outlets per spur.

Typical Ring Main Wiring Diagram ~

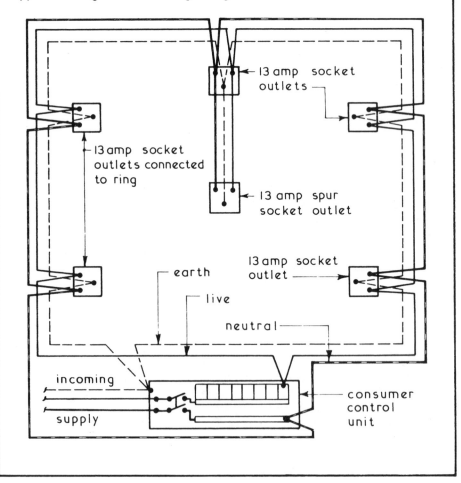

Lighting Circuits ~ these are usually wired by the loop-in method using an earthed twin cable with a 5 amp. fuse or miniature circuit breaker protection. In calculating the rating of a lighting circuit an allowance of 100 watts per outlet should be used. More than one lighting circuit should be used for each installation so that in the case of a circuit failure some lighting will be in working order.

Typical Lighting Circuit Wiring Diagram ~

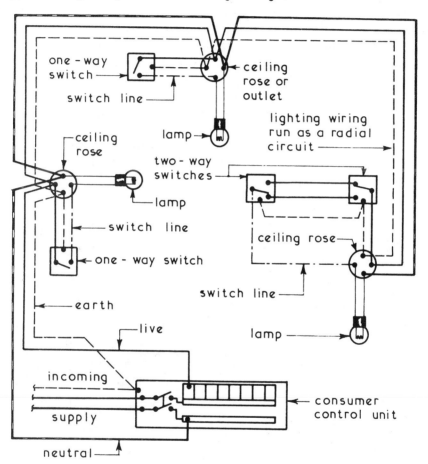

Electrical Accessories ~ for lighting circuits these consist mainly of switches and lampholders, the latter can be wall mounted, ceiling mounted or pendant in format with one or more bulb or tube holders. Switches are usually rated at 5 amps and are available in a variety of types such as double or 2 gang, dimmer and pull or pendant switches, the latter must always be used in bathrooms.

Gas Supply ~ the supplier ( British Gas plc.) under the Gas Act 1972 is obliged to supply gas to a dwelling which is within 25 yards ( 22·860) of a qualifying gas main but under the Energy Act 1976 it is not obliged to supply gas to any new premises where the amount of supply would exceed 25 000 therms per annum. The supply, appliances and installations must comply with the safety regulations made under the Gas Act 1972 and with the requirements set out in Part J of the Building Regulations 1985. The route of the gas service pipe is agreed between the Gas Region and the designer and must be situated so that it is protected from damage so as to avoid the possibility of a leak.

Typical Supply Arrangement ~ :

84

mastic sealed duct

high level external entry meter box containing control valve and meter position to be agreed with gas undertaking - requires brick opening 450 wide x 530 high

to appliances

external wall

Gas Region pays for service pipe to boundary~ maximum 9·140

boundary

pipe covered or housed in a duct if required

pipe clips as necessary

375 minimum

service pipe extends to and includes meter

gas main

service pipe - minimum diameter 25 mm - responsibility of gas region

For alternative supply arrangements see page 517

Gas Service Pipes ~

1. Whenever possible the service pipe should enter the building on the side nearest the main.
2. A service pipe must not pass under the foundations of a building.
3. No service pipe must be run within a cavity but it may pass through a cavity by the shortest route.
4. Service pipes passing through a wall or solid floor must be enclosed by a sleeve or duct which is end sealed with a suitable mastic.
5. No service pipe must be installed in an unventilated void space.
6. Suitable materials for service pipes are copper (BS 2871) or steel (BS 1387 & BS 3601).

Typical Supply Arrangements ~

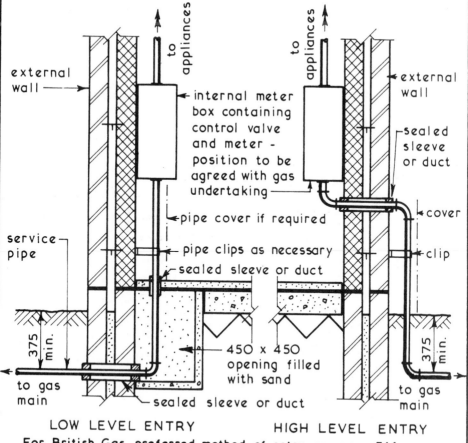

LOW LEVEL ENTRY          HIGH LEVEL ENTRY

For British Gas preferred method of entry see page 516

Gas Fires ~ for domestic use these are classified as a **gas burning appliance with a rated input of up to 60 kW and must be installed in accordance with minimum requirements set out in Part J of the Building Regulations.** Most gas fires connected to a flue are designed to provide radiant and convected heating whereas the room sealed balanced flue appliances are primarily convector heaters.

Typical Examples ~

flue blocks or lined **flue - see page 519**

shelf

convected warm air

canopy or hood

fire bars giving off radiant heat

firebrick backing

baffle

gas burner

tiled hearth

air inlet

external wall

NB. gas fires connected to a flue can be designed as a recessed fire - **see page 519**

damp-proof course

125 mm thick solid floor

consolidated hardcore

casing

convected warm air outlets

casing gives off radiant heat equal to approximately 10 % of total heat output of appliance

gas burner

internal air inlets

air drawn in via external terminal

terminal

inlet duct

products of combustion expelled at terminal via outlet duct

Gas Fire Flues ~ these can be defined as a passage for the discharge of the products of combustion to the outside air and can be formed by means of a chimney, special flue blocks or by using a flue pipe. In all cases the type and size of the flue as recommended in Approved Document J will meet the requirements of the Building Regulations.

Typical Single Gas Fire Flues ~

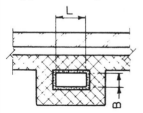

LINED CHIMNEY ON
EXTERNAL WALL

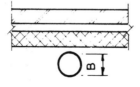

FLUE PIPE ON
EXTERNAL WALL

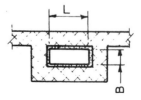

LINED CHIMNEY ON
INTERNAL WALL

**Flue Size Requirements ~**
1. No dimension should be less than 63 mm.
2. Flue for decorative appliance should have a minimum dimension measured across the axis of 175 mm
3. Flues for gas fires - min. area = 12000 mm$^2$; max. L = 6 × B
4. Any other appliance should have a flue with a cross-sectional area at least equal to the outlet size of the appliance and L = not more than 5 × B.

Flue Blocks ~

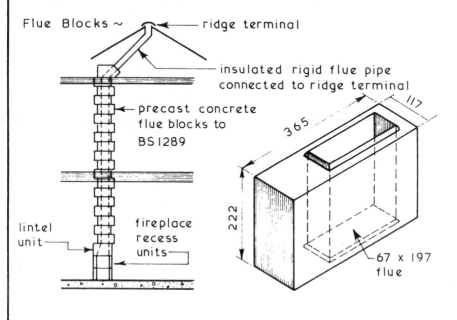

ridge terminal

insulated rigid flue pipe
connected to ridge terminal

precast concrete
flue blocks to
BS 1289

lintel
unit

fireplace
recess
units

222

365

117

67 × 197
flue

519

Open Fireplaces ~ for domestic purposes these are a means of providing a heat source by consuming solid fuels with an output rating of under 45 kW. Room-heaters can be defined in a similar manner but these are an enclosed appliance as opposed to the open recessed fireplace.

Components ~ the complete construction required for a domestic open fireplace installation is composed of the hearth, fireplace recess, chimney, flue and terminal.

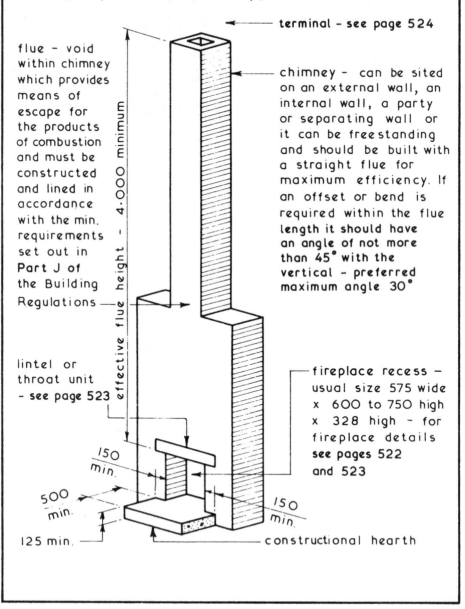

terminal – see page 524

flue – void within chimney which provides means of escape for the products of combustion and must be constructed and lined in accordance with the min. requirements set out in **Part J of the Building Regulations** –

chimney – can be sited on an external wall, an internal wall, a party or separating wall or it can be freestanding and should be built with a straight flue for maximum efficiency. If an offset or bend is **required within the flue length it should have an angle of not more than 45° with the vertical – preferred maximum angle 30°**

effective flue height – 4.000 minimum

lintel or throat unit **– see page 523**

fireplace recess – usual size 575 wide x 600 to 750 high x 328 high – for fireplace details **see pages 522 and 523**

150 min.

500 min.

150 min.

125 min. –

constructional hearth

Open Fireplace Recesses ~ these must have a constructional hearth and can be constructed of bricks or blocks of concrete or burnt clay or they can be of cast insitu concrete. All fireplace recesses must have jambs on both sides of the opening and a backing wall of a minimum thickness in accordance with its position and such jambs and backing walls must extend to the full height of the fireplace recess.

Typical Examples ~

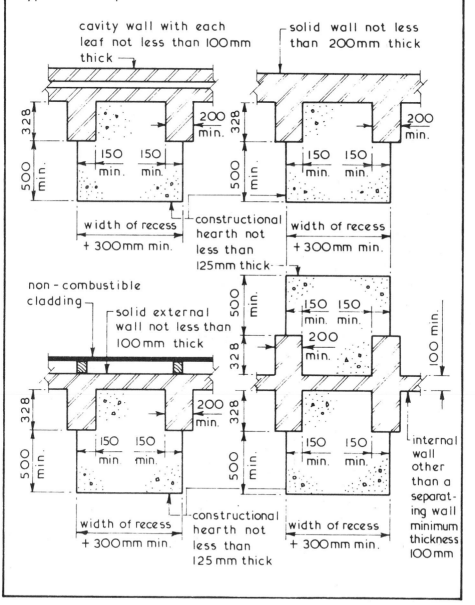

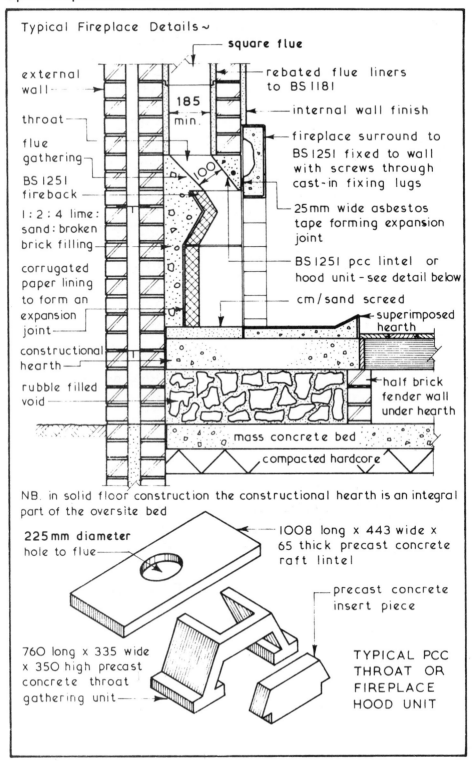

Typical Fireplace Details~

square flue

external wall

throat

flue gathering

BS 1251 fireback

1 : 2 : 4 lime : sand : broken brick filling

corrugated paper lining to form an expansion joint

constructional hearth

rubble filled void

185 min.

100

rebated flue liners to BS 1181

internal wall finish

fireplace surround to BS 1251 fixed to wall with screws through cast-in fixing lugs

25mm wide asbestos tape forming expansion joint

BS 1251 pcc lintel or hood unit - see detail below

cm / sand screed

superimposed hearth

half brick fender wall under hearth

mass concrete bed

compacted hardcore

NB. in solid floor construction the constructional hearth is an integral part of the oversite bed

225mm diameter hole to flue

1008 long x 443 wide x 65 thick precast concrete raft lintel

precast concrete insert piece

760 long x 335 wide x 350 high precast concrete throat gathering unit

TYPICAL PCC THROAT OR FIREPLACE HOOD UNIT

Open Fireplace Chimneys and Flues ~ the main functions of a chimney and flue are to :-

1. Induce an adequate supply of air for the combustion of the fuel being used.

2. Remove the products of combustion.

In fulfilling the above functions a chimney will also encourage a flow of ventilating air promoting constant air changes within the room which will assist in the prevention of condensation.

Approved Document J recommends that all flues should be lined with approved materials so that the minimum size of the flue so formed will be 200 mm diameter or a square section of equivalent area. Flues should also be terminated above the roof level as set out in the Approved Document.

Typical Examples ~

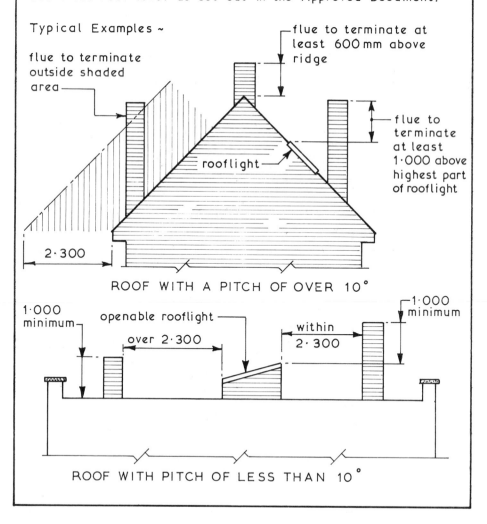

flue to terminate outside shaded area

flue to terminate at least 600 mm above ridge

rooflight

flue to terminate at least 1·000 above highest part of rooflight

2·300

ROOF WITH A PITCH OF OVER 10°

1·000 minimum

openable rooflight

over 2·300

within 2·300

1·000 minimum

ROOF WITH PITCH OF LESS THAN 10°

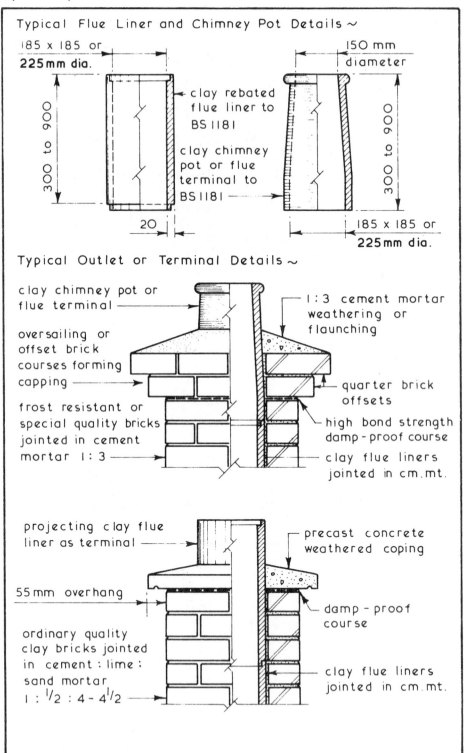

Typical Flue Liner and Chimney Pot Details ~

185 x 185 or
**225mm dia.**

150 mm
diameter

300 to 900

clay rebated
flue liner to
BS 1181

clay chimney
pot or flue
terminal to
BS 1181

300 to 900

20

185 x 185 or
**225mm dia.**

Typical Outlet or Terminal Details ~

clay chimney pot or
flue terminal

1:3 cement mortar
weathering or
flaunching

oversailing or
offset brick
courses forming
capping

quarter brick
offsets

frost resistant or
special quality bricks
jointed in cement
mortar 1:3

high bond strength
damp-proof course

clay flue liners
jointed in cm.mt.

projecting clay flue
liner as terminal

precast concrete
weathered coping

55mm overhang

damp-proof
course

ordinary quality
clay bricks jointed
in cement : lime :
sand mortar
1 : $\frac{1}{2}$ : 4 - 4$\frac{1}{2}$

clay flue liners
jointed in cm.mt.

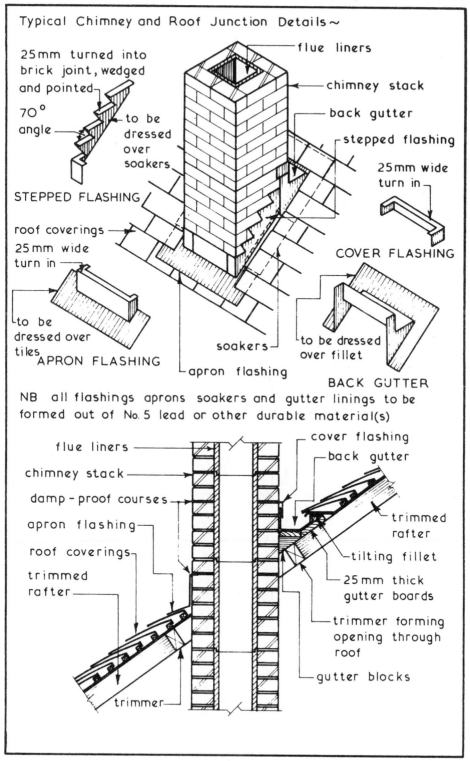

Typical Chimney and Roof Junction Details~

flue liners

chimney stack

back gutter

stepped flashing

25mm turned into brick joint, wedged and pointed

70° angle

to be dressed over soakers

STEPPED FLASHING

25mm wide turn in

COVER FLASHING

roof coverings

25mm wide turn in

to be dressed over tiles

APRON FLASHING

soakers

apron flashing

to be dressed over fillet

BACK GUTTER

NB all flashings aprons soakers and gutter linings to be formed out of No. 5 lead or other durable material(s)

flue liners

chimney stack

damp-proof courses

apron flashing

roof coverings

trimmed rafter

trimmer

cover flashing

back gutter

trimmed rafter

tilting fillet

25mm thick gutter boards

trimmer forming opening through roof

gutter blocks

Combustion Air ~ it is a Building Regulation requirement that in the case of open fireplaces provision must be made for the introduction of combustion air in sufficient quantity to ensure the efficient operation of the open fire. Traditionally such air is taken from the volume of the room in which the open fire is situated, this can create air movements resulting in draughts. An alternative method is to construct an ash pit below the hearth level fret and introduce the air necessary for combustion via the ash by means of a duct.

Typical Details ~

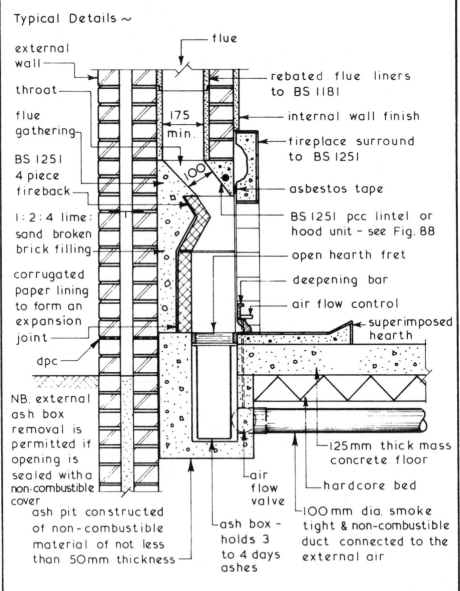

external wall

throat

flue gathering

BS 1251 4 piece fireback

1:2:4 lime: sand broken brick filling

corrugated paper lining to form an expansion joint

dpc

flue

175 min.

100

NB. external ash box removal is permitted if opening is sealed with a non-combustible cover

ash pit constructed of non-combustible material of not less than 50mm thickness

air flow valve

ash box - holds 3 to 4 days ashes

rebated flue liners to BS 1181

internal wall finish

fireplace surround to BS 1251

asbestos tape

BS 1251 pcc lintel or hood unit – see Fig. 88

open hearth fret

deepening bar

air flow control

superimposed hearth

125mm thick mass concrete floor

hardcore bed

100mm dia. smoke tight & non-combustible duct connected to the external air

Telephone Installations ~ unlike other services such as water gas and electricity telephones cannot be connected to a common mains supply each telephone requires a pair of wires connecting it to the telephone exchange. The complete service installation both internal and external is carried by British Telecom engineers.

Typical Supply Arrangements ~

underground cables must be installed when the building is constructed - they are hidden and therefore have little or no effect on the surrounding environment

extension telephone if required

lead-in terminating point

external wall

350 minimum

underground cable and earth wire

19mm internal diameter duct sealed at both ends

wall hook at or near eaves

overhead cable

19mm internal diameter duct sealed at both ends

extension telephone if required

junction box

overhead cables are smaller and cheaper than underground cables and are convenient for providing new installations

external wall

earth wire

lead-in terminating box

earth rod

527

# INDEX